Wallless Cities

Shen Hou

Wallless Cities

Cities and Nature in American History

palgrave
macmillan

Shen Hou
Department of History
Peking University
Beijing, China

Translated by
Ruitian Li
School of Translation Studies
Xi'an International Studies University
Xi'an, China

ISBN 978-981-97-7826-3 ISBN 978-981-97-7827-0 (eBook)
https://doi.org/10.1007/978-981-97-7827-0

Jointly published with Sichuan People's Publishing House Co., Ltd.

The print edition is not for sale in China (Mainland). Customers from China (Mainland) please order the print book from: Sichuan People's Publishing House Co., Ltd.
ISBN of the Co-Publisher's edition: 978-7-220-12260-6

This Palgrave Macmillan imprint is published by the registered company Springer Nature Singapore Pte Ltd.
The registered company address is: 152 Beach Road, #21-01/04 Gateway East, Singapore 189721, Singapore

To my father
Hou Shangzhi

Contents

About the Translator

Ruitian Li, Ph.D. is a seasoned translator and interpreter working between Chinese and English. A graduate of The Hong Kong Polytechnic University, his research areas include corpus-based translation studies, translation language variants, and conference interpreting. He holds translation and interpreting certificates from both NAATI in Australia and CATTI in China. With extensive experience living and studying overseas, he is currently a faculty member at the School of Translation Studies at Xi'an International Studies University, where he teaches various courses to undergraduate and graduate students. He has published multiple research papers in SSCI-indexed journals and has led several translation research projects.

CHAPTER 1

Introduction: Expanding Cities and Shrinking Continents

As one bids farewell to the vibrant and bustling Beijing Capital International Airport, embarking on a journey across the vast and solitary wilderness of Siberia, the ice-clad expanse of the Bering Strait, and the sprawling darkness of the Canadian Shield, the anticipation of entering a new country is tempered by the arduous hours spent in the cramped confines of the aircraft. Upon arrival and departure from the airport in a major American city, traversing the labyrinth of highways leading to the city center amidst the sluggish traffic of the CBD, one may find oneself yearning for sights that ignite a sense of uniqueness and Americanness. Yet, such sentiments are often elusive.

Airports worldwide share a common architectural blueprint, emblematic of the technological marvels and consumerist culture of the era. Their metallic edifices, lined with rows of duty-free shops and fast-food outlets, reverberate with a familiar ambiance, complemented by the dulcet tones of uniform music and offerings. Likewise, the arteries connecting airports to urban cores bear a striking resemblance. Flanked by orderly rows of roadside trees and punctuated by imposing power poles, they blur into fleeting shadows against the backdrop of a swiftly passing landscape. Though the vistas beyond these arboreal sentinels may appear expansive, they lack the primordial desolation witnessed from aerial heights. Instead, they present a curated tableau of monotony, a testament to the meticulous hand of human intervention in shaping the natural realm. Even the skyscrapers that adorn CBD skylines, each bearing the

S. Hou, *Wallless Cities*,
https://doi.org/10.1007/978-981-97-7827-0_1

hallmark of a visionary architect, blend into a collective tapestry when illuminated by the pulsating glow of neon lights. Whether in Beijing, Tokyo, Los Angeles, or New York, they converge to embody the global commercial ethos. The uniformity of the urban looks should come as no surprise. They are the offspring of a modern industrial ethos, employing standardized methods in pursuit of shared objectives. To seek national identity within their confines is to chase after a mirage. If one tries to seek, only the billboards, shop signs, or the vernacular of airport dining menus might offer a fleeting glimpse into the soul of a nation. Alternatively, the intrepid explorer might choose to disembark, forsaking the comforts of modern transit, and venture forth on foot. Through the labyrinthine alleys, quaint neighborhoods, and bustling markets, they may uncover the pulse of everyday life, imbued with the distinctive flavor of America. Strolling through the aged, secluded streets, communities, and quaint corner stores, he immerses himself in the essence of bygone eras. Engaging the senses of sight, smell, and taste, he delves into the ordinary yet captivating essence of daily life. Perhaps through such an exploration, he can grasp a distinctly American flavor.

To be equitable, each city possesses its own idiosyncrasies, be it in the United States, China, or elsewhere. Embedded within their fabric are unique historical legacies and cultural nuances, etched onto the facades of buildings, the layout of neighborhoods, and the contours of landscapes, and manifested in the rituals, traditions, and customs upheld by its inhabitants. Likewise, the nature of a city defines its singular allure. Whether nestled along rivers or nestled against mountainous backdrops, whether adorned with sunlit beaches or sprawling plains, whether adorned in perpetual floral bloom or blanketed in ice and snow half the year—these environmental factors intricately weave into the tapestry of local life. Climate, terrain, flora, and fauna, despite the era of globalization, persist in shaping the daily rhythms of indigenous dietary preferences, sartorial choices, architectural aesthetics, and leisurely pursuits. When juxtaposed with villages and small towns, cities emerge as colossal, bustling, and endlessly diverse entities, defying easy categorization. Their sheer scale and complexity render them resistant to simplistic labels. While cities across the globe share certain commonalities, such as dense populations, intricate social networks, imposing architectural landscapes, and intricate transportation and commercial infrastructures, they steadfastly resist being pigeonholed into generic descriptors. A metropolis stands as a vast biome, akin to the majesty of the Amazon rainforest or the

Mississippi River basin. In juxtaposition, a village or small town mirrors the simplicity of a single-celled organism under the magnifying lens of a microscope.[1] However, as the distinctive features of a city persist within its natural backdrop, historical traditions, and architectural panorama, sometimes even blossoming vigorously, the regional or local characteristics may overshadow and blur its national identity.

In hemispheres where civilizations have thrived for millennia, one can glimpse the national memories etched into the edifices of yesteryears. Yet, how does one decipher the national essence concealed within American cities—a nation that, through its imperial expansion in the twentieth century, reshaped and assimilated the global urban landscape? How can such disparate cities as Boston, Kansas City, and San Francisco be gathered under a single national label? Despite their differing environments, origins, and dimensions in terms of area, population, and wealth, they are all identified as "American cities." Compared to cities in other nations, where lies their Americanness? Boston, a city nestled by the harbor, shares its maritime vitality with counterparts like Marseille in France and Shanghai in China. Yet, what renders Boston uniquely American, setting it apart from its European or Asian cities? To untangle these queries, we must traverse the annals of history, delving not only into the architectural chronicles of the city but also into the ongoing dialogue between its culture and natural environment.

As we step into the historical realm, a vivid image emerges—the silhouette of city walls. In the modern urban tapestry, the once-mighty ramparts that once repelled adversaries and fortified power now stand as mere echoes of the past. Even where remnants persist, they serve as silent witnesses, submerged beneath layers of history or fading into collective memory. Yet, in the tapestry of yesteryears, city walls played a crucial role, shaping the contours of ancient cities of the Old World, demarcating their boundaries, and defining their relationship with the world beyond. By day, the city gates swung open, ushering in a symphony of exchange of people, energy, and ideas. But as the gates closed, the city walls stood sentinel,

[1] In the realm of urban ecology, an enduring debate persists over whether cities themselves can be classified as a biome. In cities, various species coexist, fostering intricate relationships among them. However, unlike in the Amazon rainforest or natural river basins, where human influence is also evident, humans are not the absolute dominant species. In cities, humans reign supreme as the singular and unequivocal dominant species. Consequently, many urban ecologists hesitate to categorize cities as a biome, despite their undeniable composition of multiple interconnected ecosystems.

asserting safety as the foremost concern, safeguarding lives, property, and the sanctity of ideological authority and social order. The walls stand tall, warding off a multitude of adversaries, each more perilous and disordered than the last. At the forefront of this array are those who yearn to seize the fecundity of humanity and the flourishing of civilization within the natural world. Among them are the "foreigners," clad in fur and blood, their faces etched with tattoos, and their garments draped in an unfamiliar manner; the rebels, clad in iron and resolve, their clamor echoing with the thundering hooves of their steeds; and the displaced, wandering souls driven to desperate measures, even to the unthinkable act of consuming their own offspring. Yet, amidst this throng, there exist a plethora of exotic and unorthodox entities, alongside fervent adherents of a radical creed, their heads bared and their fervor unyielding. Vigilance against these myriad threats is a constant necessity, for they encroach upon our realm with unwavering determination. While the solemn duty of protecting city walls remains unshakable, cities enclosed by towering ramparts often present as insular, cramped, and stratified environments. Even amidst the vibrant hustle and bustle of urban life, concepts of equality, freedom, openness, and progress frequently take a backseat.

The presence or absence of city walls serves as a defining boundary between modern and traditional cities. Walls of American cities have never been dismantled because no tangible walls have ever been erected. The notion of "wallless cities" epitomizes ethos of American cities, symbolizing their unwavering commitment to outward expansion and embodying a spirit of openness and inclusivity distinct from ancient cities.

From Boston to San Francisco, from Chicago to Houston, American cities sprawl and unfold within their respective hinterlands, spanning from the colonial era to the present day. Indeed, when Dutch settlers arrived in Manhattan, they raised the modest Wall Street, now synonymous with modern finance. Yet, in the seventeenth-century landscape of New Amsterdam, the rudimentary wooden walls merely hindered free-roaming livestock. In contrast, the Hudson River and the East River might function as practical equivalents of walls, albeit in a different guise. Unlike the moats encircling ancient cities, rivers simultaneously linked urban centers with the wider world while maintaining a measure of separation. During that era, colonists harbored aspirations of pioneering the construction of modern bridges, ferries, and even canals, aiming to establish themselves as global leaders. Eventually, in 1825, amidst the renaming of New

Amsterdam to New York, a fledgling United States emerged on the continent. A monumental feat came to fruition, a canal spanning 584 km, linking the thriving Atlantic coast city to the vast expanses of forests and grasslands in the midwest and west. The Erie Canal seamlessly interconnected the Great Lakes with New York, transforming the sprawling lands of the central and western regions into newfound hinterlands for the city, and putting New York on the top of American cities, surpassing Boston and Philadelphia. In contrast to historical canal projects spanning millennia, the Erie Canal was completed in a mere eight years from commencement. Meanwhile, Manhattan metamorphosed into a gateway to an expansive and uncharted continent within a mere century and a half. Here, the city burgeoned with resolute determination and unprecedented velocity. Echoing the sentiment of Emily Dickinson's verse: "The Sun Came out It changed the world's countenance, Cars hurried back and forth, like messengers, Yesterday has aged."[2]

Indeed, "yesterday has aged," serving as the fundamental creed for American cities. Despite their manifold diversity, one can intuitively discern commonalities deeply ingrained in their nascent chronicles. None of the extant American cities predates 1500, and prior to the arrival of white colonists on the continental North America, cities were exceedingly rare. In the eleventh to fourteenth centuries, Cahokia flourished on the fertile lands near the confluence of the Mississippi and Missouri Rivers. At its zenith, it boasted a population of 14,000–18,000 inhabitants. Before 1492, it stood as the largest settlement in America, north of Mexico.[3] The Cahokia people erected sacrificial platforms, established a strict order, practiced slash-and-burn agriculture, and engaged in trade exchanges. By the thirteenth century, this once magnificent city mysteriously declined and then vanished, leaving only earthen altars standing in the desolate western winds. When Europeans arrived on the continental North America, there was no legacy of urban civilization here. From colonial era to the founding of the United States, newly built cities emerged as products of a world undergoing a series of modern revolutions. Waves of immigrants arrived from across the ocean, bringing with them updated modern systems and ideologies. It is this pristine continent that bestows

[2] Emily Dickinson, *Selected Poems of Emily Dickinson*, translated by Jiang Feng, Beijing: Foreign Language Teaching and Research Press, 2012, p. vi.

[3] Please refer to Timothy R. Pauketat, *Cahokia: Ancient America's Great City on the Mississippi*. New York, N. Y.: Viking, 2009 for the history of *Cahokia*.

upon them a fresh vitality that the Old World cannot provide, enabling them to multiply and prosper. Since its inception, cities have exuded a distinctly modern temperament: capitalism, nationalism, individualism, and liberalism, which not only showcase the power of the nation state but also strive for democratic ideals. Ultimately, cities were born out of the dream of limitless natural abundance.[4] Cities in the New World inevitably become innovations of the modern world—wallless cities, synonymous with modern cities.

For this reason, the shared landscape among American cities remains perpetually new and ever evolving, leaving the relics of yesteryears behind in the shadows of yesterday. To linger on the nostalgia of ancient capitals proves futile, as only continual innovation serves as the inexhaustible wellspring of progress and the cornerstone of our journey into the future. Here, there are no grand edifices or monuments beckoning reflection on a distant past. While Boston may boast the preservation of Paul Revere's former residence, dating back to around 1680, its significance pales in comparison to the weathered temples, halls, and churches of the Old World's ancient cities. Moreover, even in Boston, one of the oldest American cities, such antiquated structures are rare, let alone in cities like Chicago, Los Angeles, and Las Vegas, which emerged on maps only in the nineteenth or even the twentieth century.

The pursuit of novelty perpetually maintains American cities in a state of flux, embodying what economist Joseph Alois Schumpeter famously termed "creative destruction." In *Capitalism, Socialism and Democracy* published in 1942, Joseph Alois Schumpeter wrote: "The opening up of new markets, foreign or domestic, and the organizational development from the craft shop and factory to such concerns as U.S. Steel illustrate the same process of industrial mutation. The mutation, if appropriate, incessantly revolutionizes the economic structure, and creates a new structure form while incessantly destroying the old structure form. The process of creative destruction is the essential fact about capitalism."[5]

[4] Donald Worster believed that it was the material abundance of the New World that stimulated and ignited a succession of modern revolutions, including the scientific revolution, the capitalist revolution, the reorganization of world power, social and political revolutions within Europe and America, and the industrial revolution. Donald Worster, *Shrinking the Earth: The Rise & Decline of American Abundance*. New York: Oxford University Press, 2016, Chapter II.

[5] Joseph Alois Schumpeter, *Capitalism, Socialism, and Democracy*, translated by Wu Liangjian, Beijing: The Commercial Press, 1999, pp. 146–147.

According to Schumpeter, capitalism doesn't stagnate in a rigescent and fixed state due to the ongoing changes in both its social and natural environments. Capitalism must adapt to these changes, yet the method of adaptation isn't merely gradual adjustment; it entails what he terms "mutation" or the revolution of structure. It's worth noting that Schumpeter emphasized the need to adapt not only to the social environment but also to the changing natural environment. American cities, serving as hubs of trade, investment, and profit, embody the essence of "creative destruction" in both their appearance and structure. Eternity and sanctity seldom define American cities. Whether it is the land, ancestors, or monumental buildings erected for a particular religion, creed, or belief, everything is subject to constant destruction and reconstruction. Cemeteries yield to highways and even renowned historical landmarks like New York's Pennsylvania Station can be replaced without significant public outcry. Certainly, there are numerous exceptions demonstrating efforts in American cities to preserve their heritage through legislation or citizen initiatives. However, compared to European cities, American cities remain in a state of flux, both in form and essence. They continuously reinvent themselves, perpetually enacting the spectacle of "urban renewal," transforming and reimagining the urban landscape with relentless fervor. For them, modernization is an ongoing process. Perhaps this relentless pursuit of "newness" is the most palpable characteristic of American cities, always seeking novelty, even amidst its multifaceted connotations.

Every generalization inevitably brings exceptions, counterexamples, and even chaos, yet the evolution of American urban history stands distinct, awakening observers to its divergence from elsewhere. These cities, often new or newly developed, perpetually exist in a state of flux, their identities amorphous and open, transient and prodigal. While they showcase a mosaic of wealth, cleanliness, and fashion, they invariably reflect inequality within their egalitarian ethos. Though they hurtle along paths of innovation and upheaval, serious observers can discern amidst their landscapes the meanderings, the pauses, and the stagnations, uncovering the intricate tapestry of their evolution. They frequently undergo explosive growth, swiftly annexing new territories and erecting new structures. Then, as the construction machinery recedes and vanishes, urban vistas stabilize. But after a period of equilibrium, the cranes roar anew, signaling another cycle of rebuilding. Each phase etches itself onto the city skyline, mirroring its ebbs and flows, measuring its pulse. Every structure tells a tale of its past and present, every community reflects the shifts, every immigrant's journey weaves into the tapestry of races.

At the heart of shaping the complex urban history are the immigrants, continuously weaving into the fabric of American urban life. Their diversity in race, region, culture, beliefs, and customs sets the tone for multiculturalism in modern cities. No pre-modern city embraces multiculturalism as profoundly as American metropolises like New York, St. Louis, and Los Angeles. While ancient Rome or Chang'an might attract denizens from across the Old World, the role of these denizens was more ornamental than the protagonists. In a sense, North American cities have been international since the colonial era. Diverse cultural groups, whether by choice or coercion, have contributed to their construction. Native Americans, Africans, Europeans, and Asians bring diverse languages, perspectives, and visions for urban life, shaping not only coastal but also inland cities. St. Louis, nestled near Cahokia in bygone eras, was initially crafted by the French before falling under Spanish dominion. Through subsequent epochs, it housed black slaves, Spanish and French-Canadian fur traders, and German brewers. Even into the twenty-first century, it resonates with the rising tide of international multiculturalism.[6] As a consequence, the diversity and differentiation of cultures have emerged as defining features of American cities. Instead of a singular "culture," these cities boast a kaleidoscope of dozens of cultural influences, colliding and intertwining in a perpetual state of transformation. Undoubtedly, not all cultures command equal standing. Like any human society, hierarchy, order, and differentiation are entrenched realities in American cities. Distinct cultural cohorts have varied interests and perspectives, each yearning for expression and realization within the urban fabric. Just as the cityscape undergoes continual renovation, so too does its cultural milieu evolve, transitioning from periphery to center, from dominance to decline. Regardless of the prevailing circumstances, these cities extend a promise to newcomers: this is a space of openness, freedom, equal opportunity, and boundless adventure.

What binds together these culturally diverse groups is a shared pursuit—capitalism. Many are ensnared by the pursuit, from slaves to

[6] The most comprehensive work on the environmental history of St. Louis is the collection of essays edited by Andrew Hurley. Andrew Hurley ed., *Common Fields: An Environmental History of St. Louis*. St. Louis: Missouri Historical Society Press, 1997.

indigenous peoples; while others fervently champion its cause. Yet, regardless of individual aspirations, all are inexorably drawn into the common marketplace where their positions are disparate and unequal. Some ascend to the pinnacles of power, while others comprise the laboring masses of the multicultural mosaic, situated at the base of the economic pyramid. In the annals of the urban development of the Old World, there has never been a place quite like the American cities where capitalism asserted its dominance from the very outset.

Compared to American cities, European cities are steeped in a deeply ingrained aristocratic tradition. Take Munich, for instance. This modern city stands at the forefront of today's industrial technology. Yet, a century ago, it was still shaped by its elite ruling class. This class constructed opulent palaces and meticulously manicured gardens, while also commissioning various statues and monuments to showcase their illustrious achievements. The nobility didn't merely see themselves as wielders and consumers of power and wealth; they also saw themselves as stewards and patrons of the arts. Even long after their political and economic influence waned, the aristocratic ambiance of European cities persisted. There, the urban dwellers uphold a traditional consensus on the city's form and aesthetic preferences, and the decider to dictate and standardize the city's character. In America, however, it is vastly different. While capitalism reigns supreme, the bourgeoisie is a fluid and evolving group. While some aristocratic dynasties have formed in East Coast cities, the rise of the nouveau riche has been relentless, constantly supplanting the old guard and seizing control over urban development. Their demands are paramount, prompting American cities to channel their wealth, architectural prowess, and enthusiasm into the construction of banks, malls, and towering edifices. Though Americans sometimes bemoan their cities' perceived lack of character and taste, attempts to replicate European urban art often rely on sponsors whose wealth is often fleeting.

Yet, within American cities, there exists a counterforce to private capital. Particularly in the late nineteenth century, during the progressive era, all levels of government began emphasizing their roles in social management and assuming responsibilities for urban public affairs. The burgeoning middle class and the public sentiment they champion converge to form a collective voice that restrains the excesses of private capital. However, neither the government nor the middle class has ever been able to completely supplant private capital and dictate the course of urban development. Moreover, in many instances, they have been complicit allies with capital rather than adversarial foes. Security once characterized the cities of the Old World, with city walls providing both a material and psychological safeguard. In contrast, modern American cities, unencumbered by walls, prioritized growth in both thought and planning from its inception. This emphasis on growth represents the true consensus among capital, government, and the middle class in American cities—growth in volume, population, wealth, and power. As cities sprawled across the land, they invariably responded to the call of capitalism, with its ethos of unlimited growth.

Confidence in and pursuit of unlimited growth have been defining features of modern civilization. Natural resources have been capitalized and commoditized to maximize profit. Capitalist culture demands that all institutions, including the state and government, support and legitimize this pursuit. However, endeavor in such a pursuit requires more than just human ingenuity and technology; it necessitates a tangible material foundation. Environmental historian Donald Worster termed this the discovery of the "Second Earth." In the book *Shrinking the Earth: The Rise & Decline of American Abundance*, Worster wrote: "Beyond the Western Ocean lies a vast treasure trove of untapped natural resources. Alongside the material and human ecological revolutions came profound shifts in concepts, values, systems and behaviors. The discovery of the 'Second Earth' ushered in an era of unprecedented material abundance. Ultimately, it would overwhelm older civilizations with new resources and the freedoms they afforded. This era has endured for 500 years."[7] Around 1500, the dawn of the modern world commenced as human civilization finally began to unravel the millennia-old curse of ecological scarcity in the Old World. This liberation extended to break the economic, political, and cultural shackles born from scarcity. The Western Hemisphere, comprising vast continents and oceans, offered a promise of material richness akin to the "Second Earth," including soil, water, minerals, precious metals, energy, flora, and fauna, all of which laid the groundwork for a New World, new values, new cultures, and new cities. The foundation of the American wallless cities lies in the abundant nature of the "Second Earth."

Conquering the "Second Earth" and exploiting its boundless abundance to satisfy the insatiable growth demands of cities epitomize the

[7] Donald Worster, *Shrinking the Earth*: p. 13.

guiding beliefs of American urban development. In the relentless expansion of cities, akin to an amoeba, a diverse array of individual, including farmers, plantation owners, whalers, gold prospectors, buffalo hunters, agricultural suppliers, miners, railroad and telegraph workers, embarked from urban centers to partake in the relentless exploitation of nature, all the while ensuring the flow of goods back to the city through an extensive supply chain. Across the vast expanse of continental North America, indigenous tribes dwindled and vanished, while the absence of hereditary nobility or long-standing gentry, as seen in Europe, allowed cities to sprawl unchecked, encroaching upon and reshaping the surrounding landscapes without restraint. Cities often appear as unstoppable power, wreaking havoc on countryside and wilderness areas. Just in a century, the United States transformed from a cluster of British colonies confined to the Atlantic coast into a continental empire that sought to control two oceans, from the Arctic Circle in the north to the Gulf of Mexico in the south. By the late nineteenth century, the United States had emerged as one of the world's preeminent powers, with its environmental influence extending globally. As the locus of capital power, the city serves as both the genesis and culmination of environmental expansion. Nowhere else has any country allowed cities to accumulate natural wealth with such rapidity and intensity as the United States.

Cities enriched by abundant nature have grown so formidable that they foster an illusion suggesting they are solely the product of human ingenuity and culture. This perception posits that cities only interact with their natural surroundings upon their initial establishment, relegating the significance of the "Second Earth" to a mere provider of resources facilitating unchecked growth. However, delving deeper into the history of cities reveals that cities cannot exist in isolation from the natural world. The history of American cities has given rise to a unique ecological paradox, reshaping their culture, politics, economy, landscape, and ecology.

This first ecological paradox found in the American urban history centers on the interplay between freedom and nature, with the wallless cities symbolizing a commitment to the pursuit of freedom. Whether it was the religious fervor of the first Puritans stepping onto the continental North America, the relentless pursuit of wealth by fur trappers venturing deep into the continent's wilderness, the insatiable hunger for land driving thousands westward, or the vivid dreams of countless individuals gazing upon the Statue of Liberty as they sailed into the Port of

New York, the underlying impetus behind these endeavors was the quest for freedom in thought, economy, politics, and culture. The pursuit of freedom became a noble cause, yet it often veiled the organized massacre of indigenous people, the systemic enslavement of African Americans, the exploitation of women and child laborers seeking meager livelihoods in factories, the discrimination against Chinese laborers toiling in gold mines and railroads, and the oppression of Latin American migrant workers drifting through orchards and pastures. Under the guise of pursuing freedom, the suppression of the freedom of other races, groups, and individuals has been rampant throughout the modern world over the past five centuries. Urban dwellers, consciously or unconsciously, have become the primary consumers of this exploitative labor, even as they themselves remain ensnared in the pervasive traps of modern society.

The most significant legacy of contemporary historical inquiry lies in exposing the suffering of the enslaved amidst the celebration of freedom, prompting a critical reflection on the absurdity and hypocrisy underlying the value of freedom and prompting a reevaluation of its cultural paradoxes. Yet, so far, many historians have overlooked or sidestepped the related but more universal and profound ecological paradox—the conflict between the pursuit of freedom and the subjugation of nature. New cities continue to emerge and flourish, offering everyone the liberty to exploit and conquer nature openly, without the need for walls. They suggest that freedom is within reach, available for the taking. Everyone has enlisted in the war against the wild and perilous nature, and in their efforts to subjugate it, none can claim innocence. As nature yields and civilization revels in its triumphs, a tragicomedy unfolds—a paradoxical dance between culture and nature, where the domination of the natural world invariably leads to the subjugation of one group of people by another. At the turn of the nineteenth and twentieth centuries, amidst the fervent urbanization sweeping across America, a palpable sense of suffocation gripped the people. In search of newfound freedoms, they turned to preserving the wilderness, seeking solace in the mountains, or imbuing their cities with a touch of the natural, hoping to fulfill the unattainable promises of freedom and rediscover aesthetic and emotional redemption.

For modern cities, this process of naturalization holds profound significance. It heralds the emergence of a new modernity that redefines our relationship with nature in a more complicated way. Yet, it has failed to stem the tide of the second ecological paradox: the paradox between multiculturalism and ecological simplification. Multiculturalism stands as the crowning jewel of civilization of American cities. Amidst the enduring power struggles of race, class, and gender, multiculturalism has been strengthened and highlighted through cycles of creative destruction, flowing through the veins of culture of American cities, and infusing it with a perpetual vitality. In stark contrast lies the relentless simplification of urban ecology and the erosion of biodiversity. Humanity finds itself confronted with a crisis of species extinction. Unlike the preceding five mass extinctions, the sixth is distinguished by being the first driven by the actions of a single biological species in the annals of Earth's history. This crisis transcends mere ecological concerns related to the safety and health of the Earth's ecosystem, it poses an ethical quandary, challenging the very entitlement of humans, as a species, to manipulate and exploit others for their own gain. This issue extends far beyond the confines of American urban history, yet its implications reverberate deeply within the fabric of urban civilization. The ecological paradox it presents is one that American cities cannot evade. The causes of species extinction are manifold, with researchers often highlighting varied factors across different species and historical contexts. However, a prevailing consensus points to the pivotal role played by the disappearance and degradation of species habitats. Cities, alongside industrial agriculture that underpins operation and consumption of cities, have become the heralds of destruction for once-thriving ecosystems, sounding the clarion call of swamps, forests, lakes, and grasslands.

The loss of biodiversity wrought by urbanization extends far beyond the city limits, encompassing the entirety of our globalized world. In this vast expanse lies a new significance for the paradox between the burgeoning tapestry of multiculturalism and the fading echoes of biodiversity. The fascination and curiosity of the former toward diverse cultures not only foster tolerance and dynamism but also ignite a passion for global consumption. Under the influence of multiculturalism, foreign cuisines, clothing, artifacts, and even the cultures themselves have emerged as the latest trends in mass consumption. In the conclusion of the world history The Human Web: A Bird's-Eye View of World History, penned in collaboration with his father William, John McNeill articulated: "Human history unfolds as an evolution from simple homogeneity toward diversity, and then towards complex homogeneity."[8] Interestingly, the contemporary trend toward homogeneity aligns seamlessly with the process of urbanization. Paradoxically, multiculturalism not only fosters the simplification and homogenization of the Earth's ecosystem to cater to global consumption but also potentially leads to the homogenization and simplification of global culture, mirroring the strikingly uniform global urban landscapes observed at the outset of this chapter.

Consequently, this trend inevitably gives rise to the third and perhaps most fundamental ecological paradox: the paradox between the expanding city and the shrinking planet. The material underpinnings of modern American cities rest upon the abundance of what can be termed as the "Second Earth." This abundance has served as the catalyst for a revolution, innovation, and disruption in the modern sense. However, with the restructuring of the global economy, politics, and ecology, the "Second Earth" has seamlessly merged with the "First Earth." As a result, the vigor, passion, and confidence once derived from the abundance of the "Second Earth" have waned, signaling the demise of the modern world crafted amidst the bountiful and pristine natural landscapes. Presently, human civilization confronts a singular reality—a shrinking earth. Even as American cities continue to expand, driven not solely by geographical and demographic factors but also by the proliferation of urban lifestyles, consumption patterns, growth ambitions, and aspirations for innovation that permeate the global consciousness, cities become increasingly congested, voracious, and poisoned. In the wake of the modern world, cities have shed the need for physical fortifications against tangible foes and dissenting ideologies. Yet, whether in historical retrospect or present-day realities, is the city really as safe and solid as we imagined? Has nature been tamed by the city, eradicating all threats like viruses and "pests," floods and earthquakes, and catering to all necessities such as energy and food, health and beauty? Or have we consistently, since the dawn of urbanization, or even until now, been meticulously maintaining a fragile yet indispensable equilibrium with nature—a power not engendered by human culture?

This book narrates the history of American cities and their intertwined evolution with the natural world, encapsulated within the ecological

[8] J. R. McNeill & William H. McNeill: *The Human Web: A Bird's-Eye View of World History*, translated by Wang Jinxin, Song Baojun, et al., Beijing: Peking University Press, 2007, p. 312.

paradox they've engendered. It stands as an environmental history of American cities. Environmental historians often weave tales centered around a specific "place." However, given the vast expanse of the United States, it's challenging to evoke what Wallace Stegner, a renowned American author of the West, referred to as "the sense of place"—a multi-faceted perception that builds up over time. Attaining this sense requires a tactile engagement with the land, weather, calamities, and abundance through personal labor. It entails understanding the nuances of a place's seasons, skies, scents, and sounds through daily life, grasping the unspoken past through tapping into ancestral memories, and capturing the subtlest changes through local poets.[9] Many environmental historians strive to unearth the sense of place when delving into the history of a place. Yet, perhaps only those rooted in a village or town for generations can fully embody it. For those navigating the currents of modern society, it remains an elusive perception. Even for urban dwellers born and bred in a specific place, such a sense of place becomes a luxury. American cities, like most modern metropolises, undergo substantial transformations within a single generation. In this rootless modern world, Wallace Stegner's sense of place may feel antiquated and distant. Hence, there arises a need for a new sense of place, one that aids environmental historians in comprehending a place's ecological journey.

The United States, though a political entity demarcated by geographical boundaries, transcends mere politics. It embodies a cultural, geographical, and ecological identity. While nature's evolution often surpasses human-made boundaries, these artificial boundaries play a pivotal role in shaping the interplay between culture and nature. In the modern world, nation-states delineate the most significant boundaries in the artificially constructed world, while the artificially constructed world also inhabits its own ecosystems, even becoming a vital power in shaping these ecosystems. Within these boundaries, American cities maintain intricate connections with their hinterlands and share certain commonalities with other cities. It is these connections and commonalities that endow them with a unique sense of place, rooted in cultural, ecological, and

[9] Wallace Stegner, "The Sense of Place", http://www.pugetsound.edu/files/resources/7040_Stegner,%20Wallace%20%20Sense%20of%20Place.pdf.pugetsound.edu/files/resources/7040_Stegner, %20Wallace%20%20Sense%20of%20Place. pdf, reprinted with permission from Random House, Inc. which was logged in on August 1, 2020.

historical cores, laying the foundation for the crafting of their collective history.

For a long time, urban history, like other historical disciplines, was merely chronicled as human affairs. However, this changed dramatically with the emergence of the environmental history of cities. Historians, such as Joel Tarr and Martin Melosi, began to focus on the environmental evolution within cities, seeking to understand the operational mechanisms of urban areas based on the relationship between the infrastructure, such as waste disposal, sewage, sanitation, and the environment. By 1991, William Cronon's seminal work, *Nature's Metropolis: Chicago and the Great West*, revolutionized the paradigm of urban environmental history research and caused a stir in the field of environmental history. In this book, Cronon shattered the confines of cities, placing them within a broader ecological and historical context for discussion, thereby forging new frontiers in environmental history research. He said: "The binary opposition between cities and countryside, and between human beings and nature, obscures our vision, preventing us from seeing the deep unity between these artificially divided concepts. Focusing solely on cities, we often perceive it as the ultimate symbol of 'human' conquering 'nature,' while overlooking the fact that urban dwellers will continue to rely on the non-human world, much like they depend on each other. For people everywhere, many aspects of their lives and relationship, whether in cities or countryside, factories or wilderness, workshops or offices, cannot be encapsulated by the simplistic notion of singular humans conquering singular nature. They're not a specific person. As we neglect them and their histories, we enclose ourselves within walls and isolate ourselves from the vast ecosystem, including our urban homes of which we are a part."[10]

This excerpt, penned nearly three decades ago, continues to offer enlightenment to contemporary environmental historians. *Nature's Metropolis: Chicago and the Great West* adeptly situates Chicago's growth within the broader context of the Great West, charting new heights in environmental history research by exploring the ecological and economic interplay between countryside and cities, a pinnacle yet unrivaled. However, regrettably, constrained by personal interests and the limitations of book-length, Cronon didn't fully integrate the diverse city dwellers with the ever-evolving ecology of the urban hinterlands. Where did they

[10] William Cronon, *Nature's Metropolis: Chicago and the Great West*, translated by Huang Yanjie et al., Nanjing: Jiangsu People's Publishing House, 2020, p. 22.

come from, how did they eke out a living, why were some impoverished, how did others thrive, and to what extent were their urban planning endeavors (in a broad sense) driven by long-term considerations and prudent assessments or forced adaptations to natural constraints? Similarly, this book didn't narrate the environmental experiences of ordinary individuals: the injustices they face and the aspirations they harbor. Is their longing for nature solely a product of Western cultural education, as Cronon suggests, or is it a universal human instinct?[11]

Despite Cronon's repeated warnings to his successors against oversimplifying the emergence of cities as a narrative of conquering nature, urging them instead to recognize that "urban dwellers will continue to rely on the non-human world," *Nature's Metropolis: Chicago and the Great West* ultimately portrays the conquest of vast plains and indigenous species including human, by urban civilization. It narrates how the "Second Nature" overwhelms and potentially supersedes the "First Nature." Cronon posits that within the "First Nature," besides climatic factors, ecology is fundamentally regional, shaped by local landforms, hydrology, and biodiversity. In contrast, the "Second Nature" embodies the logic of capital, where central cities exert enduring market influence over their peripheries, severing the link between hinterland ecology and its habitat environment. Undoubtedly, this perspective presents a formidable challenge. However, if we define nature as a power generated by non-human entities, a power to which humans inherently belong, then the so-called "Second Nature" merely represents the interactive

[11] Several works have delved into the intersection of racial issues, urban pollution, urban planning, and waste disposal in Chicago from the perspective of environmental justice. Among these, Sylvia Washington's Packing Them In: An Archaeology of Environmental Racism in Chicago stands out, exploring the roots and evolution of environmental racism in Chicago through the prism of Foucault's power/knowledge relationship. (Sylvia Washington, *Packing Them In: An Archaeology of Environmental Racism in Chicago, 1865–1954*. Lanham, Maryland: Lexington Books, 2005.) Brian McCammack's Landscapes of Hope: Nature and the Great Migration in Chicago take a fresh perspective in examining the relationship between African Americans and nature. Unlike past explorations focusing on neighborhood movements and grassroots environmental activism, McCammack's work delves into how black immigrants in Chicago seek, fight for, build, and cherish natural and scenic spaces influenced by environmental memory. This book garnered several prestigious awards in 2018, including the Turner Award (First Book Award) from the Organization of American Historians and the Marsh Award (Best Book Award) from the American Society for Environmental History. (Brian McCammack, *Landscapes of Hope: Nature and the Great Migration in Chicago*. Cambridge: Harvard University Press, 2017).

dynamics between non-human power and capital, rather than an unequivocal triumph of the capital over the non-human power. Cronon is not oblivious to this complexity. He wrote that the "Second Nature" is indeed an abstract concept, merely representing numerous entities and their more diversified relationships. However, despite this acknowledgment, Cronon hesitates to fully abstract the concept of the "Second Nature," suggesting that "abstract concepts still possess some form of entities." He pointed out that cities like Chicago and its expansive hinterlands embody a materialized form of "Second Nature," in which capital emerges as the paramount dominant power.[12] Given this, the distinction between the "Second Nature" and the artificial environment seems to blur, since all artificial environments still rely on materials and energy from the non-human world for construction, what then is the significance of the concept of the "Second Nature"?

Despite the pervasive spread of capital's logic over the past two centuries, the power of nature has persistently stood in opposition. Chicago's ecological narrative does not culminate with the "White City," nor does it succumb to the dominance of capital's logic, simplifying itself into Cronon's notion of the "Second nature." On the contrary, as an autonomous and spontaneous power, nature persists across cities, countryside, and the far-reaching hinterlands, continually shaping the ecology, economy, and individual lives within cities.[13] Furthermore, cities are not solely the grand creation of city dwellers. Countless farmers, shepherds, and miners residing far from cities contribute to the construction of cities. When delving into urban history, their experiences, labor, migrations, hardships, or triumphs warrant historians' attention. Likewise, the history of cities cannot be reduced to a narrative solely focused on the human species or the capital power they instigate. A vast, non-human power and the myriad forms of life and inanimate existence it engenders have always been integral to the evolution of the history of cities. These forces,

[12] William Cronon, *Nature's Metropolis: Chicago and the Great West*, p. 379, pp. 381–382.

[13] "White City" refers to the building groups of the World's Columbian Exposition Chicago World's Fair in 1893, whose chief designer is Daniel Burnham and the landscape designer is Frederick Law Olmsted. Cronon's narrative in *Nature's Metropolis: Chicago and the Great West* reaches its conclusion with a pilgrimage to the "White City". For Cronon, the "White City" signifies not only the splendid ascent of Chicago as a global metropolis but also the complete integration of cities and countryside.

intertwined with the city's fluctuations, cultural ideologies, and reciprocal interactions, collectively form the ecological paradox of evolution of American cities. This is the main point of this book.

Since the pioneering works of Joel Tarr, Martin Melosi and Cronon, the research into environmental history of American cities has flourished over the past three decades. Nearly every significant American city boasts multiple environmental history biographies, with some cities like New York, Boston, Chicago, and Los Angeles featuring five to six or more. However, despite these advancements, urban environmental historians have largely overlooked the United States as a geographic entity, neglecting to thoroughly contemplate the intricate relationship between cities and nature.[14] Undoubtedly, the environmental history of American cities is a comprehensive topic, offering an opportunity to reinterpret various aspects of American history from the micro to the macro, from cultural ideology to political economy, within its ecological paradox. Consequently, this book inevitably faces the task of making choices in its discourse. For instance, transportation and energy serve as pivotal factors in ushering in the era of American cities. They not only act as the fundamental driving force connecting cities to their hinterlands and sustaining urban operations but also form the cornerstone for cities to attain dominance and reshape natural ecosystems, including the atmosphere. However, lacking the expertise to offer innovative perspectives on these two topics, I opt to embed the analysis within the text rather than dedicate a separate chapter to discussion. In terms of personal research focus, a significant sacrifice of the book is its lack of discussion on the fragility and fear experienced by cities in the face of disasters. Initially envisioned as spanning two chapters, this exploration was curtailed after

[14] Currently, the comprehensive investigation into environmental history of American cities is epitomized by *Liberal Dreams and Nature's Limits: Great Cities of North America since 1600*. This book, organized chronologically, delves into the interaction between nature and the economic, political, social, residential, and occupational environments in five major North American cities: Philadelphia, New York, Chicago, Los Angeles and Toronto. However, while the book meticulously explores political, economic and social life within the city, it largely neglects the relationship between cities and countryside or wilderness environments in the process of urban development. Moreover, its examination of the ideological underpinnings of urban development and its relationship with nature is confined to economic thought, overlooking the broader ideological contexts of urban planning, urban parks and anti-urban development movements. (James Lemon, *Liberal Dreams and Nature's Limits: Great Cities of North America since 1600*, New York: Oxford University Press, 1996).

encountering Joanna Dyl's groundbreaking work, *Seismic City: An Environmental History of San Francisco's 1906 Earthquake*, published in 2019. Dyl's insights on the nature and power, her analysis of the intricate relationship between natural disasters and cities built upon technological self-assurance, and her examination of diverse reactions to natural disasters among different racial and socioeconomic groups resonated deeply with my own thoughts.[15]

After omitting the chapter on disasters, in addition to the introduction,[16] this book is organized into four main parts, spanning seven chapters in total. The first part *City Writing*, meticulously traces the evolution of the environmental history of American cities. On this basis, it reflects on existing paradigms, encountered challenges, and potential future research directions. The narrative of the environmental history of American cities, on the basis of the ecological paradox of American cities, necessitates transcending urban boundaries and forging connections between the history of cities and the broader history of ecological evolution. Firstly, it delves into the temporal dimension, uncovering the interconnectedness between the history of cities and the deeper and more enduring natural history. Secondly, it explores the spatial dimension, or the realm of energy exchange, intertwining the urban ecosystem with the hinterland ecosystem that it both consumes and changes. This exploration aims to unravel the intricate interplay between economy and ecology. Thirdly, it explores the ideological dimension, probing the diverse perspectives of different societal groups on the relationship between culture and nature in the urban age. This exploration encompasses concerns and awareness about resource scarcity in consumer societies, the endeavor to reconstruct the aesthetic imagination of the city, the reflection on the civilization orientation of cities, and the consideration of the possibility and form of nature in the urban age amidst the

[15] Joanna L. Dyl, *Seismic City: An Environmental History of San Francisco's 1906 Earthquake*. Seattle: University of Washington Press, 2019. The contents of this book, along with other works of environmental history related to San Francisco, will be further discussed in the chapter dedicated to San Francisco; hence I'll refrain from elaborating here.

[16] The contents of this part were originally published under the titles "City without Borders: From the History of American Cities to the Environmental History of Cities" and "Intricate Trajectory: Rewriting the History of Cities in Nature". They underwent comprehensive revisions and adjustments when the book was completed.

transformation of countryside and nature into the spiritual foreign land of urban dwellers.

On this basis, the second part, "Growing Cities," the third part, "Expanding Cities," and the fourth part, "Ideological Landscape of Cities," respectively, delve into the evolution of environmental history of American cities from the internal, external, and ideological perspectives of cities. It's essential to note that discussing these three levels separately adopts a reductionist narrative approach. In reality, due to the flow of population, energy, and ideas, they are in a constant interactive state, forming an inseparable whole. Each part comprises a general chapter and a chapter on urban fields, intending to offer a personalized and locally flavored comprehensive analysis of representative cities.

The second part, "Growing Cities" focuses on the internal dynamics of cities. The first chapter, "Crowded, Poisoned, and Beautiful," delves into the interactions between people from diverse cultures and various species and their material environment. It explores how new energy and technology infiltrate cities, how cities oscillate between environmental challenges and solutions, and how to seek health and beauty amidst crowded and poisoned cities. Presently, discussions about race, class, and gender, akin to other fields of American history research, dominate the mainstream discourse on the internal environmental history of American cities. These topics inevitably permeate the considerations of this book. However, even for cities whose ecosystems are dominated by human, narrating human relationships should not overshadow reflections on the relationship between human and the rest of nature. Environmental justice should not extend beyond highlighting the differences in environmental experiences among different groups but also incorporate environmental ethics concerning other species. Guided by such reflections, this chapter endeavors to explore the environmental history of American cities from a perspective distinct from Martin Melosi's *Sanitary City*, reviewing the tensions and transformations of freedom, health, and morality throughout the history.

The urban field associated with this chapter is Boston, a city nestled atop a mountain. Situated on the east coast of the United States, it stood as one of the earliest cities established in North American colonies. It served as an early hub of commerce and industry, a cultural epicenter of the nation, the cradle of significant intellectual trends such as transcendentalism and pragmatism, and the birthplace of influential intellectual movements that are most closely associated with American urban thought.

From the colonial era to the contemporary era, the historical trajectory of Boston is intricately linked with key milestones such as the American Revolution, industrialization, and progressivism. Throughout its history, Boston has grappled with profound reflections on the relationship between culture and urban environments. After a concise overview of Boston's environmental history, this chapter aims to delve into Boston's engagement with various forms of nature within the city, seeking to redefine the concept of civilization in the urban age.

The third part, "Expanding Cities," focuses on the interplay between the cities and the world beyond the borders of cities. It reviews how wallless cities have influenced the natural world outside the confines of cities, while remaining subject to the constraints and influences of nature. The first chapter of this part commences with an exploration of urban population movements, highlighting the phenomenon of ecological teleconnection and delving into the ecological ties between the migration patterns of the American urban population and the environments from which they originate. Urban dwellers rely on essential resources such as water, food, and energy for both urban functionality and personal sustenance. Moreover, the urban lifestyle fosters additional consumer desires, shaping the dynamics of urban consumption. The population movements spurred by urbanization entail not only numerical and symbolic changes or changes in family and personal destiny but also transformations in the interconnected ecosystem that is inseparable from cities. The process of urban expansion is also a process of constructing its hinterland. I call this process as "making bubbles." In the relationship between a city and a hinterland in a traditional society, the size of a city is largely determined by the ecological carrying capacity of its surrounding hinterland. In the process of "making bubbles," capital is added to the production and consumption chains between the city and the hinterland. By virtue of modern transportation technology, the hinterland of a city is no longer just a region closely connected with it in the geographical sense, but expands to the entire continent, even the entire planet. During the expansion of the city hinterland in all directions, the ecological independence of the city itself is also lost day by day, thus relying more and more on the ecological system of other distant places to support their own operations. However, the "bubbles" are fragile, and the hinterland is not stable. With its limits, shrinkage, and drastic changes, the nature constantly challenges the structure of the city and its hinterland, forcing the urban culture to change and adapt to the new environment. Meanwhile, the alienation caused by the continuous expansion of the city stimulates the emotional desire for the natural world. I call this situation as "emotional rift," which once again urges urban people to protect those places far away from the city.

This section of the urban fields will be conducted in Kansas City, also known as the "City in the Sea of Grass." This city is located in the center of the United States. When we are observing the map of the United States, we can find that this city is situated in the center of the country's geographical location regardless of south-north direction or east–west direction. However, American people generally believe that this city is located on the edge of their country. I have ever lived and studied at the University of Kansas in Lawrence for six years. Lawrence is known as a college town. It took me 30 minutes to drive there from downtown. In my heart, this city is like my hometown. More importantly, this city is not so universal as Chicago and other metropolises. Its development and its relationship with nature represent the common characteristics of those central and western cities booming in the agricultural hinterland. Moreover, it reflects the country's ambition, confidence, success, and frustration while developing itself in the plentiful Western natural environment. For the above reason, I hope to continue this chapter by focusing on the relationship between Kansas City and its agricultural hinterland. Based on this relationship, I reflect on the paradigm established by Cronon in *Nature's Metropolis* to review the relationship with nature after the founding and growth of cities.

Ideological landscape of cities is the highlight of the fourth part. This part focuses on the discussion of the relationship between the city and nature, rather than all broad thoughts about the city. Since the founding of the United States accompanied by a generation of intellectuals and political elites such as Thomas Jefferson, it has been amid the dispute over encouraging or suppressing city development. Successors of anti-urbanism thoughts are never absent from the ideological spectrum of the United Stations. When cities develop to different stages, the theme has changed. Since Frederick Law Olmstead, urbanism has become a new and unstoppable trend of thought, and cities have been given new meaning and value. City and city life are unavoidable. Under such circumstances, urbanists shift their focus on how to make cities better. In fact, there is no clear distinction between urbanism and anti-urbanism. Behind their emergence and evolution is always the ecological paradox between city and nature, and the paradox absolutely means dialectical and mutual shaping.

The thinking over anti-urbanism always forces city proponents to adapt to the idea of a beautiful city in a subtle way. In turn, the voice and action of city reformers stimulate people's imagination of wilderness and rurality.

Urban fields mentioned in this part will extend westwards to San Francisco on the Pacific coast. If Boston is always facing and thinking about the natural wealth of the Atlantic Ocean and the European civilization on the other shore, Kansas City is the blueprint for conceiving American cities on the continent of North America under the vast western sky, and the people of San Francisco breathe the humid air from the Pacific Ocean, looking out at the widest waters on the earth and the unfamiliar Asian civilization on the other shore. The environmental history of this city includes all conceivable elements, from the influx of people from eastern, western, northern, and southern regions, to the mining development brought by the gold rush and the irrigation agriculture formed by the river of the Jimi Empire, to the confrontation of modern cities supported by technology against various natural disasters such as earthquakes and wildfires, to the vigorous development of urban green politics, and the construction and protection of urban parks and national parks. Each of the perspectives can stimulate the ambition and imagination of urban environmental history writers. Fields are part of my thoughts. My stories about San Francisco linger in the poetic imagination of its magnificent bays, rugged shores, continuous mountains, gentle morning fog, and undulating urban landscape. Such imagination is a response to the natural landscape that attracts people here, an inheritance of and rebellion against the rigid, mainstream eastern urban civilization, and a meditation on the unique taste of nature and life in exotic Asian cultures.

American cities may still be on this shrinking planet for many centuries. They may become greener and more ecological. They may not do so. However, some of them are pulsating, growing, or dying. Fundamentally, they are communities consisting of different kinds of life, such as forests and estuaries, which consume energy and realize metabolism. American culture is keen to think about the future, and cities of tomorrow have become the theme of every turning point planner. Carl Sandburg is the most faithful successor to Whitman and a passionate proponent of American cities. He wrote the following words in 1918 before the upcoming first high moment of America in global history:

I talk about new cities, new people/I tell you the past is a bucket of ashes/I tell you yesterday is a ceased wind/A sunset that has descended in the west/I tell you there's nothing else in the world/But a vast ocean full of tomorrows/A vast sky full of tomorrows.[17]

No one knows exactly what the scenario will be, but history is not over, and the city that breathes again in the seas and skies of tomorrow must confront the ecological paradox it creates.

[17] Carl Sandburg, "Prairie". The translated text is excerpted from Zhao Yiheng's compilation: *Anthology of Modern American Poetry (Volume I)*, Beijing: Foreign Language Teaching and Research Press, 2019, p. 175.

PART I

City Writing

CHAPTER 2

Rewrite the History of Cities in Nature

From 1833 to 1836, Thomas Cole, the founder of the Hudson River School in the United States, focused on the completion of a large group of paintings—"The Process of Empire." In the first painting with the shaded sky, dense forest, and steep mountains, Cole named it *Savage State*. In the second painting named *Arcadia or the Arcadian or Pastoral State*, the spring sun is rising and the grass is green. In the third painting, the summer sun is shining brightly over brilliant buildings, so it is called *the Consummation of Empire*. The fourth painting is full of stern autumn sceneries and dense smoke of wars, so it is named *Destruction*. In the last painting, we can observe the winter moon, mournful wind, and broken walls, so it is named *Desolation*. Immigrated from England to the United States, Cole, like most of the early immigrants, longed for the vitality of the new continent, hoping to cast off the dust, disease, and decaying morality of cities in the Old World. He settled in New York State, but not in New York City, which was already expanding at that time. He lived in the upper state of Catskill. Till now after nearly two centuries, it is still very much like what he painted in his group painting *Arcadian*, with fragrant grass falling everywhere, fertile fields, and beautiful pools constituting idyllic scenery. In his view, such an intermediate state is the most brilliant and fresh period of civilization and the most harmonious and intimate period with nature. In the city depicted in *the Consummation of Empire*, the land is covered by bricks and tiles, the mountains are shaded

S. Hou, *Wallless Cities*,
https://doi.org/10.1007/978-981-97-7827-0_2

by tall buildings, corruption, anxiety, and violence are surging under the prosperity and civilization is declining.

1 City and Nature: An Unbreakable Tie

Cities seem to be ubiquitous and extremely common today. For a long time in human history, they have not been the place where most people live in. At that time, human homes were the caves dug on the loess slopes on one side of farmland or the trails opened by nomads on the land where they picked and hunted for food. Sometimes, their dwellings were a crop-land completed after great hardships in pioneer work and changed with the seasons. Sometimes they were a small village with several brick houses, isolated from another village on the other side for a long time, hearing the sound of chickens and dogs with rare face-to-face interactions. Cities exist, but they are too far away from the daily lives of most people.

The nineteenth century was the first exclusive era for cities in human history. Suddenly, massive people rushed into cities from rural areas to seek a new living space and lifestyle. By the turn of the twentieth century, the urban population of most developed countries had far exceeded the rural population, and the vast majority of the urban population lived in metropolises. In the twenty-first century, various inventions of the modern industry have liberated more people from agriculture. Whether in developed or developing countries on every continent of the earth, a large number of people continue to migrate to cities, leading to an irreversible trend. Meanwhile, more and more food, water, and other resources were transported into cities from rural areas to support the urban operation. However, the urban population is basically not involved in the production of these materials they depend on for survival. They rely on everything endowed by the earth in an indirect way mediated by layers. An ordinary urbanite has almost no direct contact with the earth on which they live, or with the natural forces and resources that sustain their lives.

In less than a century after the completion of Cole's group paintings, the world civilization experienced the rise and fall of empires, the rise of cities, and finally the outbreak of World War I. The German Oswald Spengler, who was at the center of the war, published the book *Decline of the West* on the impasse of civilization in his eyes after the war. Featuring a distinct mysticism tendency, Spengler was unequivocally opposed to the theory of evolution, so his outlook on history was significantly different from the view of progressive history that emerged and gradually became the popular outlook during that era. This difference lies not only in his challenge to western centralism, but also in his analogy of civilization as a kind of superorganism, which goes through the process of birth, growth, and death, and consummates the cycle of history in the sequence of spring, summer, autumn, and winter. Spengler's description of the course of civilization echoes the group paintings created by Cole a hundred years ago and can almost become a textual annotation of the paintings. For him, agriculture gives birth to a new emotion, "hostile nature becomes a friend, and the land becomes earth mother", and "culture itself grows in turn from the earth's landscape, constantly renewing and strengthening the intimate relationship between human race and the land... Only in a civilization with great cities can we once again show contempt for these roots and let ourselves free from these roots." Spengler believed that every city generated in the adolescence of every civilization is not a real city, but a derivative of rural areas, because the two are the same in spirit and are rooted in the land; only the cities in the late civilization can develop the independent intelligence and soul of cities, making rural areas become their own subsidiaries, constantly exploiting these rural areas, and eventually leading to the complete bankruptcy of the latter. Both nature and rural areas have become the alien land of cities: "The emergence of cities worldwide marks the end of the era of organic growth and the beginning of the inorganic, uncontrolled and infinite accumulation process. The contemporary world also sees artificial and mathematical products fully isolated from the land, i.e. the cities designed by urban architects, which are purely a kind of mental satisfaction."[1] Finally, a civilization ends in the carnival of urban intelligence.

In Spengler's argument, the reason why the "cities of late civilization" are separated from the land is that the intimate relationship between cities and nature collapses at the spiritual level. Therefore, the buildings in cities no longer grow out of the land, but only stand on the land. From this point of view, although Spengler biologicalized civilization, the basis of his historical theory is completely idealistic, ignoring the materiality of history. However, his historical interpretation is more accepted by later historians than his historical cycle theory. Since the publication of this highly controversial book, his outlook on civilization history has

[1] Written by Oswald Spengler, translated by Wu Qiong: *The Decline of the West*, Perspectives on World History, Volume II, Shanghai: SDX Joint Publishing Company, 2006, pp. 78–79, p. 88.

produced a profound impact, whereas his historical cycle theory has been criticized now and then. However, whether its followers or critics focus on the rationality of its civilization cycle theory or the pros and cons of his individual civilization discussion, they seldom discuss the relationship between the two core concepts of his argument, "city" and "nature." This pair of relationships has not attracted the research attention of most historical researchers. In the opinion of most city history scholars, cities are purely the product of civilization and nature is beyond cities. Whether such separation will lead to the final collapse of cities and even civilizations is a question for discussion. However, the separation itself seems to be a self-evident prerequisite for city history and urban criticism. It reflects the general perception in the era of urbanization.

In fact, whether it is the "homesickness" that pervades the public feelings, or the insight of "urban–rural estrangement" that penetrates every piece of paper from scholars, it is just the psychological construction of the so-called natural "alien land" by urban people. To some extent, they narrate the changes in cities, rural areas, and nature at the level of spirit and shallow landscape, but they cannot be fully reflected in the deep ecological and historical levels. An inextricable and unbreakable tie exists among cities, rural areas, and nature. The new study of urban environmental history requires that cities should be put back into the history of vast ecosystem extension, and that the power change, economic rise and fall, and ideological ups and downs of cities should be reviewed on the basis of material flow, resource competition and ecological change, as well as the history of interaction between a city and its rural hinterland. Only in this way can we more clearly realize a basic fact about cities: cities have never been just the product of human civilization, but the result of the interaction between civilization and natural forces.

On the issue of exploring the historical relationship between cities and rural areas, Marx initiated a more material proposition, the "metabolic rift" summarized by the ecological Marxists later.[2] Marx believed that after the second agricultural revolution based on chemical agriculture, the metabolism between urban and rural areas had an "irreparable rift." While discussing the relationship between industry and industrialized agriculture in Volume I of *Capital*, he expounded that "Capitalist production

[2] John Bellamy Foster, "Marx's Theory of Metabolic Rift: Classical Foundation for Ecological Sociology," *The American Journal of Sociology*, 1999 (2): pp. 366–405; refer to Foster, *Marx's Ecology: Materialism and Nature*. New York: Monthly Review, 2000.

makes the urban population gathered in the major centers more and more dominant. On the one hand, it gathers the historical power of society but destroys the material exchange between human race and land; in other words, the components of land consumed by people in the form of food and clothing cannot return to the land, thus destroying the eternal natural conditions of the lasting fertility within the land."[3] Here, I will not discuss the rationality of ecological Marxists' efforts to "green" Marx's thoughts, because no matter whether Marx was sympathetic to the sacrifices made by nature in the process of capitalism expansion or not, he clearly realized that the change of urban–rural relationship is not only dependent on the change of human identity but also can be explained by the change of productivity and class structure. Behind the change in this pair of relationships lurks a major shift in the ecological relationship, with the material exchange between urban and rural areas changing from the former organic metabolism to the unilateral energy exploitation in rural areas by cities. The "waste" produced by the consumption of food in the city is fed back to the land in the form of fertilizer through the original metabolic process, and then the fertilizer is absorbed by the land and converted into food; with the replacement of organic fertilizer by chemical fertilizer in chemical agriculture, the potential fertilizer becomes real waste, and the chain of material exchange between cities and rural areas is broken.

"Metabolic rift" brings self-evident inspiration to the understanding of the evolution of urban and rural ecological relations. It is worth noting that although Marx devoted himself to the dual relationship between urban and rural areas, emphasizing the double oppression of land and labor in the process, the thinking over metabolic rift is also beneficial to studies on the environmental history of cities. For example, after the rise of bacterial science, the transformation of the modern city from an organic city to a sanitary city (sanitary city), the transformation of urban basic services such as water supply, sewage, and garbage treatment, from the physical process of transportation and transfer to the chemical process

[3] Written by Marx, translated by the CPC Central Compilation and Translation Bureau for the Works of Marx, Engels, Lenin and Stalin: *Capital* Volume I, *the Production Process of Capital, the Complete Works of Marx and Engels* Volume XXIII, Beijing: People's Publishing House, 1972, p. 552.

of disinfection and sterilization, can be used for theoretical reference in the proposition of metabolic rift.[4]

Although there is a great difference between the "metabolic rift" and Spengler's "end of the era of organic growth" in terms of materiality, Marx still concentrated on the rift. When he pointed out that capitalism disenchanted nature, he neither intended to restore the divinity of nature nor was he interested in reviewing nature with a holistic and organic view. In Marx's analysis of cities and nature, nature is reduced to land and resources on the land, and the relationship between the two is simplified to how people use the land and resources and how they are used. Therefore, Marx's metabolic rift theory cuts off only the organic exchange between cities and rural areas at the agricultural level, but it does not reflect the connection between them and the ecosystem in which they live together. Although the feet of urbanites were just plucked from the soil soaked with urban manure, their bodies were still closely related to the air they breathed, the food they ate, the energy they consumed, and the organisms they lived with. The power flow and hierarchical order in urbanized society are largely based on the allocation of resources, and the lack and abundance of resources play a pivotal role in defining human relations. Even the thinking of the urbanized era over the reflection of civilization or the inspection of power is more or less intentionally or unintentionally subject to and influences the changes of the surrounding environment.

2 Evolutionary Urban Ecosystem

While Marx focused on human social relations, another intellectual community became interested in the relationship between living things and their environment. In 1866, just a year before Marx published Volume I of *Capital*, a new term appeared on the map of human knowledge, ecology. In several decades since its publication, this term developed into a mature discipline. Like all other disciplines, "ecology" has been constantly refined and alienated in the process of developing itself and establishing the legitimacy in the scientific community. Meanwhile, various branches and interpretations have appeared, and various new concepts have been initiated.

[4] Martin Melosi, *The Sanitary City: Urban Infrastructure in America from Colonial Times to the Present*. Baltimore: Johns Hopkins University Press, 1999.

An important contribution made by ecology lies in observing evolution in nature, thus historicizing nature.[5] Paleogeologists and paleontologists are exploring the history of the earth in nature, from Archaean era to Mesozoic era and then to Cenozoic era, from unicellular organisms to dinosaurs and then to Homo sapiens. However, such natural and historical changes seem to stay in the distant past, also the "prehistoric era" is often referred to by historians. In the era studied by historians, also 10,000 years after the emergence of agriculture or thousands of years after the emergence of writing for more historians, the evolution of nature ended and history became the forbidden domain of human civilization, the time of nature and the time of civilization became two separate dimensions, and the environment became the static structural existence behind civilization.[6] The emergence of ecology broke the cognitive deadlock over the time of nature, while the emergence of environmental history reintegrates the time of nature and the time of civilization into a whole, making the former redefine our understanding toward the latter to a great extent. For example, the traditional Chinese time sequence follows the rise and fall of dynasties, but after the time of nature intervened in the range of thinking, the historical narrative with dynasties as the basic time unit was no longer a matter of course. The big issues include climate change while the small ones include the reproduction of germs, and the issues relating to humans such as the diversion of the Yellow River, the domestication and spread of American species, the soil loss, the imbalance of flood and drought, both tend to make the evolution of history out of the wishful control of dynastic civilization system, and rushed in the natural and cultural entanglement full of time turbulence and uncertainty.

The evolutionary history of cities also requires a review in this new time dimension. Before the rise of cities, any piece of land had a long history of evolution. Its geographical location, geological structure, soil, hydrology, rocks, plants, animals, and microorganisms have gradually formed in the ancient world where Homo sapiens species have not yet appeared. After a long process of evolution, when human beings began to settle down and build cities, they determined to a large extent whether cities could appear,

[5] Donald Worster, *Nature's Economy: A History of Ecological Ideas*, translated by Hou Wenhui, Beijing: the Commercial Press, 1999, pp. 485–486.

[6] Fernand Braudel & Immanuel Wallerstein, "History and the Social Sciences: The Longue Durée," *Review*, 2009 (2): pp. 171–203. The original French text was published in the yearbook in 1958.

the form of cities (a commercial hub or a military fortress, a mining base or a distribution center for agricultural products, a political center or a cultural center); the scale of its growth requires the mobilization of financial, material, and human resources; they also determined whether the population was mainly accumulated through immigration or through the natural growth of the local population.

The overall ecological history of a region and the history of a city are indispensable when the foundation of the city was laid, but this does not mean that it is only the curtain of the city. With the development of technology, the rhythm of the nature seems to be no longer clear in the operation of cities: the invention of trains and airplanes broke through the traffic restrictions imposed by the seasonal changes and the high water period and low water period of rivers, and various effective power generation methods blurred the difference between day and night. However, the history of nature still exists continuously and forcefully or is forced to change direction or actively participate in the historical process of cities. In the process of taking by force and encroaching upon the land by cities, countless original ecological habitats disappeared, countless indigenous species were replaced by alien species such as humans and their domesticated creatures, and the original process of nature was disrupted, but its evolution did not stop because of these changes. Bacteria and those insects and animals that could be called as small but strong creatures evolved together with the city's environment, making the city population unable to relax just a little. Although the use of fossil energy has saved human beings from the cold of winter, the smog of London, the halo on the impressionist canvas, and the dark sky of Beijing in winter are reminders of the transformation of the city's population and seasons. A catastrophe that caused social unrest like a revolution also occurred: the reason why the year 2005 became one of the most important historical nodes in New Orleans of the United States was not the election of a certain mayor, nor was it due to ethnic conflicts or ideological confrontation. Actually, the most significant historical node was Hurricane Katrina that year. From urban planning to rescue and reconstruction, it completely challenged the system, economy, and belief of cities in the United States, the most powerful country in the world. For nature, this hurricane was a natural phenomenon in its evolution; for the city and the people living in it, it was a disaster. Different races and classes had different experiences in this disaster, but its driving force originated from nature, and then it became a clear collision between nature and civilization in the city, redefining the history and memory of New Orleans.[7] The outbreak of COVID-19 pandemic in 2020 was called "a biological uprising" by historian Xia Mingfang and "would" eventually become an indelible symbol and mark of the future.[8]

For a long time, there was no direct connection between ecology as a natural science and history as a discipline of humanities. As the fundamental theory of ecology, the theory of evolution has been flattened and rounded by social Darwinists, serving as a weapon to break the historical cycle theory and establish a progressive view of history based on western centralism. Therefore, it was cursed widely. Although many concepts of ecology, such as ecological niche, top community, competition, and balance, have entered the studies of humanistic ecology and some historians, and have been used to explain the establishment and subversion of human social orders, the studies fail to really put human history into the evolution of living ecosystems to review the history shared by culture and nature. The attempt to consciously apply ecological theory to the study of history and use it as an important theoretical basis to explore the history of the interaction between nature and culture began with the rise of environmental history in the 1970s. In the studies relating to this discipline, the concepts of ecosystem, evolution and adaptation, dynamic equilibrium, and chaos are no longer the nouns of similar phenomena in the human society, but the material explanation about the history of interaction between nature and history. City, as the physical existence shaped by human forces and non-human natural forces, needs the inspiration and help of ecological thoughts to interpret its history.

Donald Worster defined environmental history as the study of the relationship between human society and culture as well as the natural world in history. As he wrote in his article *Transformations of the Earth*, environmental history is the study to analyze the role and position of nature in human life. It studies the interaction between the non-human world, a world that is not created by humans at all, and all the societies in the past. Here, he referred to "nature" as the basis of human survival and development of all fundamental biological, geographical, and physical elements. Nature is the collection of all non-scientific or non-technological forces; at

[7] Ted Steinberg, *Acts of God: The Unnatural History of Natural Disaster of America*. New York: Oxford University Press, 2006, the 2nd edition.

[8] Xia Mingfang: *The Biphasic Civilization: The Entanglement of Disaster and History*, Guilin: Guangxi Normal University Press, 2020, pp. 4 and 18.

least for the most time of history, these forces include the cycle of climate and weather, the geology of continents, the energy of the sun, the evolution and distribution of plants and animals, and the form and operation of ecosystems. All these forces constitute a world that we humans must learn to live with and adapt to, even as we try to escape from it. In recent centuries, human influence on the natural world has been growing, and even today the Earth's climate is being altered by human activities. But human beings only change nature and cannot create nature. Nature never runs in a fixed order. On the contrary, it is always in constant change. This state of change existed long before the birth of mankind. Undoubtedly, nature will also survive the changes we impose, and it will remain in some form long after the mankind has gone extinct as a species.[9]

So what is the position of cities in environmental history, an ambitious discipline in its attempt to revolutionize the way people understand the past? In other words, what role does nature play as a combination of power and existence created by non-human factors in cities as the product of culture? Such a question may lead to many different answers and opinions. Undoubtedly, the city is the environment of human life, but the early urban history rarely regards a city as an environment in which man and nature interact with each other. When American urban historians studied the history of cities for a long time, they were often indifferent to or even completely ignored the broader and ancient experience of human beings, the experience in the era of picking, nomadism, and farming, their relationship with the history of a city and the natural resources which the city depends on for survival. Cities are regarded as a unique creation of human beings, showing no contact with nature. The story about cities always highlights the relationship between people, not the relationship between people and the non-human world. Cities are seen as a series of

[9] Donald Worster, "Transformations of the Earth," *The Wealth of Nature: Environmental History and the Ecological Imagination*. New York: Oxford University Press: pp. 46–48. Worster is often criticized by urban environmental historians for excluding cities from the study of environmental history. However, Worster does not deny the position of cities in environmental history. What he criticizes is the study of urban environmental history that stops within the artificial buildings of cities. Worster has always emphasized the interaction between mankind and the natural world, and he also believes that the history of urban environment should emphasize the exchange and role between cities and non-artificial environments and forces. As another evident fact, there are several studies on the history of urban environments in the doctoral thesis written under the guidance of Worster.

buildings, streets, political gatherings, factories, and their internal labor conflicts, where people of different ethnic groups meet and collide with each other.

The human ecology introduced by the Chicago School in the 1920s and 1930s when studying urban issues was called ecology. However, it is just a model of ecosystem in nature and applied to human society. The concepts of ecosystem, community, ecological niche, micro-habitat, competition, and balance in ecology are used to explain the changes, confrontation, and cooperation of various groups and classes in urban communities. Human ecology has a profound impact on urban historians who actively use sociological theories and methods. However, few of them regard a city as a place, where rivers flow, plants grow, microorganisms spread, energy consumption takes place, and material information swaps, or a place, where the ecosystem health is intertwined with the health of human beings. This legacy tendency artificially isolates cities from nature. In other words, the original studies on urban history are still faltering, environmental historians are bringing cities back to nature, or bringing nature into cities, treating cities as organisms in which humanistic ecosystems and a natural ecosystems are intertwined, acting and evolving together, thus giving the studies on urban history a new direction.[10]

In the early twentieth century, the biology professor Lawrence Hendelson from Harvard University initiated "the fitness of the environment" based on Darwin's theory of species evolution. In his opinion, the theory of survival of the fittest applies to both organisms and the environment. While organic matter adapts itself to the environment, the environment is also changing itself in order to become the most suitable habitat for biological survival. He wrote: "Since the initiation of Darwin's theory, biologists have developed a habit of studying adaptation, considering only the adaptation of organisms to their environment. For them, the environment has always been an independent variable in the past, present and future. Any modern study has not considered whether the material universe also obeyed the laws essential to organic evolution. However, adaptation must exist not only in organic species, but also in the environment." In his opinion, the evolutionary process must include

[10] For the application of human ecology in American urban history, see *the Theory of Human Ecology in American Studies of Urban History* written by Jiang Peng, *Historiography Bimonthly*, 2001, No. 2, pp. 105–117.

a two-way relationship between environmental evolution and biological evolution. This means that they adapt to each other and change each other. As a biochemist, Henderson focuses on the increase and decrease of energy in the environment. He pointed out that maintaining the "complexity" and "sustainability" of life requires a stable environment, and maintaining such stability requires continuous automatic adjustment of the environment. Moreover, "an organism must be active, therefore, it needs the nourishment of matter and energy and must have a corresponding exchange of energy and matter with the environment."[11]

So far, no scholars have reviewed the history of urban development from the perspective of environmental adaptation. However, if this theory is introduced into the study of history, it can be seen that the city has not only formed its own unique environment but also an individual that has changed in the evolutionary system. When this large system consisting of cities, rural areas, wilderness, and other various individuals is evolving, they depend on each other and compete with each other.[12] Using the perspective of ecology to investigate the history of cities means to study the ecological connections between individuals and between them and larger systems in an objective evolutionary material environment; ecology is also applied to observe how a city maintains the complexity, sustainability, and vitality of itself and the organisms on which it depends through the exchange of corresponding energy and material in the process of co-evolution.

Specifically, urban environmental history on the basis of environmental evolution should be written in three aspects: first, the study of urban environmental history must focus on the evolution of the vast overall ecosystem that has an impact on the city and is also affected by the city.

[11] Lawrence J. Henderson, "The Fitness of the Environment, An Inquiry into the Biological Significance of the Properties of Matter," *The American Naturalist*. Vol. 47 (554), 1913: pp. 105–6.

[12] The concept of ecosystem, which formally appeared until 1935, was completely contrary to Henderson's theoretical basis of treating the environment as a superorganism. The founder of ecosystem, A. G. Tansley, aimed to cut off the connection between the early ecology and the organic theory of Romanticism, remove the content that could not be easily quantified and analyzed, describe all the connections between organisms as mere material exchanges, and study them as a physical system rather than an organic community. (Donald Worster: *Nature's Economy: A History of Ecological Ideas*) However, the use of ecosystems in the subsequent development, especially in the development of non-mainstream ecologists, still surpassed Tansley's original intention and closely integrated with biological philosophy and environmental protection.

Second, in order to study the urban environmental history, we must see the diversity, lamination, and alternation of urban ecosystems. In other words, urban ecosystems are neither single nor static; in cities, there are not only a number of different ecosystems but also different ecosystems changing in different times. Finally, in order to study the urban environmental history, we need to see that the complexity and vulnerability of urban cultural systems are closely related to the complexity and vulnerability of urban ecosystems. It is their influence on each other that restricts to a large extent, and even determines the stability and disorder of the city.

In *the Culture of Cities*, a book published just before the cold winter of 1938, Lewis Mumford reviewed the close tie established between regions and cities based on ecology. In his view, administrative boundaries determined in disregard of objective geography, economy, and society were arbitrary and unsustainable. The above action attempts to sever the organic connections within a region based on crude and exclusive consistency, turning a vibrant network of life into a kind of monotonous and crude political symbol with a monopoly tinge. However, boundaries with such absolute meaning are inevitably blurred and invalidated by the growth of the organism itself, including human beings. Although the region has gone beyond its original geographical state through human activities, it has always been restricted and influenced by nature. Mumford wrote as follows:

A region is a unit area, an original state formed by geographical structure, soil, surface undulation, water system, climate, vegetation, and animal life. Several of the following factors, including human settlements, domestication of new species and adaptation to environment and climate, rural and communities, controls determined by science and technology on several land, energy, climate, and movement, have transformed and redefined this original state to some extent.

All in all, the human domain is a complex of geographical, economic, and cultural elements. Neither can it be regarded as a natural product nor is it only a creation of human will and Utopia. The region, like a corresponding man-made city, is a collective work of art.

However, what role should the city play in this organic formation of the region? Mumford believed that the city naturally forms an attractive but open center because it is a regional distribution center for people, energy, and materials. A region changed its areas but also expressed the "personality" of these areas. A cooperative relationship existed between man and nature in the city from the beginning. "People accept and promote nature, even if human beings transform nature into the structural language of the city." "Even if the rough silk threads of geographical conditions are not noticeable, their quantity, quality and density will affect the final woven fabric in the network of urban civilization. The combination of diverse effects caused by nature to a city as well as complexity and individuality of such effect is actually an eternal guarantee to counteract the human tendency of oversimplification."[13]

Mumford is an urbanization scholar with deep historical thinking. As he embedded cities in the ecosystem of his region theory, he also wove the history of cities into the evolution of regional geography and the web of life. When urbanization and machine civilization developed rapidly from 1930 to 1970, he believed that cities had entered a stage of disorderly development as "giant cities," becoming an amoeba that lacked personality and stretched pseudopods everywhere, and also a monopoly force that plundered the "homeland of the earth" with the help of big machines. On the one hand, he keenly saw that the city's grab and consumption of resources was further intensified, indicating that the city's ecological hinterland was becoming wider and wider, and its material dependence on the overall regional ecosystem was becoming more and more diverse and complex. He cited the demand for water and expansion of pollution as an example to elaborate on the so-called "deformed vastness."[14] On the other hand, he warned that in the expansion of power and technology, the city officials and dwellers have lost the insight of understanding the interdependence. As a consequence, the city is devoid of individuality in the mandatory unity. Souls of the urban people are desolate, and the ecological hinterland on which the city depends is devastated. The reconstruction of urban regionalism is not retrogression to the relationship between urban and rural areas in the agricultural era, but to help the city out of the control of big machines and become the real controller of machines. Moreover, biotechnologies are used to establish an ecological regionalism for the future.

[13] Written by Lewis Mumford, *the Culture of Cities*, translated by Song Junling et al., Beijing: China Architecture Industry Press, 2009, p. 404, p. 354, p. 357, p. 360.

[14] Mumford: *Culture of the Cities*, pp. 271–279.

3 Environmental Turn of Urban History Studies

Contemporary American history was born in an era of urbanization. Most contemporary American historians grew up and lived in towns and suburbs. For them, the desire to understand the history of cities is an extremely natural response. Meanwhile, we must also admit that modern history is the era of cities. If intending to thoroughly interpret modern history, historians must care about the history of cities and its role in promoting the development of modern history. Compared with the relatively peaceful and monotonous rural history, the urban history that brings together different ethnic groups and cultures seems to be more colorful and climaxes appear repeatedly, and the historical research on this theme is more diverse and extensive.

American historians have begun to write the history of cities as cities spread rapidly across the American landscape. However, as an important branch of history, "urban history," really appeared in the mid-1960s only. This new branch began to study the historical characteristics of cities and the historical role of urbanization. These outstanding new "urban historians" tended to regard the city as a "process" rather than a specific "place." By referring to the method of sociological "quantitative research," they begin to focus on the people in the middle of the city who have lost their voice to resist the elite research of traditional historiography. The themes of the new urban history include social and political changes, origin of class relations, and racial, ethnic, and gender conflicts. This new branch of history also presents interdisciplinary characteristics, and a large number of theories and methods from social history, architectural history, and urban sociology are introduced into the research.

The pioneering work about the new urban history of the United States is the classic work of Stephan Thernstrom, *Poverty and Progress: Social Mobility in a Nineteenth Century City*. It was published in 1964. The book used a large number of census data to study the social changes in Newburyport, Massachusetts, between 1850 and 1880. It had a great impact on the narrative of American urban history in the next 10 or even 20 years. New urban historians use relevant data to further develop a "bottom-up" narrative perspective, thus studying the knowledge and experiences of people at the bottom of society and comparing the upward mobility of different ethnic groups as a social pattern.[15]

Some other influential works in this branch include *the Milwaukee of the Immigrants 1836–1860* written by Kathleen Conzen (1976), *Class and Community: Lynn's Industrial Revolution* written by Alan Dawley (1976), *the People of Hamilton, Western Canada* written by Michael B. Katz (1975), *Perilous Classes: Sin and Poverty in Columbus, Ohio, 1860–1865* written by Eric H. Monkkonen (1975), and *Social Change in an Industrial Town: The Progressive Model in Warren, Pennsylvania, from Civil War to World War I* written by Michael P. Weber (1976).[16]

Compared with the previous new urban history, the historical narrative of urban "biography" that appeared in the same period was less influenced by sociology. For the public, it may be the most common way to study urban history. Urban "biography" attempts to give a historical narrative about a particular city. Similar to biographical historians, urban biographical historians try to connect various complex aspects of the city, such as its origin, leaders, economic foundation, transportation, municipal administration, geographical expansion, demographic characteristics, schools, and various other cultural facilities, to describe the experience of the city. Each particular city has won a certain special comprehensive trait, and no longer exists only as a social phenomenon. As a result, the city has once again become a specific but special "place" with sufficient details and a penchant for narratives. One of the most representative works is *Gotham: A History of New York City to 1898* written by Edwin G. Burrows and Mike Wallace and published in 1999.[17]

[15] Stephan Thernstrom, *Poverty and Progress: Social Mobility in a Nineteenth Century City*. Cambridge: Harvard University Press, 1964.

[16] Kathleen Conzen, *Immigrant Milwaukee, 1836–1860: Accommodation and Community in a Frontier City*. Cambridge: Harvard University Press, 1976; Alan Dawley, *Class and Community: The Industrial Revolution in Lynn*. Cambridge: Harvard University Press, 1976; Michael Katz, *The People of Hamilton, Canada West: Family and Class in a Mid-nineteenth-century City*. Cambridge: Harvard University Press, 1975; Eric Monkkonen, *The Dangerous Class: Crime and Poverty in Columbus, Ohio, 1860–1885*. Cambridge: Harvard University Press, 1975; Michael P. Weber, *Social Change in an Industrial Town: Patterns of Progress in Warren, Pennsylvania, from Civil War to World War I*. University Park: Pennsylvania State University Press, 1976.

[17] Edwin G. Burrows and Mike Wallace, *Gotham: A history of New York City to 1898*. New York: Oxford University Press, 1999.

Therefore, urban history became an important branch of history whether at the public level or in the academic circle about 30 years ago, but its development since then is quite anticlimactic. However brilliant it may have been, the discipline seems to have declined. One of the most striking signs is that many universities in the United States have stopped offering courses about this branch. The University of Kansas is a typical state university in the United States, so we can take it as an example. In its history department, there are nearly 40 teachers, more than 100 graduate students, and nearly 500 undergraduates, but the department offers almost no undergraduate or postgraduate courses about urban history. In the annual job market in the United States, there are almost no new positions for the graduates majoring in urban history or focusing on urban history. Why did this happen? Why did this so promising discipline begin to lose its appeal? Cities are still expanding day by day, but why did the study on urban history stagnate after earlier successes?

The answer to this question is very complex, and to a large extent, we can regard it as a necessity for the development of American history. When a discipline develops to a certain level, it will no longer be satisfied with the discipline norms formulated at the beginning of the discipline. In turn, it will be eager to collide, blend, and then differentiate with many other disciplines and thoughts. American cities are expanding day by day, and their problems are becoming more and more complex and specific. In order to study a city from a specific view, or conduct a case study of a specific city, it is more attractive than considering urbanization as a unified process without personality no matter in terms of accuracy, feasibility, or the attractiveness of historical narrative. On the one hand, writing a biography for a single city is a popular option for a large number of new urban historians. On the other hand, urban history itself begins to split or merge into other branches of history, such as social history, economic history, studies of race, class or gender, or environmental history.

Since the twenty-first century, urban historians have been discussing the future of this branch. Clay Mcshane provided a rather "pessimistic conclusion" on this branch in his discussions. Based on an exhaustive study of the syllabus and awards relating to urban history, Mcshane pointed out that the branch was no longer the favorite branch of historiography and lagged behind the overall development of historiography. In his opinion, urban history is devoid of a clear definition and core as well as discussions on the changing social phenomena. Besides, this branch has also lost its methodological precision. In the further analysis of Mcshane,

"[Stuart] Blumin pointed out long ago that the works of urban history are devoid of theories or methods and the characteristics of this discipline. This is still an accurate view nowadays. Exceptions still exist, but they are not the mainstream thoughts. Such disinterest in theory and method will most likely make us despised by our peers, especially social scientists."

Yet the problems in urban history seem to go deeper than the absence of methodological precision. Mcshane also admitted that the most influential works in this branch are outdated. In other words, there are not many new books to promote the progress and increase the influence of this branch. After he listed and compared the most frequently cited books at present, he pointed out that, in fact, the degree of acceptance for all these works failed to make them jointly recognized classics, and those four works with the highest citation are not even academic works. In the end, Mcshane concluded "it seems that there is a very significant rift between urban history and the remaining branches within the discipline of history. The following questions are asked: is urban history still a discipline even if it doesn't produce a classic, even a controversial one?"[18] This pessimistic evaluation of urban history at least shows that urban history, as an independent discipline in the United States, has been difficult to maintain its consistency or continue its lasting influence on historians. It did not have the power to attract creative thinkers from the new generation, nor did it produce books that would keep the discipline going. While cities are growing up, urban historians are dilatory with a weak will.

However, we review this branch from a more optimistic perspective. Perhaps urban history has not declined but has changed. It has gradually broken away from the close connection with sociology but in an unpredictable way for urban historians. As a consequence, it has begun to revitalize in some unexpected places. Urban history is not dying. It has simply changed its focus, perspective, and approach. For example, its rebirth occurred at the moment when urban history meets environmental history, and from that moment on, urban history discovered a new definition. In the United States, environmental history has injected new thoughts and vitalities into the study of urban history at least, and such convergence has produced breathtaking books and articles.

Urban–suburban landscapes began to emerge in the branch of environmental history in the 1980s, took hold in the 1990s, and now become the

[18] Clay Mcshane, "The State of the Art in North American Urban History," *Journal of Urban History*. 32 (2006): pp. 594–5.

most popular new frontier for the research in this field. Historians have realized that the city has a profound impact on the utilization or protection of wilderness areas farthest away from it and also exists as a place where nature reserves its power and leaves some indelible marks. Nature is not just a distant prairie or forest. It also includes the air and water that flow around our homes, the energy that makes urban machines bustle all the time, and all the plants, animals, and microorganisms that find their niches in the middle of the city. Scholars such as Kathleen Brosnan, Andrew Hurley, Martin Mellosey, Adam Rome, and Joe Tarr have greatly broadened our thinking on the interface between man and nature and proved that human settlements, like human physiological systems, are a collection of a series of existence to be replenished and excreted. The operation of urban metabolic systems is like a farm or factory, proving that the city is rooted in nature even if it is to a large extent the combined result of human creation and natural force.[19]

In a review associated with urban environmental history, Joe Tarr, one of the pioneers specialized in the urban environmental history of the United States, summarized the development of the country's urban environmental history into five aspects: first, study the impact of artificial environment and human activities on the natural environment; second, study the efforts and role of social response to alleviate environmental problems; third, explore the impact of natural environment on urban life; fourth, analyze the relationship between cities and their expanding hinterland; fifth, study the relationship of races, classes and genders with environmental problems in cities.[20] In these five aspects, environmental historians have been making new works in the recent 30 years. From the

[19] Kathleen Brosnan, *Uniting Mountain & Plain: Cities, Law, and Environmental Change along the Front Range*. Albuquerque: University of New Mexico Press, 2002; Andrew Hurley, *Environmental Inequalities: Class, Race, and Industrial Pollution in Gary, Indiana, 1945–1980*. Chapel Hill: University of North Carolina Press, 1995; Martin Melosi, *Garbage in the Cities, Refuse, Reform, and the Environment, 1880–1980*. College Station, TX: Texas A & M University Press, 1981, Adam Rome, *The Bulldozer in the Countryside: Suburban Sprawl and the Rise of American Environmentalism*. New York: Cambridge University Press, 2001; Joel Tarr, *The Search for the Ultimate Sink: Urban Pollution in Historical Perspective*. Akron: University of Akron Press, 1996.

[20] Joel Tarr, *Urban History and Environmental History in the United States*: Complementary and Overlapping Fields," http://www.h-net.org/~environ/historiography/usurban.htm, accepted on August 25,2010, indexed into Christoph Bernhardt ed., *Environmental Problems in European Cities of the 19th and 20th Century*. New York/Muenchen/Berlin: Waxmann, Muenster, 2001: pp. 25–39.

early comprehensive studies on the urban environment by Melosi et al. to the attention of many scholars to a specific city, much attention has been paid to the activities and thinking of human beings in this city, and the interaction between the city and its surrounding nature. So far, the history of the urban environment has become more and more rooted in the land. Like their counterparts in environmental history, the new generation of urban environmental history scholars pays attention to the changes in the details of the urban landscape, seeks for the characteristics of each specific city in the sense of environmental history, and rewrites the "biography" of a city.[21]

4 Nature's Metropolis

Provided that the early urban environmental historians still focused on the interior of cities to review the impact of human activities, policy implementation, and use of technology, such as garbage and sewage treatment, energy utilization, and industrial pollution, on the urban environment and human health, like their counterparts in urban history, the new study model breaks the artificial boundaries between cities in both time and space dimensions, by combining the evolution history of the city with the longer lasting history of agriculture and animal husbandry experienced by the development of human society. As a result, they have achieved a fundamental expansion over the territory of urban environmental history in space and conducted an overall organic investigation on the environment of the city and its hinterland. The study began with William Cronon's book, *the Nature's Metropolis: Chicago and the Great West*.[22]

The book was published in 1991. It not only created a new paradigm for the writing of urban biographical history but also opened up a new situation for the development of urban environmental history. Chicago, a metropolis booming in the west of the United States has long been the focus of studies on the history of American cities. The emergence of

[21] Joel Tarr, *Urban History and Environmental History in the United States*: Complementary and Overlapping Fields," http://www.h-net.org/~environ/historiography/usurban.htm, accepted on August 25,2010, indexed into Christoph Bernhardt ed., *Environmental Problems in European Cities of the 19th and 20th Century*. New York/Muenchen/Berlin: Waxmann, Muenster, 2001: pp. 25–39.

[22] Publication information of the first edition in English: William Cronon, *Nature's Metropolis: Chicago and the Great West*, New York: W. W. Norton & Company, 1991.

this phenomenon not only lies in that the city has nurtured an important theory of urban history, and human ecology but also its own colorful and magnificent rise and development history. Its landscape is various, its politics are treacherous and changeable and the contradictions and compromises among races, classes, and genders are sharp and distinct. Both traditional political historians and new social historians have discovered the space to make full use of their talents in history. However, Cronon's work expands the study of historians on Chicago to the entire western region of the United States. In the vast land, he thought about the reasons for the formation of this nature's metropolis.

Just before the publication of *Nature's Metropolis*, Cronon had already established his status as a new generation of leader in environmental history with his book *Changes in the Land: Indians, Colonists, and the Ecology of New England* published in 1983. When *the Changes in the Land* was published, Cronon was not yet thirty years old. This book with a text of less than 200 pages fundamentally challenged the long-standing ideological line in American history and also the progressive view of history, which was deeply influenced by Whig history. It was replaced with a tone of decay theory with deep respect and sympathy for Indian culture and New England ecosystems. It won Francis Parkman Prize (a great prize for history in the United States) the next year after the book was published, helping this new historian gain a widespread reputation. Along with Worster's *Dust Bowl: The Southern Plains in the 1930s*, *Changes in the Land* also established the morale and academic position of this nascent discipline about environmental history.[23] Nearly ten years later, Cronon shocked the academic circle again and published the 540-page *Nature's Metropolis*, which won the Bancroft Prize the next year.

Tarr and Melosi had built up landmarks for this new field in the map of environmental history in ten years before *Nature's Metropolis* explored and expanded the development of urban environmental history. However, the publication of *Nature's Metropolis* is of great significance in the following two aspects.[24] First, in the 1980s, famous environmental

[23] William Cronon, *Changes in the Land: Indians, Colonists, and the Ecology of New England*. New York: Hill and Wang, 1983.

[24] Domestic scholars have discussed in detail the contributions of Tarr and Melosi to the urban environmental history. In addition to the studies of Bao Maohong and Gao Guorong, refer to *Wang Yue: Study on Urban Environmental History by American Environmental Historian Joe Tarr*, *Journal of North Minzu University*, No. 1, 2009,

historians such as Worster, Cronon, and Roderick Nash were famous for their studies on agricultural and wilderness environmental history. When environmental history was still a new discipline, urban environmental history was much more marginal. As one of the founders of environmental history morale, Cronon devoted himself to this field, helping urban environmental history enter the center of American environmental history research, and greatly promoting the development of this discipline. As the second aspect and also a more important point, Cronon conducted a double study on the city and its hinterland through commodity flow and pursued the interaction between the city and nature on the overlapping between the ecosystem of nature and the economic system of city, thus building a new research model of urban environmental history.

Cronon's new model is based on an ecological historical interpretation of Thünen Circle created by Johann Heinrich von Thünen. In 1826, von Thünen published *the Isolated State*, in which Thünen devised a simplified model for the economic and geographical relationship between the central city and its hinterland. A central city is like an independent country. Production activities on its hinterland have a direct connection with the market of this city. By calculating land rent and transportation costs, farmers rationally pursue the maximization of profits, and thus determine the circular distribution of agricultural production on the city's hinterland. The production circle closest to the city is mainly engaged in fruit, vegetable, and dairy activities; the circle second closet to the city is the forestry circle; the circle third closest to the city is the extensive grain production circle; the outermost circle is engaged in animal husbandry production. Cronon admitted that although this assumption was too simple and abstract compared to the complex and changing actual relationship between the city and its hinterland, it could still effectively explain Chicago's rapid development in the nineteenth century. With regard to the difference, the forestry circle of "Thünen Circle" is located in the second ring on the periphery of the city, following the "effective use" of scientific forestry developed by Prussia since the seventeenth century, while the American continent was still addicted to the myth of inexhaustible resources when Chicago emerged, so unrestrained deforestation was located in the outer ring for animal husbandry. The outermost ring was scattered with wild areas and picking, hunting, and

pp. 132–136; Mao Da: *Analysis of Important Academic Phenomena in the Development of Urban Environmental History*, *World History*, No. 3, 2011, pp. 37–45.

fur trading of the western region of the United States in the nineteenth century. Undoubtedly, the use of "Thünen Circle" by no means stops at local corrections. More importantly, he made contributed to the development of ecology and history for an economic and geographical theory, so that the solidified theory was fused into a dynamic evolution process.[25]

From the perspective of ecology, Cronon pointed out that the "Thünen Circle" is not only a process of economic system expansion but also a story of ecosystem succession for the extension of cities in their hinterland. The invincible power of capital has not only changed the landscape of Chicago, the former "wild garlic land," where tall buildings replaced the riverside clusters of wild garlic, but also stimulated changes in the economic system and ecosystem of the entire Great Plains. It was here that Cronon walked out of the artificial boundary that divided the city and nature, and discovered the connection of the city with wheat chaff, cattle hissing, sheep baaing, and wood debris outside the city in the orderly operation of the city's capital. Every new expansion of the "Thünen Circle" is a process of transition from the old to the new ecosystem. Straightened wheat fields replace native grasslands, raised cattle and sheep occupy the ecological niche of bison herds, and forests in the Great Lakes Region disappear rapidly in the unrestrained deforestation. Cronon emphasized that this is a process for the second nature to substitute the first nature, and a process for the ecosystem shaped by economic, technological, and political power to substitute the ecosystem formed by the power nature.

The concept of "second nature" was by no means the first of its kind by Cronon, but the introduction of such a concept into the study of environmental history began with Cronon. As early as in his book *Changes in the Land*, Cronon vaguely initiated this concept through the use of land in New England by original inhabitants. In *Nature's Metropolis*, Cronon began to unswervingly promote thinking over this concept. The concept of the second nature is the reconciliation of the relationship between nature and artificial environment, which has been discussed by urban environmental historians for a long time. Its original intention is to emphasize the inseparable connection between the artificial environment and the ecosystem of nature and between the city and rural areas. Environmental

[25] Johann Heinrich von Thünen, *The isolated state in relation to agriculture and political economy*, translated by Wu Hengkang, The Commercial Press, published for the first time in 1986, reprinted in 2010.

history exactly happens on this connection. However, under the guidance of Cronon, the discussion about "the second nature" no longer stays in the field of urban environmental history, and attracts a large number of postmodern historians, who begin to carry out cultural deconstruction of "nature" with great interest.[26]

In the opinion of some environmental historians, the term "artificial environment" was originally too emotionless, while "second nature" was more friendly and tender. Since then, severely affected by the wave of postmodernism, some environmental history studies have also actively participated in the torrent of "linguistic turn." Under the postmodern deconstruction, "nature" is regarded as a pure cultural construction rather than an objective material existence. Such a wave is noticeably complementary to the postmodern "cultural analysis" with race, class, and gender as the main line. For any new historian, postmodernism first means challenging authority, representing a process that is extremely risky, yet full of fun and intelligence. The sudden emergence of Worster, Cronon et al. in the history world was a challenge to the paradigm of Western history since Lanke. In the process of further developing the environmental history, the rebellion in that era has become the current authority, and new rebels have emerged accordingly. This new wave of challenges came from the middle-aged and young environmental historians, who were deeply immersed in social history research.

They believed that the older generation of environmental historians regarded the "human" in the relationship between nature and human beings as a "class," and therefore ignored the different relationships between people of various ethnic groups, classes, and genders in this interaction. The presupposition here is that nature is the product of human thoughts and behaviors, and there is no ecosystem without human interference under the boundless sky. In this way, the logic of postmodern environmental historians is formed. Since nature is a cultural construction and there is no difference between artificial environment and natural environment, the emphasis of environmental history on nature

[26] In the collection of essays named Uncommon Ground: Toward Reinventing Nature and edited by Cronon in 1995, three years after the publication of *Nature's Metropolis*, Cronon and other authors deconstructed the concepts of nature and wilderness and promoted the thinking towards the "second nature". William Cronon ed., *Uncommon Ground: Toward Reinventing Nature*. New York: W. W. Norton & Co., 1995.

becomes aimless, and all research returns to the discussion about artificial environment or cultural power.

Worster was sensitive to the arrival of such transformation in the mid- to late-1980s, so he stressed the existence of nature as an independent force, so as to resist the deep-rooted arrogance and ignorance of nature in historical research and even human thought. In the passages often criticized by urban environmental historians, Worster wrote as follows:

According to the typical concept, we think of "nature" as a non-human world, a world that is fundamentally not created by us. The state in which human beings influence each other in the absence of nature is not included in the "social environment." The same is true of constructed or artificial environments, which are clusters of things made by human beings. They are so common that they constitute the "second nature" around them. As humans continue to stamp their marks in forests, gene banks, and polar ice caps, it seems that there may be no actual difference between "natural" environment and "artificial" environment. However, it is necessary to preserve the difference because it reminds us that there are different forces at work in the world, not all of them are from human beings and some forces have always been self-contained. The constructed environment is a highly expressive culture, and the study of constructed environments has been well-developed in the history of architecture, technology, and cities. But in phenomena like forests and water cycles, we encounter spontaneous forces that are not driven by us.

These forces impinge on human life, provoking a certain reaction, a certain resistance, and a certain ambition. As we cross the world of human self-care and encounter non-human regions, environmental history finds its theme.[27]

Since the beginning of the development of environmental history, it has not neglected the marginal or vulnerable groups in society. From the sympathetic description of Worster about the ecological refugees who had left their homes in *Dust Bowl* to the emphasis of Carolyn Merchant in *the Death of Nature* on the special relationship between women and nature,

[27] Written by Donald Worster, translated by Hou Wenhui: *Three Levels of Environmental History Research*, *World History*, No. 4, 2011, pp. 98–99. This article was originally extracted from a collection of essays edited by Donald Worster and named *the Ends of Earth: Perspectives of Modern Environmental History*, New York: Oxford University Press, 1988), long postscript.

the questioning of Crosby about the superiority theory of white civilization in *Columbian Exchange* and the yearning of Cronon in *Changes in Land* for the state of Indian people getting along with nature, the environmental historians got rid of the fetters of elite history and political history and discovered different forces and relationships in the historical structure in their early years.[28] However, unlike the founders of social history in the same era, the above scholars had a broader and deeper vision, thereby making the care of historiography penetrate all kinds of human affairs and reach a world closely related to us, but not the same one with ours. Nature has become an important theme of history study, constituting the greatest contribution of environmental history to historiography. As reflected by the current development of this discipline, environmental history is likely to return to the traditional principle that "human affairs" is the only theme of historiography four decades after its setup under the impact of this transformation.

Although Cronon tried his best to promote the thinking toward "second nature," such a development momentum may not be his original intention. Whether it is his preference for the harmonious land relations of the Indians revealed in *Changes in the Land* or his insistence on animal rights and moral economy in *the Nature's Metropolis*, Cronon always expressed respect and sympathy for the existence of nature (or the first nature as he said). In the study on urban environmental history, he focused on the connection between the first nature and the second nature rather than the two together. In this regard, Cronon and Worster shared the same expectation for the development of environmental history. The two great environmental historians had the same spiritual affinity and became good friends just in the process of "crossing the world of human self-care and encountering with non-human regions."

Returning to the study of urban environmental history, the wave of "cultural transformation" in the environmental history is surging in this field with extremely subtle differences between artificial and natural systems. After the publication of *Nature's Metropolis*, the reputation was

[28] Donald Worster, *Dust Bowl: The Southern Plainsinthe1930s*. New York: Oxford University Press, 1979; Carolyn Merchant, *The Death of Nature: Women, Ecology and the Scientific Revolution*. San Francisco: Harper & Row, 1980; Alfred Crosby, *Ecological Imperialism: The Biological Expansion of Europe, 900–1900*. New York: Cambridge University Press, 1986.

mixed with its own dissonance of criticism. One of the most powerful criticisms focuses on Cronon's intentional avoidance of the understandings and needs of different urban populations toward nature in such a book on urban history. The story of Chicago told by Cronon shows the circulation of natural energy between the city and rural areas driven by the power of capital, rather than the interaction between people with very different cultures and experiences and the environment in the city. Undoubtedly, the above situation is in disagreement with the ascendant cultural analysis methods. Although urban environmental history had become popular due to the participation of Cronon, only a few middle-aged and young historians really followed the urban environmental history model constructed by Cronon.[29]

When the capital influence of cities is getting weaker and weaker nowadays, not only going beyond their own administrative boundaries but also crossing national and continental boundaries, the ties between urban people and nature unreachable for the capital influence are becoming more and more complex. Chicago cited by Cronon in his book only consumed the grasslands of the Great Plains and the forests of the Great Lakes, while the Chicagoans at that time only got food and energy from the western region of the United States. Today, people in any natural metropolis may eat fruits from California, vegetables from Shouguang in China's Shandong Province, soybeans from Brazil, and beef from Australia. The energy to support the functioning of cities may come from distant Africa, turbulent Central Asia, and desolate Siberia. The garbage discharged by cities pollutes the vast coastal areas and territorial seas, and all kinds of industrial and domestic waste gas reach up to the space. Due to the above facts, environmental historians need to think more deeply about the research model established by Cronon.

[29] After Cronon, the most famous scholars who tried to break through the urban boundaries and study the ecological and economic relations of the city and its hinterland include Mike Davis, *Ecology of Fear: Los Angeles and the Imagination of Disaster*. New York: Metropolitan Books, 1998; Kathleen Brosnan: *Combining Mountains and Plains: Cities, Law and Environmental Change on the Frontier*; Stéphane Castonguay, *Metropolitan Nature: Environmental Histories of Montreal*. Pittsburgh: University of Pittsburgh Press, 2011.

5 Post-Cronon Era

Robert Gottlieb's new book *Forcing the Spring: The Transformation of the American Environmental Movement* came out two years after the publication of Nature's Metropolis.[30] Since its publication, the book got both praise and censure, yet nobody could deny its challenge to the narrative of previous American environmental movements. Since the publication of Samuel Hays' classic work *the Gospel of Resource Conservation and Efficiency*, the environmental movement or reform in the United States in the late nineteenth century has often been described as an activity carried out outside the city, whether it was a resource conservation reform based on scientific forestry or a natural conservation movement based on national parks. In the narratives of early environmental historians in the United States, it seemed that they always tried to realize their ambitions outside fields.[31] Undoubtedly, this narrative has not only narrowed a much more complex and comprehensive environmental reform movement greatly but also made the environmental justice movement that has been in full swing in the inland cities of the United States since the 1980s a tree without roots. Gottlieb pointed out that the American environmental protection movement is fundamentally an amendment to the various environmental problems brought about by industrialization and urbanization in the late nineteenth century, and the movement occurred just in the metropolises that emerged during this period, which was reflected in various urban health and medical reforms during the progressive period. Driven by *Forcing Spring*, race, class, and gender entered the territory of the early American environmental movement, and cities rose impressively during the environmental movement.

It is really hard to define Gottlieb's book as a work of urban environmental history. However, it has still made a significant contribution to urban environmental history. For a long time, environmental movement and policy have been important themes in the study of environmental history in the United States, but cities have always been ambiguous in the study of environmental reform movements during the progressive

[30] Robert Gottlieb, *Forcing the Spring: The Transformation of the American Environmental Movement*. Washington, D. C.: Island Press, 1993.

[31] Samuel Hays, *Conservation and the Gospel of Efficiency: The Progressive Conservation Movement, 1890–1920*. Cambridge: Harvard University Press, 1959.

period, which occupies half of the history of the environmental movement. In this regard, Gottlieb's work has surely eliminated the ambiguity. Gottlieb focused on the marginal groups of the environmental movement and the close tie between human health and environmental health, thus inspiring his peers of the urban environmental history to expand the study on the relationship between the urban environment and different social groups. However, it is worth noting that Gottlieb's study on the origin of the American environmental movement is not an attempt to restore the panorama of a movement that has been narrowed by historians, but to narrow it again from another angle. As an activist of the environmental justice movement, he has his own political and social demands. When he relocated the front of the early environmental movement from the countryside and wilderness to the city, he made the public health reform of industrial urban communities become the core element for the development of the environmental movement. According to his writings, the movement that transcended the narrow self-care of human beings in many aspects and pursued the balance, beauty, and health of ecosystems including human beings seemed to be a measure for the health of human beings, especially a certain group of human beings. If not completely hidden, nature will be marginalized again according to his narrative. Meanwhile, this tendency is becoming more and more intense among the new generation of environmental historians who use cultural analysis techniques.

After the publication of Forcing Spring, the social historian Andrew Hurley's book Environmental Inequality: Class, Race, and Industrial Pollution in Gary, Indiana, 1945–1980 was sent to the press. This book can be regarded as a model of the combination of urban environmental history and social history research. This city chosen by Hurley was not a metropolis like Chicago, New York, or Pittsburgh, but a small industrial city rarely known by others and suffering from severe pollution in Indiana Hurley's research reveals that in Gary, the victims of environmental pollution are mainly the poor black and white working class living in the inner city. The vulnerable groups in society are the vulnerable groups in environmental pollution, and environmental inequality is the reflection of social injustice. Like most social historians, Hurley expresses in his research a deep sympathy for marginalized groups and a harsh critique of environmental inequality. However, unlike most social historians, Hurley's research focuses on air, water, and energy outside of human culture, politics, economy, and society.

Hurley's work was a great success among environmental historians and soon made it into the required reading list of most environmental history study courses, while Hurley's personal orientation was interesting. In both public interviews and private dialogs, Hurley sees himself as a social historian, denying the label of an environmental historian. Hurley was also rarely seen at the annual meeting of American Society for Environmental History. In an interview with Hurley by Dr. Gao Guorong, a Chinese environmental historian, a conversation made people think deeply. When asked about his current research, Hurley replied that he was not currently engaged in any environmental history research, but was working on the issue of historic relics preservation. After the reply, he began to talk about his friend, the environmental historian David Stradling because David believed that this was still the study of environmental history, and emphasized that problems involving ground landscape, whether man-made or natural ground landscape, are part of the environmental history study. However, Hurley didn't agree with him and retorted that in his opinion, "non-human nature can be defined as the result of environmental history only when it becomes an important part of the study. As for me, environmental history has such a threshold. Nature should be the center of environmental history study.[32] Hurley's denial inevitably makes the pioneer postmodern environmental historians have the suspicion of wishful thinking, which is a bit embarrassing. However, we should further think deeply about the following question: From the perspective of the study object, how does a work about urban history with the seclusion of nature find a special view for environmental history and how does it surpass the previous generation of urban history studies? From the perspective of the study method, can the urban environmental history study still have the significance of independent existence once the urban environmental history is dominated by simple "cultural significance?".

Fortunately, not all middle-aged and young environmental historians prefer to dwell on the traditional state of human affairs and cultural analysis, and even within the urban environmental history, there are always scholars with ecological consciousness and care. In 2000, Adam Rome published *the Bulldozer in the Countryside: Suburban Sprawl and the Rise of American Environmentalism*, which made a great contribution to the

[32] Gao Guorong: *Public Historians Focusing on Environment and Cities: An Interview with Professor Andrew Hurley*, *Clio at Beida*, Beijing: Peking University Press, 2012, pp. 362–387.

study of urban environmental history.[33] First of all, the book has made an undeniable contribution to help explore and expand the development of urban environment history. The basic model of urban development in the United States after the 1950s was the suburban sprawl marked by mass housing, and the suburbs at the junction of urban and rural areas became the basic habitats of urban populations in the United States. In the study of urban history, there are many excellent works on the study of suburbs, such as *Crabgrass Frontier: The Suburbanization of the United States*, which is still popular now. It reviews the process of suburbanization and its impact on American society from the perspectives of politics, economy, urban development, and social population mobility patterns.[34] However, in the field of urban environmental history research, Rome is the first scholar to open up the atmosphere of suburban research. However, as the more important contribution made by the book to urban environmental history for the first time, it makes the study of urban environmental history different from the previous works and truly based on the land community composed of soil, water, animals, and plants, air and human beings, to investigate the various contradictions between artificial environment and natural environment in the process of suburban spread. It also investigates people's understanding toward these contradictions and reflections on their impact on society at the physiological level, material level, aesthetic level, moral level, scientific level, and ecological level.

Equally important, as the clarion call for cultural analysis rings through the sky, Rome is able to maintain a serious dialogue with the natural sciences for which environmental history was founded. As a consequence, *Bulldozer in the Countryside* not only retains the textual analysis and historical narrative that historians are familiar with and good at but also skillfully uses a large number of meticulous scientific data and theories. Its understanding toward the land community casts off the fetters of cultural politics. And the interdisciplinary communication helps humanists realize that nature is not a narrow vocabulary for deconstruction, nor is it a landscape that exists only through human aesthetic imagination. In essence,

[33] Adam Rome, translated by Gao Guorong et al., *the Bulldozer in the Countryside: Suburban Sprawl and the Rise of American Environmentalism*, Beijing: China Social Sciences Press, 2011. For information on the publication of original works in English, see Note 33.

[34] Kenneth T. Jackson, *Crabgrass Frontier: The Suburbanization of the United States*, translated by Wang Xu, Beijing: The Commercial Press, 2017.

nature is an objective independent force and material reality. While calmly and objectively analyzing the historical facts, Rome did not try to hide the profound ecological care. It is this kind of ecological care that makes his review of history go beyond the narrow human affairs, thereby becoming more profound and extensive.

Rome's work was published in 2000, which became a new watershed in the study of environmental history to some extent. In the following 20 years, the study of urban environmental history has been a great sight and new works have been published one after another. According to the urban environmental history bibliography hosted by Melosi and Tarr and published by the University of Pittsburgh Press, we can get a glimpse of the continuous efforts on the depth and breadth of this field over the past two decades.[35] Among these works, there are works not only reviewing the environmental problems of American cities in different themes or aspects but also generating more environmental biographies about every single city. In the first category of works, special attention should be paid to the encyclopedia-type work named *The Sanitary City: American Urban Infrastructure from Colonial Times to the Present* and written by Melosi, the founder of urban environmental history. Harold L. Platt explored the relationship between organic cities and modernity in different regions from a comparative perspective using his work *Building the Urban Environment: A View of Organic Cities in the United States, Europe, and Latin America*. Dawn Biehler and Zachary Falck were aware of the unwelcome creatures in the city respectively: pests and weeds, and reviewed how these creatures that urbanites tried their best to eliminate tenaciously occupied the ecological niche in the city system, and how their existence affected the relationship between different classes and ethnic groups in the city. Melanie Kiechle's *Taste Detective* has a unique style, allowing the taste to permeate the history of American cities in the nineteenth century and then reviewing the transformation of urban environment and sanitation concepts.[36]

[35] For the series books published by the University of Pittsburgh Press on urban environmental history, please visit the website: https://upittpress.org/series/history-of-the-urban-environment/, logged in on November 29,2020.

[36] Martin Melosi, *The Sanitary City*; Harold L. Platt, *Building the Urban Environment: Visions of the Organic City in the United States, Europe, and Latin America*. Philadelphia: Temple University Press, 2015; Dawn Day Biehler, *Pests in the City: Flies, Bedbugs, Cockroaches, and Rats*. Seattle: University of Washington Press, 2015; Zachary J.

In the biographies of every single city, environmental historians of the twenty-first century showed more vigorous creativity, their topics for discussion were more diverse, and their thinking was deeper. Besides, they were sensitive to the power structure formed by the differences of race, class, and gender, and their discussions permeated their reviews over the relationship between city and nature. Meanwhile, more and more scholars of urban environmental history realized that nature, a force not created by human beings, runs through the evolution of cities and often declares its existence through disasters. These two aspects do not appear as two parallel clues in the study of urban environmental history. On the contrary, they have always been intertwined, urging environmental historians to realize that the non-nature of natural disasters is hidden in social injustice. As a city suffering frequent natural disasters and serious air pollution but enjoying a colorful culture, Los Angeles is particularly noticeable. Since the 1990s, Mike Davis started the environmental narrative of Los Angeles with two environmental biographies about this city. Although disasters lingered when the discipline of environmental history was first established, it really began to become an important theme of urban environmental history, which was initiated by Davis' fear and imagination around the fragile ecology of Los Angeles.[37] Among the writings about the inextricable relationship between cities and disasters, the outstanding study named *Seismic City* and conducted by Joanna Dyl about San Francisco's 1906 Earthquake calls for deep thought.[38] New Orleans is favored by urban environmental historians because of its unique and charming southern experience and the doomsday impact

S. Falck, *Weeds: An Environmental History of Metropolitan America*. Pittsburgh: University of Pittsburgh Press, 2016; Melanie Kiechle, *Smell Detective: An Olfactory History of Nineteenth-Century Urban America*. Seattle: University of Washington Press, 2019.

[37] See Mike Davis, *City of Quartz: Excavating the Future in Los Angeles*. London: Verso, 1990, and *Ecology of Fear: Los Angeles and the Imagination of Disaster*. New York: Vintage, 1998. From a more macroscopic perspective, disasters have always been one of the motifs of environmental history. As a cornerstone work of environmental history research, Donald Worster's *Dust Bowl* tells one of the biggest ecological disasters in the history of the United States. After that, Theodore Steinberg wrote *Acts of God*, which had a great impact on the study of environmental history about disasters. Theodore Steinberg, *Acts of God: The Unnatural History of Natural Disasters*. New York: Oxford University Press, 1st ed. 2000 and 2nd edition 2006 with a new chapter on Katrina.

[38] Joanna L. Dyl, *Seismic City: An Environmental History of San Francisco's 1906 Earthquake*.

of Hurricane Katrina. In Andy Horowitz's new book named *Katrina: A History, 1915–2015*, he put this disaster that shocked the world into a period as long as a century to review and reflect on the relationship among disaster, infrastructure, and social structural injustice in American cities represented by New Orleans. Earlier environmental history books on Seattle focused on the confidence of New Orleans in the conquest of nature by technology and their forgetfulness toward the pain caused by disasters.[39]

Undoubtedly, the study of urban environmental history over the past two decades involves not only the scene of disasters but also the traditional issues of urban environmental history, such as garbage, water supply, and pollution. These issues have flourished for two decades. In addition to these issues, more roles have stepped onto the stage of urban environmental history. Taking different cities as an example, such as New York, there is not only the new book *Flesh Kills: A History of Consuming and Discarding in New York City* written by Melosi about Fresh Kills and published in 2020, the largest landfill in the United States but also the overall study on the fundamental deconstruction and reshaping of the city's history. Among them, the most challenging work is Theodore Steinberg's masterpiece *Unrestrained Gotham*, which is discussed in more detail later. Another environmental biography of New York City named *Taming Manhattan* published in the same year reflects on the relationship between man and nature in New York City within the framework of race, class, and gender. Moreover, both Matthew Gandy and David Stradling set their sights on New York to further analyze and explain the binary opposition between city and nature.[40] The same is true of Seattle's environmental history. Frederick Brown and Matthew Klingle

[39] Andy Horowitz, *Katrina: A History, 1915–2015*. Cambridge, Mass.: Harvard University Press, 2020; Ari Kelman, *A River and Its City: The Nature of Landscape in New Orleans*. Berkeley Cal.: University of California Press, 2003 1st edition and 2006 2nd edition with a new chapter on Katrina.

[40] Martin Melosi, *Fresh Kills: A History of Consuming and Discarding in New York City*. New York: Columbia University Press, 2020; Ted Steinberg, *Gotham Unbound: The Ecological History of Greater New York*, New York: Simon and Schuster, 2014; Catherine McNeur, *Taming Manhattan: Environmental Battles in the Antebellum City*. Cambridge, Mass.: Harvard University Press, 2014; Matthew Gandy, *Concrete and Clay: Remaking Nature in New York City*. Cambridge: MIT Press, 2002; and David Stradling, *Making Mountains: New York City and the Catskills*. Seattle: University of Washington Press, 2007.

involved themselves in the study from totally different points of penetration. Frederick Brown focused on the transformation of relations between human beings and animals in Seattle while Matthew Klingle told a green story about this city.[41] The study on western desert cities also presents a thematic update, such as Phoenix, which emerged as a new force in the second half of the twentieth century. In his comparative study of Phoenix and Tucson, Michael Logan focused on water availability and cultural differences between the two cities. The book *Transmission Lines*, which won the 2016 Book Award of American Society for Environmental History, rewrites the history of the rise of Phoenix in the games among coal mining, thermal power generation, energy consumption, and social justice.[42]

Unfortunately, most of the above works fail to extend the model of urban environmental history created by Cronon. In the history about the rise of Chicago, Cronon realized that the artificial boundaries of the city not only restricted the imagination of historians but also caused one-sided interpretation of the evolution of the city itself. Therefore, he called for breaking through the boundaries of cities to study the urban environmental history. Despite the impressive achievements made by urban environmental historians in just 40 years, most of the books, with the exception of a few urban biographies, still stop within the boundaries of cities. They discussed the internal metabolism of urban organisms, the close connection between urban populations and environmental health, the different perceptions of different groups toward the urban landscape, and the logical relationship between environmental justice and social justice. Undoubtedly, such study is an important part of urban environmental history, which is not only necessary but also will still maintain its vitality.

[41] Frederick Brown, *The City Is More Than Human: An Animal History of Seattle*. Seattle: University of Washington Press, 2019; Matthew Klingle, *Emerald City: An Environmental History of Seattle*. New Haven: Yale University Press, 2007.

[42] Michael F. Logan, *Desert Cities: The Environmental History of Phoenix and Tucson*. Pittsburgh: University of Pittsburgh Press, 2006; and Andrew Needham, *Power Lines: Phoenix and the Making of the Modern Southwest*. Princeton University Press, 2015.

6 Unrestrained Gotham

However, besides the prevalent narrative paradigm of urban disasters, depicting nature as a victim has become a cliche in the writing of urban environmental history. This clearly contradicts the original intent of environmental history—to explore the historical interplay between culture and nature rather than portraying a one-way relationship. Moreover, it diverges from the ecological reality of cities. Fortunately, in 2014, Theodore Steinberg's groundbreaking work, *Gotham Unbound: The Ecological History of Greater New York*, reintroduced the force of nature into the narrative of urban ecological history. While undoubtedly influenced by *Nature's Metropolis*, Steinberg's book surpasses its predecessor, emerging as a seminal work in contemporary urban environmental history. In *Gotham Unbound*, nature is seamlessly woven into the developmental history of New York, and its profound influence on the city is not just limited to the moment of disaster. Originally an estuary city, New York's geographical features have shaped its ecosystem, landscape, and ongoing interactions with the natural world characterized by cooperation, compromise, and confrontation. As New York gradually transforms its ecosystem, other species and ecosystems continually strive for survival. Despite often appearing as losers, they persistently inhabit certain ecological niches within the city, poised for opportunistic shifts. As a global hub of commerce and fashion, teeming with diverse languages, cultures, and vibrant urban life, New York relies on essential resources such as water, air, food, and various energy sources. It engages in a perpetual struggle against pathogens and grapples with the challenge of waste disposal. Steinberg convincingly argues that exploring our natural environment not only illuminates our present but also offers valuable insights into our past. New Yorkers have continually attempted to harness the untamed natural forces coursing through the Gotham Estuary with cultural order, yet the latter's unbridled power has consistently challenged human endeavors.

Therefore, the study of urban environmental history demands historians to reinterpret the history of urban evolution within the broader context of evolution and ecosystem dynamics, thereby enriching urban history with greater materiality. However, this doesn't imply that the study of urban environmental history will neglect the ideological shifts within cities. Regarding materiality, Steinberg meticulously analyzes the transitions in the ecosystem of urban-nature coexistence. Nevertheless, he falls short of integrating the voices of thinkers into the flowing currents of the Hudson River. In fact, it poses a significant challenge for scholars in both fields to amalgamate urban environmental history with intellectual history. However, it is this amalgamation that transforms urban environmental history into more than just a presentation of the physical environment. It becomes a stage for the interaction and dialogue between the diverse ideas expressed by individuals and groups and the ecosystems they physically inhabit.

Exploring what urban populations think about the relationship between nature and the city is not a new topic for historians. Intellectual giants such as Marx, Spengler, and Mumford have extensively analyzed this relationship, providing endless research topics for future historians. Additionally, artists and writers have expressed and written about the connection and disconnection within the relationship in various forms, attracting the attention of numerous historians.[43] All beings in cities also harbor their own desires and expectations for nature and the city, which have long been the focus of historical studies.[44] These studies are undoubtedly important and necessary, advancing our understanding of urban thought. However, it is now time for ideas to become grounded, allowing intellectual history to seamlessly integrate with environmental history.

Such a fusion implies that scholars of environmental and intellectual history must not remain stagnant within the realm of ideas, despite its already astonishing richness and diversity, which can be overwhelming. As historians venture into the intricate tunnels of historical thought, they must emerge to witness the unfolding changes in the ecosystem of the world where ideas manifest. It was the fertile land saturated with fertilizers that sparked Marx's contemplation on metabolic disruptions. The asphalt-covered, skyscraper-laden, and concrete-dominated landscapes provoked Spengler's concerns about urban civilization. Additionally, the polluted air, pesticide-infused food, and monotonous cityscape fermented the urban populace's yearning for the countryside and wilderness. When ecological catastrophes such as earthquakes, tsunamis, hurricanes, and pandemics strike, their profound impacts on the psyche of urban dwellers linger for generations. Similarly, when cities are situated in inhospitable

[43] Raymond Williams, *The Country and the City in the Modern Novel*, translated by Han Ziman, et al.: Commercial Press, Beijing: 2013.

[44] Peter Schmitt, *Back to Nature: The Arcadian Myth in Urban America*, New York: Oxford University Press, 1969.

ecological zones, the fear of impending disasters often shapes their collective consciousness. Urban environmental historians must grasp the dynamic changes occurring within the ecosystems they study, thereby tracing the psychological and intellectual evolution of the inhabitants.

Furthermore, environmental intellectual history must explore not only how human cognition is molded by environmental shifts but also how people's perceptions of cities and nature have reshaped urban environments and their surrounding ecological landscapes. The desires for urban growth have spurred ecological transitions; dissatisfaction with urban landscapes and fear of urban alienation have driven the establishment of urban parks, fostering the emergence of diverse ecosystems. Concerns about resource and energy consumption by urban society have transformed into conservation movements, while the yearning for natural beauty and wilderness, despite residing in urban areas, has propelled the construction of national parks. Additionally, studies on animal diversity in urban laboratories have laid the theoretical foundation for nature reserves. All these issues have been explored by environmental historians, yet there's a need for more bidirectional and multi-directional research to enrich urban environmental history studies.

Breaking the boundaries of the city doesn't imply the disappearance of cities from environmental history; rather, cities remain central to urban environmental history research. Nor does it suggest blending the city with the countryside and wilderness. As Worster emphasizes, differences exist and are worth respecting. However, it is essential for urban environmental historians to focus on the connections between these differences. Such connections are primarily established along the time dimension. Urban environmental history narratives should not be isolated from the more ancient and enduring histories of nature, agriculture, and pastoralism. Urban environmental historians should contextualize the city's development within the overall evolution of the region. Furthermore, historians can uncover spatial connections by linking the evolution of the urban physical environment with the distant and vast hinterlands it consumes and exploits, exploring the economic and ecological interactions between them, and examining the environmental and social changes resulting from these interactions. Moreover, such connections also exist in the imaginative realm of urban dwellers. Combining their understanding and imagination of the relationship between the city and nature with wild and free landscapes involves addressing concerns about the scarcity of natural resources in consumer-driven societies, extending urban aesthetic imagination, interpreting biodiversity, and ultimately, sustaining the health, beauty, and longevity of ecosystems on an increasingly urbanized planet.[45]

Over half a century ago, Aldo Leopold wrote in an abandoned farm not far from the college town of Madison: "People without a farm may entertain two possible illusions, one being that breakfast comes from the grocery store and the other that heat comes from the furnace."[46] Few contemporary historians own farms. Like most urban dwellers, we consume nature in illusion. However, as historians, we must not allow ourselves to write our past in the fog of illusion and blindness. Instead, we must trace the ecological footprint of the egg that appears on our breakfast table and retrace the ecological path that provides warm air for us on cold winter days. These elements are not only fundamental to the functioning of the cities we inhabit but also the basic units of a vast interconnected life network. Reflecting on their trajectories reveals more than just the changes occurring at a certain moment in a city and its ecological hinterland; it also unveils a series of social, ideological, economic, and ecological evolutionary processes that ultimately lead to these changes. With this understanding, historians will reassess our definition of the city and its history.

[45] My book *The City Natural: Garden and Forest Magazine and the Rise of American Environmentalism*. (Hou Shen, *The City Natural: Garden and Forest Magazine and the Rise of American Environmentalism*.

[46] Aldo Leopold, *A Sand County Almanac*, translated by Hou Wenhui, Beijing: Commercial Press, 2016, p. 6.

PART II

Growing Cities

CHAPTER 3

Crowded, Poisoned, and Beautiful

1 Prologue: "Hell with the Lid Taken Off"

In 1866, the second year after the end of the American Civil War, cities and countryside in the South were still searching for a new order on the ravaged land, while the dark clouds of war had vanished in the northern sky. Steamboats shuttled back and forth on the Ohio River, and hundreds of trains ran daily between cities in the northeast. In that year, James Parton traveled by train to Pittsburgh, the most active city in the industrial landscape of the United States, which had amassed a great deal of wealth during the Civil War due to its developed military industry. At the age of 44, Parton became the most famous biographer of his time. His celebrities' biographies focused on not only political figures such as Franklin, Jefferson, and Jackson but also cultural elites such as Voltaire and Greeley. Unlike most male writers of his time, Parton paid special attention to women, especially those with strong intellectual charm, and wrote *Eminent Women of the Age: Being Natives of the Lives and Deeds of the most Prominent Women of the Present Gentlemen*. For Parton, it was not surprising, after all, his wife, who was ten years older than him, was Fanny Fern, the most outstanding female columnist in the United States at that time. She was not as beautiful as flowers but a feminist with incomparable intellect. Parton's trip to Pittsburgh left such a deep impression on him that he temporarily withdrew from his interactions with those varied and graceful historical figures and wrote a vivid biography of the city called *Pittsburgh*, which was published in the *Atlantic*

S. Hou, *Wallless Cities*,
https://doi.org/10.1007/978-981-97-7827-0_3

Monthly in January 1868.[1] It is in the urban biography that he wrote this famous phrase to describe Pittsburgh: "hell with the lid taken off," which, although not as famous as his wife's famous quote "the way to a man's heart is through his stomach," was still widely quoted by later writers and scholars who describe air pollution.

In the article, Parton's depiction conveys a message of not only worry and regret but also optimism and pride. As an East Coast journalist known for outstanding figures documenting Western civilization, rather than a marginal person like Thoreau who focused on thinking and writing about nature, the information Parton provided may better reflect the ambivalence of his time about urban and industrial civilization. At the beginning of the article, he openly praised, "On that low point of land, fringed now with steamboats and covered with grimy houses, scarcely visible in the November fog and smoke, modern history began." For Parton and many of his contemporaries, modern history meant industrial and urban civilization. It was in Pittsburgh, a city located at the confluence of such three rivers as the Allegheny River, the Monongahela River, and the Ohio River, that a new chapter of American civilization was opened. Because it is not just the confluence of great rivers, where there were roaring sirens and thriving trade, and the starting point of Lewis and Clark expedition, but also the gateway to the western fertile hinterland—every Midwest city calls itself this. Most importantly, in the mountains, there was coal, the black gold that was formed from the remains of organisms millions of years ago; it is the fuel that ignites the light of modern civilization. The coal was so rich and good so that it could satisfy all the industrial ambitions of modern people. "All those hills that frown down upon Pittsburg, and those that rise from the rivers back of Pittsburg, have a stratum of coal in them from four to twelve feet thick. … It is simply a great cake of coal, … uniform, compact, as though this region had once been a lake of liquid coal, upon which mountains had been tossed, pressing it solid." Parton wrote, "The 'great fact' of Pittsburgh is coal!" The steelmaking industry can be brought to coal regions, while gains cannot make up for losses if coal is brought to the steel industry. Starting from the second half of

[1] James Parton, "Pittsburgh", *The Atlantic Monthly*. Vol. 12 (Jan. 1868). *The Atlantic* Online Archive, https://www.theatlantic.com/magazine/archive/1868/01/pittsburg/536130/, its last login date is September 22, 2020. All citations in the following text, if not otherwise specified, are from the article and such citations will not appear any longer.

the nineteenth century, when the siren of industrialization sounded, the demand for coal increased sharply and coal production soared. The four counties around Pittsburgh produced only 500,000 tons of coal in 1850, but it had increased to 5.5 million tons by 1870. It was 11.6 million tons in 1881 and reached 66 million tons in 1911.[2]

Of course, even better than that, Pittsburgh didn't just have coal. In Parton's view, it's also mixed with various natural treasures needed by industrial civilization. Faced with them, Parton completely unleashed his passion and satisfaction for conquering nature:

Brave and steadfast men have battled nobly here with the substances that offer the greatest resistance to our control, and which serve us best when subjugated; and in the hills and valleys round about [Pittsburg], nature has stored those substances away with unequalled profusion. Besides placing a thick layer of excellent bituminous coal halfway up those winding heights, besides accumulating within them exhaustless supplies of iron, besides sinking under them unfathomable wells of oil and salt water, nature has coiled about their bases a system of navigable streams, all of which form themselves into two rivers, the Alleghany and Monongahela, and at Pittsburg unite to form Ohio, and give the city access to every port on earth. It is chiefly at Pittsburg that the products of the Pennsylvania hills and mountains were converted into wealth and distributed over the world. The wonder is, not that Pittsburg is an assemblage of flourishing towns of 230,000 inhabitants, but that, placed at such a commanding point, it is not the *most* flourishing and the *most* populous city in America.

Such a joyful and material tone runs through the evolution of nineteenth-century America. At that time, Americans praised not only democracy, freedom, and ambition but also the materiality of nature. In the transition stage of American society from agriculture to industry and cities they saw more clearly than later citizens in the United States what the material abundance of the Second Earth meant for the young country and the emerging cities it deliberately built. The correct spirit, system, and capital shall rely on the material resources of nature to create more wealth, and nature shall also become efficient and orderly in the hands of these brave and resilient people.

[2] Edward K. Muller & Joel Tarr, "The Interaction of Natural and Built Environments in the Pittsburgh Landscape," in *Devastation and Renewal: An Environmental History of Pittsburgh and Its Region*, ed. by Joel Tarr. Pittsburgh: University of Pittsburgh Press, 2005: pp. 11–41.

It is precisely because of such a perfect combination that Parton sees the victory of urban industrialization and the awe-inspiring beauty in huge black smoke. "Those who approach it in the night see before them, first of all, a black hill, in the side of which are six round flaming fires, in a row, like six fiery eyes." But if people imagine that behind the hills and mountains, there will be a hot labor scene of "the furnace fire lights up earth and sky, red sparks fly pell-mell into purple smoke," they will be absolutely wrong. The era of manpower is fading away, and it is the coke oven that tirelessly does its duty day and night, converting "waste coal-dust" into service-able coke. When he reached the coal mine, the scene before him made him irresistibly ride on the cable car to enter the hill, explore the depths of the mine, and experience the power of civilization. His praise of industrial landscapes is full of quasi-naturalistic analogies, just like Whitman's attitude toward nature, industry, and cities in *Leaves of Grass*. Parton's description shows Whitman's style, after all, his wife has a fanatical passion for Whitman's poetry. As he was about to end his day of wandering in Pittsburgh, he was conducted to the edge of the abyss and looked over the "abyss" (interestingly, he used the word "abyss"), where he saw "the most striking spectacle." The entire space lying between the hills was filled with blackest smoke, from out of which the hidden chimneys sent forth tongues of flame, while from the depths of the abyss came up the noise of hundreds of steam-hammers, and he wrote, "It is an unprofitable business, view-hunting; if anyone would enjoy a spectacle as striking as Niagara, he may do so by simply walking up a long hill to Cliff Street in Pittsburg, and looking over into—hell with the lid taken off." In his cognition, the greatest natural wonders can be equated with the greatest industrial landscapes.

If this is hell, what are those who are punished and destroyed in it? Obviously, it is not this great and vigorous city, or those huge and icy machines, nor the people living proudly in the dense smoke in Pittsburghers live—although only the most fashionable young girls occasionally wear gorgeous dresses, while other pragmatic and thrifty men and women dress in simple and dark clothes because any clothing will be covered with a layer of black and yellow coal ash after walking through half a block. Parton expressed temperate compassion for the miners who had to enter the "abyss" every day to work. In addition to the large number of Irish people who went to the New World to find a new life after failure of the potato crop, there were also thousands of Germans and Welshes. However, Parton did not care as much about the workspace of these workers as the progressives in the late nineteenth century who believe that it is one of the key components in raising workers' pay to improve the working conditions. In 1904, when left-wing novelist Upton Sinclair published *The Jungle,* Chicago's urban smoke and the unpleasant odor, sewage, and noise of the slaughter houses had become the focus of criticism. The latter's cognition stems from the study of the relationship between the environment and respiratory health in modern medicine. In Parton's time, although he had cautious doubts and instinctive aversion to dense smoke, but it is only a certain perceptual cognition When he was on a one-day tour in a dark and smoky city, he repeatedly wrote, "Smoke, smoke, smoke—everywhere smoke." During the day, they can only read in natural light for half an hour in the afternoon, and at any other time, the city lights up all its gas lights. When Parton nervously asked the locals if they would detest such black smoke, they told him that smoke would have its inconvenience, but those outsiders were influenced by their prejudice against smoke in which they have been educated for a long time. Once they transcended their prejudice, they would find that the smoke of bituminous coal can kill malaria and save their eyesight. "The local philosopher" discoursed meaningfully, "The really pernicious atmospheres ... convey to man no intimation of the poison with which they are laden, and we inhale death while enjoying every breath we draw; but this smoke is an evil only to the imagination, and it destroys every property of the atmosphere which is hostile to life." Parton was skeptical of the intelligent view, but speaks out in his own defense, although he sarcastically stated that a more obvious fact was that the fashion of living a few miles out of the smoke is beginning to prevail among the people of Pittsburgh. In the end, Parton prophesied, "In coming years, the smoky abyss of Pittsburgh will be occupied only by factories and 'work', and that nearly the whole population will deny themselves the privilege of living in the smoke."

Alternatively, for Parton, the flames, ovens, black smoke, and abyss only awakened his visual imagination of Dante's *Divine Comedy*, without pondering what was tormented in the "hell with the lid taken off." His sight was captured by the artificial light of the city; he had no intention of exploring life and things in hell although he took off the lid of hell. He was not without regret for the natural beauty lost in the bustling industrial achievements. When he took a train from Cincinnati through long and short tunnels to Pittsburgh, looking north from the south bank of the Monongahela River, he lamented, "Was there ever such a dismal lookout anywhere else in this world?" Those hills, once so beautifully rounded and in such harmony with the scene, have been cut down, sliced off, pierced, slanted, zigzagged, built upon, built under, until almost every trace of their former outline has been obliterated, without receiving from man's hand any atoning beauty. As he gazed at the black Pittsburgh under the smoke screen, he, comparing the past with the present, lamented that scarcely whites lived here over a hundred years ago, but now "we have found better uses for it since. But these better uses have played havoc with the striking beauties of the landscape." However, soon his sighs were overshadowed by the black smoke of Pittsburgh and eventually disappeared into the celebration of industrial victory. There comes an urban age.

The natural landscape, social landscape, and ideological landscape of the United States will all be dominated by cities in the new age. If we define the urban age from an ecological perspective, it has the following characteristics: firstly, cities become the distribution, production, and consumption center of various natural resources and agricultural products while the national market and transportation network are formed. Meantime, even the most remote and isolated countrysides and wastelands have become to some extent the hinterland of cities, and the nature's economy is intricately linked to the human economy. As nature is capitalized, more and more regional ecosystems are disintegrating. Secondly, specialization, a special product of the urban age, has become the most important means for human beings to manage nature. Previously, human activities in nature often relied on personal experience and indigenous knowledge. However, in the late nineteenth century, a large number of specialized professions began to emerge, with their own norms established and their connotations and boundaries defined. The entire era was permeated with a spirit of specialization and reverence for professional knowledge. While applying professional knowledge to put society in good order, many professionals at that time, such as various types of engineers, landscape designers, city planners, horticulturists, forestry managers, and agronomists, began to use it to plan and remake nature. Thirdly, the government has begun to intervene in the management of nature through acts, the establishment of specialized agencies, and other means. While striving to change the laissez-faire and the model of individualistic economy, it also emphasizes the government's responsibility to utilize natural resources and preserve natural beauty. Fourthly, the most prominent feature of the urban age is that during this period, a large number of population breaks away from the labor on the land and enters factories and cities, thus cutting off their direct connection with the means of livelihood they absorb nourishment from and consume. These changes signify not only a change in the mode of social production but also the subversion of the traditional relationship between man and land.

The emergence of urban age had a tremendous impact on American society and its natural environment, including the enormous moral impact brought about by changes in the relationship between man and nature. It is not a unique phenomenon in the United States, but a transformation that any country will experience during the transition from an agriculture society to an industrial society. At that time, European and American countries on both sides of the Atlantic entered the urban age, and people broke away from rural areas and agricultural activities to seek survival opportunities in cities, thus shaking the original moral norms of agricultural society and causing people of the time to have moral concerns about the urban age. However, the social response or shock caused by the change in the United States is particularly strong. After all, the agricultural republic complex that lasted for over a hundred years still lingered in the dreams of many people from all walks of life, while the industrial dust and urban noise were still rampant on the land which was barren three hundred years ago. The core of the American agricultural republic concept is built on the yeomen's code of ethics, which not only contains universal moral norms of agricultural society such as diligence, frugality, and honesty but also has distinct American characteristics, namely, emphasis on the ownership of land and the freedom and democratic spirit that are guaranteed by labor on land. Therefore, when the foundation on which the code of ethics relies is threatened, and the bond between man and land is severed, people begin to worry that the society, which is striving in the process of industry and urbanization, may face the risk of moral decay, thereby shaking America's commitment to freedom and democracy. People are beginning to endow freedom and morality with a new definition of the urban age, a definition whose extension and connotation are constantly changing.

2 The Redemption of Coke City

As environmental historian David Stradling once said, "throughout the nineteenth century, Americans linked all the positive changes brought about by smog and coal, namely production, prosperity and progress.

For many urban dwellers relying on coals, the smog was like a banner of civilization, rising from factory chimneys, locomotives and steamships."[3] Popular novels described it with a passionate tone, Broadway Theater made its actors praise it loudly with songs. The towering chimneys and the swirling black smoke that were painted on the covers of postcards, posters, and brochures were used to showcase the charm and vitality of the city. Parton is just one of the masses participating in the industrial civilization carnival. Although as intellectual elite on the East Coast, his education gave him a transcendent appreciation for the beauty of nature, he preferred the progress of civilization and the interests of the United States. In the article, he expressed his loyalty and pride to his country more than once. Coal and industrial growth arising from coal brought him and millions of Americans like him the hope to surpass the declining British Empire and become the first world power. After all, the Industrial Revolution ignited by coal created the glory of the British Empire, and now, the coal reserves there may only last for 80 years. But here, at Parton's feet, the thick and wide Pennsylvania coal seam can be used for thousands of years—at least that's what people told Parton.

The United States was indeed the Persian Gulf of coal, and Pittsburgh indeed became a prosperous industrial city. The coal-based steel industry and other heavy industries related to coal found the best production and distribution center here. When Parton visited Pittsburgh, the city had a population of less than 100,000, but by the time that coal consumption in the United States reached its peak in 1920, its population had soared to nearly 600,000, including a large number of Southeast European immigrants and black people who had migrated to northern cities during the Great Migration. But at that time, it was no longer the undeterminable, open, and free land of the "Second Earth." People who came to the New World were faced with places similar to the world they had left, equally crowded cities and distinct classes, and in a remote corner of the west, there might still be a piece of land waiting to be developed. However, most new immigrants were destined to find their living space in the thriving industrial city. Pittsburgh became one of the most popular cities because it had coal, steel, and Andrew Carnegie. Steel king Carnegie and steel city Pittsburgh gave each other a boost created the first great

[3] David Stradling, *Smokestacks and Progressives: Environmentalists, Engineers, and Air Quality in America, 1881–1951*, translated by Pei Guangqiang, Beijing: Social Sciences Academic Press, 2019, p. 19.

miracle of American industrial civilization and another American myth that went from a poor wretch to a tycoon. The myth that Pittsburgh and even American cities have always woven is that everyone can become the next Carnegie. Although most people are still Bob, Tom, and Joe all their lives. They have a breakfast of fried eggs with bacon in a narrow apartment filled with soot and then go through the crowded streets filled with smoke to the mines, workshops, or small offices that are also blackened by soot. In the evening, they drink a glass of inferior Bourbon Whiskey under dim and yellowish lights, and the women at home scrub without rest the coal ash that will never be washed away day and night. Perhaps they are querulous, tired, ill, poor, or perhaps don't have as magnificent imagination of civilization vision as the intellectual elites like Parton do, but as they say, they have today's bread when they see the black smoke rising from factories.

These people and those coal miners are the true source of Carnegie's billions of assets. However, Carnegie did not stay in Pittsburgh to continue enjoying the peculiar flavor of Pittsburgh's atmosphere with his employees and admirers. On the contrary, he went in the opposite direction and returned to Scotland, his hometown, where he built a huge castle—Skibo. There was lush green grass in front of the castle which was surrounded by forests, and wild but gentle highland scenery outside the castle. He reinforced the castle with steel and connected himself with the land where his wealth that sourced from and he had left in Pittsburgh through telegrams, telephones, railways, and cars, but he chose to breathe the atmosphere soaked in dew in the early morning in the Scottish countryside.

However, Pittsburgh's naive satisfaction did not last long, or at least did not continue in the way Parton had expected. It is proven that not only Pennsylvania has rich coal seams, but also West Virginia, Ohio, Kentucky, Illinois, Missouri, Kansas, Colorado, Wyoming, and Montana in the west of Pennsylvania are all rich in coal, they are waiting for brave and resilient people to arrive in. The world's coal reserves are 860.9 billion tons, of which the United States accounts for 237.2 billion tons. Although it is out of the question that the coal can be mined inexhaustibly for thousands of years, there is no doubt that it can be mined for at least 200 years according to the current coal production. However, the issue is not whether coal is available, but whether coal should continue to be used.

At the beginning of the founding of the United States, Thomas Jefferson insisted on keeping dirty cities in Europe and turning the world shrouded in soot on the other side of the ocean into a factory for the clean and fresh agricultural land of the United States. His strong rejection of soot and the moral corruption it contained was already very evident. And the doubt was a response to the trend of anti-urbanization thought in contemporary Britain. As early as the thirteenth century in London, when charcoal was in short supply and people were forced to turn to sea coal, there were already voices criticizing the "corrupted air."[4] Starting from the reign of Elizabeth I, the dirty air in London was criticized by people throughout early modern times and became an important reason for the rustic complex of the English gentry. Many historians doubted to what extent the criticism of air pollution at that time was related to health, and to what extent reflected the interpretation of morality and power by the people of the time. Indeed, critics of the time often viewed dirty air as a hindrance to the majesty of the reign, but as William M. Cavert, the author of *The Smoke of London: Energy and Environment in the Early Modern City* put it bluntly, "early modern science and medicine can support the claim that smoke is harmful to health, but did not clearly indicate the degree and nature of its harm.... It is generally found that smoke poses medical risks, but it does not seem to be understood from a precise proof and perspective that is applicable to everyone.".[5] In the eighteenth century, many scholars joined the discussion on air pollution, but too many mussy voices actually made people's understanding more ambiguous. Carver wrote, "The new developments in Common Law and natural science that are related to nuisances provided influential language for the early modern British to condemn urban smoke, but were also restricted by both. Both law and science don't provide clear support for the idea that smoke is harmful, and the contradiction makes it difficult for

[4] William H. Te Brake, "Air Pollution and Fuel Crises in Pre-Industrial London, 1250-1650," *Technology and Culture*, Vol. 16(July 1975): pp. 337–359.

[5] Carl Sandburg, "Prairie". The translated text is excerpted from Zhao Yiheng's compilation: *Anthology of Modern American Poetry (Volume I)*, Beijing: Foreign Language Teaching and Research Press, 2019, p. 175.

the anti-smoke army to defeat the growing importance of coal in people's minds."[6]

Until the nineteenth century, for ordinary people, the common sense that polluted and unpleasant air was harmful to health still came from direct physical sensations, it was free of medical endorsement based on rigorous research. Charles Dickens could depict a Coketown in *Hard Times* to heighten the difficulties of current events, but London continued to allow black smoke to emit from its chimneys. When common sense encountered the challenge of experience—the local knowledge of Pittsburghers, Parton didn't know how to cope with. More importantly, like the UK and other industrial countries on the other side of the Atlantic, coal surpassed the significance of pure fuel and got more connotations. If the pollution it created before was considered a stain on monarchy, then the enormous wealth it brought after the Industrial Revolution gave it moral legitimacy. For capitalists, it meant growth, for patriots, it represented strength, for ordinary people, it provided warmth and livelihood. Even in the first half of the twentieth century, the shadow brought by the two world wars was far more chilling than soot when the widespread harm of soot to health was no longer a controversial issue. Once again, Pittsburgh's coal and steel industry driven by it gained a solid foundation of existence. After all, everything could be temporarily sacrificed for victory. Moreover, the city had reached its peak in both economic status and population by the end of World War II. However, the pollution there also reached its peak, making "the hell with the lid taken off" a well known metaphor. Both the general public and the government felt the necessity to stop the black smoke and know the source of pollution at that time.

In the national census table, Pittsburgh, which had reached its peak in urban GDP, began to decline. Starting from the 1960s, the population continued to emigrate, and in the 1980s, steel companies closed down one after another. The steel city which once shone became a dull rivet on the rust belt. Numerous scholars contemplated the reasons for the decline of genteel industrial cities, including Pittsburgh, during the period, either from the perspective of the adjustment of regional economic structure at home or the transformation of international economic pattern, and

[6] William M. Cavert, *The Smoke of London: Energy and Environment in the Early Modern City*, translated by Wang Qingjiang et al.: Beijing, Social Sciences Academic Press, 2019, pp. 126 and 149.

created considerable achievements.[7] However, if the transformation was examined according to the real situation of the city, it was not a decline, but a renaissance as they themselves called it. Pittsburghers began to redefine the identity of their city and choose its development path. To a large extent, Pittsburgh's transformation was the result of self-reform.[8] Compared to the civilization and cities that Parton hoped to see in 1866, the Americans' understanding of civilization and cities fundamentally changed over the past one and a half centuries. The development of modern medicine made it impossible for natives to continue interpreting the relationship between soot and health on the basis of their desired local knowledge. Their own demands for life and the environment also underwent subtle changes. In other words, more and more people no longer hope to trade their health for a living, let alone accept the sacrifice of their health for growth. Thousands of Pittsburghers who hope to live on the windward side yearn for the place where Carnegie dwelt in his native place in seclusion or the lite place like that. On the path of environmental reform, Pittsburgh was undoubtedly a success. Education and high technology became its new economic pillars, among which environmental design was the highlight of its economy. Although the city was still nagged by air pollution, looking from the cliff where Parton once discovered the "hell with the lid taken off," people could clearly see a green and vibrant city.[9]

[7] Domestic scholars have conducted extensive research on the rust belt and the decline of American industrial cities. In historical studies, please refer to Chapter VIII of the *Patterns of American Urban Development from Urbanization to Metropolitanization* by Wang Xu, Beijing: Tsinghua University Press, 2006; and several articles by Han Yu, American historian, on the decline of industrial cities in the northeastern United States.

[8] For environmental reform in Pittsburgh, see Joel Tarred, *Development and Renewal* and others; In addition, see Joel Tarrand Karen Clay, "Pittsburgh as an Energy Capital: Perspectives on Coal and Natural Gas Transitions and the Environment" in Joseph A. Pratt, Martin V. Melosi, and Kathleen A. Brosnan eds., *Energy Capitals: Local Impacts, Global Influence*. Pittsburgh: University of Pittsburgh Press, 2014: pp. 5–29; and Joel Tarr, Tim Collins, and Edward K. Muller, "Pittsburgh Rivers: From Urban Industrial Infrastructure to Environmental Infrastructure" in Christ of Mauch and Tom Zellereds, *Rivers in History: Perspectives on Waterways in Europe and America*. Pittsburgh: University of Pittsburgh Press, 2008: pp. 41–62.

[9] Because the urban environmental history of Pittsburgh is so representative, its attention to the environment surpassed many other cities not only in the prosperity of environmental design but also in many aspects such as its municipal budget and urban concept. It was also at the University of Pittsburgh (PITT) that the most important

The redemption of Pittsburgh seems to be a story of victory, as the city did not truly declined, but became healthier, more livable, and even wealthier. However, the conclusion of such a victory is based on two prerequisites. One prerequisite is not to look back at the lives of individuals, groups, and other species that have been tormented and sacrificed in the "hell with the lid taken off," as well as the wild beauty that Parton expressed regret at. The other prerequisite is not to look forward to the future of Pittsburgh and not to think about where the energy and food that sustain the city's vitality come from. Historians like to tell stories, which inevitably have a beginning, an end, and a climax. However, the real world still runs and does not pause owing to the endpoint chosen by historians. Pittsburgh, like hundreds of American cities on the rust belt, the fundamental solution to air pollution lies not in the invention of new technologies, but in the conversion of fuel from coal to petroleum and the transfer of heavily polluting enterprises from domestic to overseas. North America, which used to be the hinterland of the Old World, returned to the Eastern Hemisphere in the 1950s, went to the Southern Hemisphere, and developed its own hinterland there. What will happen in the future when these hinterland areas cannot continue to support the operation of American cities like Pittsburgh?

The issue of hinterland and cities will be discussed in the next section. In the chapter, we will temporarily stay within cities and examine the changes and reforms in the internal environment of cities.

Fighting in the Name of Freedom Against in the Name of Health.

A city is never an enclosure space. As a vast, open and constantly changing ecosystem, it is porous and permeable, and to some extent, it resembles an organism.[10] Soot pollution is simply a manifestation that

books on the research into environmental history of American cities was published. Joel Tarr, a founding member of urban environmental history, was teaching at Carnegie Mellon University (CMU). Inspired by the environmental history of his city, his personal academic interests shifted to one of the pioneers of urban environmental history from a political historian studying the politics of urban "moguls".

[10] For a discussion on treating cities as an ecosystem, see Thomas R. Detwyler and Melvin G. Marcus, eds., *Urbanization and Environment: The Physical Geography of the City*. Belmont, California: Dexbury Press, 1972;Spenser W. Havlick, *The Urban Organism: The City Resources from an Environmental Perspective*. New York: Macmillain, 1974; and Richard T. Forman and Michel Gordon, *Landscape Ecology*, New York: John Wiley and Sons, 1986. The reflection on porous and permeable cities mainly stems from Rachel Carson's discussion in *Silent Spring* and Brett L. Walker's reflection on pain in *Toxic Archipelago: A History of Industrial Disease in Japan*, especially William Cronon's fine

urban operation infiltrates into the atmosphere and returns to organisms, including the porous human body, through the flow of air. Whether in Pittsburgh, New York, Boston, Detroit, or Los Angeles, modern cities of all sizes also suffered from various forms and properties of infiltration. Through gas emissions, industrial production, human excretion, and various other forms of waste, they, with water, soil, and air as media, ultimately returned to the bodies of organic matter living in cities. The connection between man and nature in cities was mainly perceived through their bodies. It was the existence, mechanisms, health, and illness of the bodies that constantly reminded people who had wrapped themselves up and armed themselves with various civilized materials, that transcending nature was just the big-headed fantasy of modern people. Toxin producing and poisoning are the common characteristics of modern cities and the biggest environmental difference between traditional cities and modern cities. All cities, traditional or non-traditional, have their hinterlands for producing food and energy, but they are different in the size of the hinterlands, the pressure they bear, and the materials they provide. All cities are also ecosystems that are not only simplified and highly domesticated but also have an impact on the habitats of diverse organisms, but their impact on pathways and degrees vary greatly. All cities are always in danger of natural risks and restrictions, such as floods, earthquakes, hurricanes, wildfires, and pathogens. Modern cities have greatly enhanced their ability to resist natural risks and rescue functions through technology, which has also exacerbated and even created more natural disasters from another perspective.[11] However, compared

preface to the book. Brett L. Walker, *Toxic Archipelago: A History of Industrial Disease in Japan*, translated by Xu Jun, Beijing: China Environment Science Press, 2012. (The original version in English was published in 2010).

[11] The latest classic example is COVID-19 Pandemic in 2020. On the one hand, modern epidemic prevention measures based on big data and the growing bacteriology since the twentieth century, compared with the plague in the traditional era, have significantly reduced human mortality. On the other hand, COVID-19 Pandemic is a link in the entire modern epidemic chain, and the causes, modes, and speed of its transmission are closely related to the survival way of people who live in a crowed environment and rapidly travel worldwide. See *Donald Worster, Another Silent Spring*, translated by Hou Shen, Chinese Reading Weekly, April 29, 2020, 13th edition; and the *Reflections: Environmental History in the Era of COVID-19* published in the 4th issue of Environmental History in 2020. The reflections include articles from 20 environmental historians from all continents in the world, among which John R. McNeill's article made a novel discussion on the formation of bat-like modern people—living in crowded environments and global

with modern industrial cities, traditional cities do not have the ability to produce toxins and therefore do not need to withstand the attack of toxins. It does not mean that traditional cities are safer than modern cities, as most cities in the world have stayed away from the threat of war only for less than 100 years. Moreover, city walls cannot often withstand huge natural disasters although they can to some extent block the military attacks of invaders. Even plague is very closely associated with disasters, which is common in human history. In extreme times, its destructive power is tantamount to massacre of all the inhabitants of a city. In addition, ancient cities in Europe and Asia were troubled to varying degrees by sewage and excrement. However, when a city could operate well, the pollutants produced by organisms would be recycled by rural areas to achieve a new energy metabolism between urban and rural areas.

The relationship between biological and physical energy exchange, all of these old ways of urban operation, was disintegrated and reconstructed upon the advent of new industrialized cities. The old biological and physical exchanges didn't disappear, but the ecological order they established was completely reorganized through the infiltration of new chemical variables. The toxins which are brought about by the use and processing of fossil fuels as well as the maturity, growth, and mass production of the chemical industry, cannot be naturally degraded or require a long time to degrade. They can be suspended in the air, soaked in soil, stored in wild and domesticated animals and plants, or flow in water and settle in food, then ultimately return to the human body. It is the new capabilities of cities, which are endowed with the powerful new technology. Most paradoxically, the invention of each new technology is to solve the problems faced by cities, but each solution creates new and larger problems until modern cities are completely immersed in various artificially synthesized toxins, and citizens and other organisms are swimming in an invisible toxic soup.

The emergence of these problems is common in every place that voluntarily or forcibly enters the Industrial Revolution and where modern technology is used to transform urban environment and life. They may break out earlier or later, heavily or slightly, but it is universal that urban environment will response to such problems. However, the reasons for

air travel—and the relationship between the occurrence and spread of COVID-19. John McNeill, "Bats, Battiness, and the COVID-19 Pandemic," *Environmental History*. Vol. 25(Nov. 2020): pp. 635–9.

these urban problems in different countries and regions are not entirely the same, and their solutions vary greatly as well. In the nineteenth century, the United States was a country where laisser-faire capitalism was highly developed. People could freely buy and sell land, mine mineral resources, cut down wood, burn coal, and even emit pollution. Any regulation of these freedoms would arouse strong resistance. If the strongest weapon against environmental regulation in other countries is the demand for economic growth, it is a collusion of freedom and growth in the United States. In the country, freedom is an invincible pass, and ultimately, environmental reforms at different levels will compete with this long-standing American pass, and redefine the boundaries and connotations of freedom. In cities, health becomes the ultimate barrier to the universal pass. However, as Martin V. Melosi pointed out in his encyclopedic masterpiece *Sanitary City: Urban Infrastructure in America from Colonial Times to the Present,* health is also a concept of change. In addition to the health in terms of directing at human body, the relationship among human health, public health, and environmental health changes as well.[12]

In the book, Melosi mainly discusses three aspects of sanitary cities: water supply, sewage systems, and solid waste disposal. He divided the sanitary service in the United States into three stages, the first stage, which he referred to as the "age of miasma," means the period from the colonial era to 1880. In this age, people's knowledge of health hazards, especially their understanding of infectious diseases, came from the traditional miasma theory. It is believed that dirty and stinking smell pose a threat to health, and these odors come from either water or garbage.[13] How to provide clean drinking water and treat sewage became the core issue of discussion during the period. During the period, the water supply

[12] Martin Melosi, *The Sanitary City.*

[13] Melanie Kiechle uses her sense of smell to perceive nineteenth century urban American and its environmental reforms in the book *Smell Detectives: An Olfactory History of Nineteenth-Century Urban America.* Ordinary people still adhered to the inference based on miasma theory that diseases were caused by foul odors, in order to urge Board of Health to promote reforms and address the health problems caused by odor pollution from garbage and water sources. During the period, germ theory, a new medical theory, was encountered. The interpretation of pollution by scientific expertise largely conflicted with the existing common sense although both sides were hoping for a reform of the urban environment. Melanie Kiechle, *Smell Detective: An Olfactory History of Nineteenth-Century Urban America.*

and sewage system in the United States was a transplant and transformation of the expertise of British public health and engineering science. It was during the period that the water-works, labeled as the "protosystem" by Melosi, emerged in some relatively mature cities in the United States and became the foundation of modern systems later on.

The second stage refers to the period from 1880 to 1945, during which the germ theory fundamentally changed people's cognition of the source of diseases. The previous miasma theory was gradually replaced by the germ theory, and at the level of public health, it could be seen that there was a broader discussion on solid waste, wastewater, and the supply of clean water. However, due to the emergence of the germ theory, professional discussions and goals on health shifted toward medical methods. On the other hand, Melosi wrote, "Although the conflict between the germ and filth theories of disease had long been settled, from an environmental standpoint there continued to be a preoccupation with biological forms of pollution at the expense of a better understanding of chemical sources, especially industrial pollutants."[14]

The third stage, from the end of World War II to the year 2000, was referred to as the "New Ecology" by Melosi.[15] However, the focus of Melosi's thinking was not on the overall development of ecology and the impact of the initiation of urban ecology on the transformation of the sanitary city during the period. He was more interested in following two aspects: on the one hand, the new environmental pressure and the demand for sanitation facilities and services, which were caused by suburban sprawl. On the other hand, the relationship between the new pollution and the tight budget of the city, which were resulted from the fact that the existing sanitation facilities were outdated and degraded. The combination of both aspects made the city increasingly inclined to accept funding from the federal and state governments in sanitation management, and the exchange condition was naturally to accept their deeper and wider interference. He wrote, "By the 1970s, federal and state government funding was a basic component of the municipal budget of almost every city." However, Melosi did not neglect the impact of the environmental protection movement that emerged during

[14] Melosi, *The Sanitary City*, p. 13.

[15] An interesting question is whether there was a fourth stage if Melosi republished the book in 2020? If there is the fourth stage, does COVID-19 or global warming become his focus of concern?

the period on the formation of new sanitary cities. From the perspective of funding and regulation, the federal government was increasingly involved in urban sanitation affairs. However, the demand for "environmental justice" and "equality" by environmental movements, especially grassroots environmental movements, became a topic that urban governments had to directly face.[16] From the perspective of the evolution of environmental knowledge and sanitation cities, Melosi pointed out that the new ecology had brought about recognition of "environmental risks", which had also led to a shift in people's concerns from "pure biological forms of pollution to chemical/industrial pollutants and pollution from municipal sewage systems."[17]

In terms of the evolution of the policies and infrastructures of the sanitary city, Melosi's observation was without doubt comprehensive and thorough. If we shift our concern from the focus on the relationship between health theory and sanitary society to the reflection on the particularly prominent confrontation between health and nature in American cities, then there is another division way. Taking the year 1945 as the watershed, in the first stage, the focus is the consideration between human health and public health. In the second stage, emphasis is placed on the connection between human health and environmental health. Both the stages are not quite separate, they heavily coincide but are also mixed with a lot of consideration for other social and environmental issues. The environmental knowledge formed in the later period of the first stage and the environmental and social issues accumulated at that period are the key triggers for the transition to the second stage.

In the first stage, the issue of public health was not just about health, but more importantly, about morality. Therefore, although whether to maintain an individual health was personal freedom, he/she didn't have the freedom to harm public health, as the latter was almost equal to morality. When the miasma theory was prevalent, the filthy smell and corrupt morality were inseparable, and the arrival of plague was often

[16] Melosi, The Sanitary City, pp. 358, 366–9. Robert Gottlieb discusses in more detail the development of environmental movement towards environmental justice movement and the transition of the home of environmental movement from wilderness to city in *Forcing the Spring: The Transformation of the American Environmental Movement*. However, the best work on the relationship between environmental equality and industrial pollution is still Andrew Hurley's discussion of the industrial city of Gary, Indiana (*Environmental Inequalities*). See the Part I of the book.

[17] Melosi, *The Sanitary City*, p. 14.

explained as a punishment from God. That's also why far more aborigines, compared with whites, were killed and whites were deeply convinced that God was on their side when viruses arrived in the New World from the Old World. As the famous Puritan priest Cotton Mather witnessed aborigines dying from smallpox, he sincerely believed, "God cleared the woods where there are dangerous creatures in order that living things better grow." Religious beliefs and the moral interpretations of health were intertwined with their inheritance of the nuisance law in British Common Law, thus forming the ideological keynote of strategies to deal with garbage, sewage, and air pollution in the colonial period and early America. Because public morality held such a high position, Boston, the city that adhered to the Puritan code of conduct, established the earliest Board of Health to constrain people's freedom to cause public nuisance.[18]

This way of thinking was not interrupted by the emergence of the germ revolution. Throughout the progressive reform, the germ theory was once again subtly transformed into issues with regard to morality and public health among middle-class reformers. However, at that time, the moral issue was not only a conflict between the white race and colored race as well as the middle class and the working class but also manifested in the game between new immigrants and old immigrants about what kind of natural environment cities needed to ensure public health. New immigrants mostly came from rural areas in Europe and the United States, so the conflict between the two sides was also a contradiction in the understanding of urban and rural health, nature, and freedom.[19] The period from World War I to World War II was a time when national interests and public spirit had the largest supporters in the country. In the first half of the twentieth century, scientists knew, without a shadow of a doubt, that the environmental problems faced by American cities, and even the entire United States, were no longer just about how to wisely and effectively protect and utilize the physical properties of resources and economic issues in the agricultural era, and health issues were not only biological problems caused by bacteria but also chemical problems created

[18] Melosi, *The Sanitary City*, p. 20.

[19] Adam Rome was the first to explore the contradictions and conflicts at the level in his article *Nature Wars, Culture Wars: Immigration and Environmental Reform in the Progressive Era*, thus showing the complexity of the issue during the Progressive Era. Adam Rome, "Nature Wars, Culture Wars: Immigration and Environmental Reform in the Progressive Era," *Environmental History*. Vol. 13(July 2008): 432–53.

by industry and cities. For example, Alice Hamilton, a doctor, chemist, and health reformer, published *Industrial Poisons in the United States* in 1925 and *Industrial Toxicology* in 1931. However, war and the Great Depression first brought about economic considerations, which were the sacrifices that individuals, driven by the moral code of patriotism, had to make.[20]

It was during the period that a large amount of knowledge about the relationship between health and environment began to fundamentally sever the mystical connection between disease and morality: dirty environments did not lead to moral decay, while degenerates were highly likely to have a healthy and beautiful body. However, that didn't make diseases become a purely private matter. On the contrary, a new generation of professionals found that diseases caused by modern industrial pollution, like plagues previously resulted from bacteria, were closely related to environmental health. If the previous reforms emphasized the obligation of cities to provide public facilities and manage public open spaces, people clearly realized that it was impossible to hinder the movement of the atmosphere and the spread of invisible dangerous toxins by cleaning up the garbage of a city and building even water and sewage pipelines which could service for more than 10,000 years after the discovery of the chemical source of air pollution. Air, like water, became a matter of national affairs. In reality, there are two key reasons why cities accept state intervention in their sanitation construction. Firstly, air pollution is not restrained by urban boundaries and its components are too complex, but broader powers and more comprehensive professional knowledge are required to cope with it. Secondly, the pollution behavior of interstate or even multinational corporations is not something that any city or even state government can restrict, it needs the regulatory power of the state. In the name of health, the constraints at that time are no longer only individual freedom or the freedom of one enterprise, nor is it the freedom of a city to arbitrarily shift its environmental pressure, the freedom of the entire capitalist economy and culture. Most importantly, human beings indiscriminately destroy the freedom of our planet in order to enjoy their own health, safety, and comfort. The focus of public health is still dependent on human beings, although it has shifted from individuals to groups of people. In the new stage, environmental health becomes a crucial link

[20] Alice Hamilton, *Industrial Poisons in the United States*. New York: McMillian, 1925; *Industrial Toxicology*. 2nd ed. New York: Scribner, 1934.

in the entire urban reform. It means that the connection between human health and the health of the environment in which they live is rediscovered, explained, and strengthened by modern science. Thus, a new morality that transcends human-centered ethics springs up. The traditional Puritan morality didn't completely vanish, but it was changed, transformed, and fused into the environmental morality established in the new urban age. Stage division is only for narrative purposes, but with the aim to truly understand how American cities entered the new morality, we shall go back to the moment when the urban age began—the Progressive Era—and explore the efforts of progressives to transform cities into public open-spaces at a critical moment of American urbanization.

3 Cities as Public Open-Spaces

From the late nineteenth century to the early twentieth century, the United States was in a transition period from an agricultural society to an industrial urban society. In the 1780s, the total population of the United States, at the time of its founding, was approximately 4 million, of which 95% were agricultural population, including a large number of slaves working on plantations. Even in the first half of the nineteenth century, namely, at the beginning of the Industrial Revolution, the country was still in the agricultural stage. Even as late as the 1850s, when the United States was about to cross the so-called threshold of modern economic development, its total population was 23 million, of which 90% were still agricultural population. However, 50 years later, namely, in 1900, the total population of the United States surged to 76 million, while its agricultural population decreased to 60%. In the 1920 Federal Population Census, more than half of Americans became urban dwellers for the first time in American history, with a total population of 106 million.[21]

If there were still large tracts of publicly owned, mostly natural land preserved in the vast western region, the East Coast of the United States had fully entered urban society in the late nineteenth century. In 1800, New York, the largest city in the United States, had a population of 60,000, followed in descending order by Philadelphia with a population of 41,000, Baltimore with a population of 26,000, and Boston with a

[21] The data source from the online U. S. Census summaries, http://www.census.gov/population/www/censusdata/files/table-4.pdf., which is logged in on August 20, 2010. In the United States, a town is defined as a place populated by more than 2,500 people.

population of only 25,000. By 1850, New York and Baltimore ranked first and second with the population of 516,000 and 169,000 respectively, while Boston became the third largest city in the United States with a total population of 137,000. By the end of the nineteenth century, Boston had dropped to fifth place, but its population had reached 561,000, while New York had a population of 3.4 million and the population of Philadelphia was 1.3 million. In addition, two emerging cities had risen from the territory of the United States, namely Chicago with a population of 1.7 million and St. Louis with a population of 575,000.[22] In just 100 years, the population growth rate of these cities all reached tens of times, and the entire territory, style and ethos of cities changed fundamentally.

Although technological development, abundant material resources, and booming knowledge have all brought unprecedented confidence in the progress of civilization and admiration for human capabilities. But on the other side of such confidence, the new society is disturbed by anxious and uneasy voices. This kind of voice is not a hidden undercurrent, but active appeals to social change in various forms, with the aim to establish new norms for cities and industries, or to restore the original order of the yeoman age for some people. The latter represents the farmers who still stay on the land, especially the large number of farmers in the central and southern regions who have been reduced to tenant farmers. They were once landowners and firmly believed that they were the backbone of American democracy, a symbol of American freedom and property, and "most moral people" as Jefferson called them. Therefore, as their land and property were encroached upon by cities and capital, they sought to reverse the raging tide which would overturn the virtues that the country should possess in their eyes so as to maintain the original value system of agricultural society, especially the unique American yeoman society, such as laissez-faire political and economic concepts, Protestantism-based moral standards, and the man-land relationship that man take what he wants from the land, thus setting off a "Populist" Movement. However, despite the initial grand and spectacular

[22] See U. S. Census Bureau, "Population of the largest 100 cities and other urban places in the U. S., 1790 to 1900," http://www.census.gov/population/www/documentation/twps0027/twps0027.html., which was logged in on August 20, 2010.

scale of the movement, its inherent contradictions and deviations from the development of the times destined it to come to a miserable end.[23]

A reform movement that later became known as "progressivism" was more tallied with the rhythm of the times. In the 1880s, some middle classes in major cities such as New York, Boston, and Chicago began to respond to various problems that arose during the urbanization process in hopes of seeking new order for the transitional society through innovative means. Compared with the Populist Movements of the time, they were greatly different in social foundation, pursuit of goals, and basic connotations. The Progressive Movement laid stress on the government functions in the management of society and natural things, adjustment of labor relations, improvement of worker welfare, advocacy of women's rights, promotion of universal education, amelioration of urban environment, protection of nature and its resources, and so on. It, like the Populist Movement, had a strong moral appeal. However, the latter had larger interests in the spirit of individualism, while the morality of the Progressive Movement put more focus on the dependence and coexistence between social groups. The reason for such differences was twofold: Firstly, people, compared with that in countryside, were inevitably required to closer collaborate with others while living and working in cities. Secondly, the limited physical space in cities constrains individual behaviors. Therefore, dualism was the moral tendency of progressive reformers. Under the basic premise of freedom and democracy, they required individuals to make a compromise on collective will, especially on the public interest.

At that time, progressive reformers generally attributed the moral problems of cities to the noisy and dirty physical environment as well

[23] For the protests and failures of populism, see Michael Kazin, *The Populist Persuasion: An American History*, Ithaca and London: Cornell Univ. Press, 1998; and Catherine M. Stock, Rural Radicals: *Righteous Rage in the American Grain*, Ithaca and London: Cornell Univ. Press, 1996. For the classic discourse on how the so-called "island" civilization, which emerged from various regions of the United States after reconstruction, connect the federal government and monopolistic enterprises into a whole to establish a new social order, see Robert Wiebe, *The Search for Order, 1877–1920*, New York: Will and Wang, 1967. In the book, Wiebe points out that populism, fundamentalism and other forms of agrarian radicalism are all resistance to the new order, and the root cause of their failure is that these rebels attempt to know about a larger world that has already been incorporated into a broader system on the basis of the small environment they are familiar with, and to control a non-private world on the basis of the experiences and customs they use in their acquaintance society.

as the cold and dull social environment. Therefore, they believed that to establish a moral order in the urban age, it was necessary to improve the urban environment and teach new urban immigrants how to adapt to such an environment which was quite different from the countryside. They were mainly concerned about the areas where the urban poor gather, the focus of urban moral issues, most of which are the living and working places of new immigrants. The most significant reform of the slum and factory environment at that time was a "municipal housing movement" led mainly by women. Many of their representatives came from big cities in the Eastern and Central United States, with the most famous being Jane Addams, who lived in Chicago.[24] In the eyes of these female reformers, the crime and moral issues in slums were not, as many social Darwinists at the time said, racial issues, which should be addressed through eugenics to select the superior and eliminate the inferior among social groups. On the contrary, in the "municipal housekeeping movement," they mainly regarded it as an environmental issue. Therefore, they appealed for the changes in the environment of slums, especially in hygiene and education, and the establishment of a more conducive place for the growth of people who couldn't help but live there, especially teenagers, in order to enhance their moral cultivation and ultimately achieve progressive reform of the entire society.

These female reformers were not just housewives who had received good traditional talents and humanistic education. Although they found municipal housekeeping, a combination of municipal and domestic affairs on the path of reform, in order to make their call more approachable and easily understood by ordinary women, they are not only passionate about encouraging women to join social reform but also calling on women to extend their family roles in public life. Their understanding of modern cities, environment, and health went beyond the scope of traditional morality. In fact, some of them were active readers of the most cutting-edge natural sciences at that time, and even directly involved in scientific research.

The representative among them is Ellen Swallow Richards, the first female graduate and also the first female teacher of Massachusetts Institute of Technology (MIT). Compared with her successor Alice Hamilton,

[24] There is a wealth of research on Jane Adams, but the most valuable one is her autobiography: Jane Addams, *Twenty Years at Hull House*, with Autobiographical Notes New York: The Macmillan company, 1937.

Richards's career as an urban environmental historian was far from receiving the attention she deserved in her writings. However, it was the female chemist who began to reinterpret people's daily lives, especially the daily necessities that women often used, with her professional knowledge.[25] The home economics she advocated was based on her own research into sanitary chemistry, which incorporates personal and household food, water, and nutrition into the environment. While public health was still in the stage of germ theory, she began to explore the hazards of chemical pollution of water and air to the human body, in a sense, she started early discussions on environmental health. At the beginning of *Air, Water and Food from a Sanitary Standpoint* (the first edition) she published in 1900, she wrote: "Air, water, and food are three essentials for healthful human life. Sanitary Chemistry deals with these three commodities in their relation to the needs of daily existence: first, as to their normal composition; second, as to natural variations from the normal; third, as to artificial variations—those produced directly by human agency with benevolent intention, or resulting from carelessness or cupidity."[26] On the basis of the analysis of the changes that humans bring to air, water, and food, Richards explored "euthenics" that she called. In 1910, she published *Euthenics, the Science of Controllable Environment; A Plea for Better Living Condition As the First Step toward Higher Human Efficiency*, which truly combined the private space of the body and family with the public space of the city into an interconnected and inseparable environment in the discourse of modern science.[27]

It is worth noting that progressivism is a social and political movement known as "Atlantic Crossings" by historian Daniel Rodgers, rather than an isolated event full of American exceptionalism.[28] Driven by the

[25] For example, in the *Sanitary City: Urban Infrastructure in America from Colonial Times to the Present*, Richards never appeared. In Robert Gottlieb's *Forcing the Spring: The Transformation of the American Environmental Movement*, Hamilton had played the core role of American environmental reform since progressivism, while the contents about Richards was only two pages.

[26] Ellen Richards and Alpheus Woodman, *Air, Water, and Food: From a Sanitary Standpoint*. 1st edition, New York: John Wiley and Sons, 1900, p. 1.

[27] Ellen Richards, *Euthenics, the Science of Controllable Environment; A Plea for Better Living Condition As the First Step toward Higher Human Efficiency*. Boston: Whitcomb and Barrow, 1910.

[28] Daniel Rodgers, *Atlantic Crossings: Social Politics in Progressive Age*, translated by Wu Wanwei: Nanjing: Yilin Publishing House, 2011.

invincible power of capitalism, the United States at the turn of the nineteenth and twentieth centuries was deeply involved in the reconstruction of the North Atlantic world, and its cities became experimental fields for political and social reform in the transatlantic world. Rodgers' academic interest concentrates on the transformation of political conception and ideas. In his view, cities have two wonderful sides, the explicit side is the indulgence of extreme individualism, and the other side hidden in the city is the interdependent network. The network is referred to by Rodgers as the "formless capitalist collectivism of the city," which "forced collective considerations into political consciousness." He keenly noticed that the most pressing issue of reform in cities in the transatlantic world is "health": "Much of the history of social politics' formation might, indeed, be written as the story of the battle for a concept of public health, rooted not only in personal discipline and hygiene but also in the social environments of labor and shelter." It is the theme that shapes the reform of the "common elements" shared by these big cities at the time. They are "the mundane, material utilities of urban life: water, streets, trolley cars, public baths, gasworks, and housing." However, the fundamental issue they refer to is whether to continue to tolerate the rampant laissez-faire principle, especially the exploitation and expansion of capitalist enterprises disregarding the public good. Therefore, "[progressives] try to realize the implicit collectivism of the city on conscious and public lines."[29] Municipal services became the responsibility of the government at that time. As a public service, it should no longer be a channel for private enterprises to make profits. Undoubtedly, it is a long process of reform, but as Rodgers said, "Municipalization was the first important Atlantic-wide progressive project. In borrowed experience and transnational example, in a shared language of germs and sewage ..., municipalizers across the North Atlantic economy began to construct a modest, alternative social economy to that of the thoroughly commercial city."[30]

Today, when we read the history of both sides of the North Atlantic at the turn of the nineteenth and twentieth centuries, we are amazed by the frequent and close exchange of knowledge and thinking in the pre-aviation (let alone the Internet) era. At that time, politicians, reformers,

[29] Rodgers: *Atlantic Crossings: Social Politics in Progressive Age*: pp. 114–117.

[30] Rodgers: *Atlantic Crossings: Social Politics in Progressive Age*: pp. 162.

and intellectual elites shuttled between countries, and books and knowledge quickly spread with the Atlantic as a medium. American travelers went to Europe, England, or admired the grandeur and elegance of Paris designed by Haussmann, marveled at the neatness and efficiency of cities after German unification, or found their spiritual home in the garden cities in England which realized Ebenezer Howard's ideal.[31] When they returned to their hometowns on the other side of the ocean, many felt ashamed of their cities which were rough, dirty, and lacked unity and completeness, while others were full of ambition, they found their love in the dazzling European urban style and proposed one blueprint after another for transforming urban public open-spaces.

However, the United States is not just a mediocre recipient of European urban aesthetics and social ideas. In Rogers' view, "No other planners in the North Atlantic economy devised bolder civic blueprints than the Americans between the 1890s and the eve of the First World War; nowhere else was the planning of cities pushed with such evangelical zeal".[32] More importantly, they began to create modern cities, the ones that could express their own characteristics. When the skyline of Chicago began to be outlined by skyscrapers made of reinforced cement in the 1890s, and the island of Manhattan rose to prominence on the other side of the Atlantic with towering buildings, American cities began to provide new directions of aesthetics for Europe and even the world. Whether they are appreciated or not, they redefined the urban landscape of the world. The contribution of American progressivism, which is equally as important, perhaps more enduring, as it, lies in the provision and preservation of the wilderness, an opposite aspect of urban civilization. Starting from the establishment of Yellowstone National Park, the first one in the world, in 1872, the most prominent aspect of American reform, in the eyes of progressive reformers in the United States, was the establishment

[31] *Garden Cities of To-morrow*, Ebenezer Howard's masterpiece, is a classic work in the history of urban design. Howard hopes to combine the opportunities and convenience of cities with the health and beauty of countrysides through the new garden cities so as to control the disorderly urban sprawl. The book, with the titled of *Tomorrow: A Peaceful Path to Real Reform*, was first published in 1898. When it was reprinted in 1902, it was renamed *Garden Cities of To-morrow*. The book had a large number of supporters on both sides of the North Atlantic, and cities or communities were designed on the basis of Howard model in Blighty, Germany, Austria, and the United States. Ebenezer Howard, *Garden Cities of To-morrow*. London: S. Sonnenschein & Co., Ltd. 1902.

[32] Rodgers: *Atlantic Crossings: Social Politics in Progressive Age*: p. 168.

of a complete set of laws, institutions and systems to protect the wilderness by using the state functions. In the next century and a half, people's understanding of the wilderness evolved with the evolution of ecological thought and American society. The original nationalism gradually faded, the established single aesthetic principle gradually became lax as well, and the meaning of wilderness became far more complex and diverse. The existence right of nature ultimately defined the value of wilderness. However, interpretations derived from the Progressive Era, in the relationship between wilderness and city and in bridging the "emotional gap" between urban populations and nature, have not undergone fundamental changes to the present day. The interpretation will also cross mountains and seas to every corner of the world.[33]

In terms of the interior of the city, it was through such exchanges that the major cities on both sides of the North Atlantic began to trend toward high homogeneity, the basic characteristics of modern urban society in the Progressive Era. If the urban morphology in traditional societies, just like the natural systems it is located in, is defined and shaped in the local ecology and culture, then cities entering modern industrial societies begin to be integrated into a homogeneous global ecological and social order. Paris, Berlin, London, and New York all sought the direction of modern cities in the midst of industrialization. They all faced a large influx of immigrants from rural areas, and the volume of cities rapidly expanded due to the sudden increase in population. The existing infrastructure, municipal services, and energy supply were in straitened circumstances and couldn't meet with the mushroom growth. They all faced inevitable disorder and chaos in the transition from agricultural society to industrial and urban society, and intense conflicts not only between workers and their employers but also between the expansion of monopoly capitalism and government functions as well as rights and obligations. More importantly, a full set of common political and scientific discourse was adopted in the more frequent exchanges between the two sides of the Atlantic due to new technological means. The discourse, in the urban age that requires maintenance through new technology and order, successfully entered political life and formed a "technocracy" that manages society and nature with professional knowledge.

[33] For specific discussions on the relationship between wilderness and cities, see Chapters IV and VI of the book.

At the same time, they also faced many common obstacles: the confrontation of enterprises and the different voices within the government, which were to be expected, and a relatively weak but stubborn resistance coming from ordinary people's defense of the boundaries of their private lives. Their customs, morals, and local knowledge that were handed down through generations, as well as the ways of handling their bodies and property were further incorporated into the discipline and adjustment of public health and morality at that time. In the eyes of progressive reformers, all obstacles were stubborn resistance to "progress" from conservatism, backwardness, or greed, but in the interpretation of each group of resisters themselves, their resistance had different connotations. For regulated enterprises, especially those in the United States and theorists who supported laissez-faire capitalism, progressive reforms not only interfered with their interests but also adhered to the principle of freedom that capitalism had upheld since its inception. The response of progressive reformers to those was clear and firm: there was no doubt that freedom constituted the fundamental moral connotation of capitalist culture, but its premise was not to harm the freedom of others. In cities, profit-seeking behaviors that disregarded or even destroyed public health pursued profit rather than freedom, as their behaviors went against the premise of freedom.[34] However, more complexity emerged at the level of private space (including the body) against public norms. It might manifest in the differences in customs and attitudes between cities and countrysides, or in the differences in knowledge and experience between opposing social classes. It was also possible that the complexity was intertwined among different genders, religions, especially Catholicism and Protestantism in English-speaking countries, and the misplaced understanding and cognition of health and morality. Such complexity ran through the progressive reforms of the most representative progressive countries on both sides of the Atlantic, such as the UK, France, Germany, and the United States on the other side.[35]

[34] There are a lot of relevant discussions in *Atlantic Crossings: Social Politics in Progressive Age*. They can also be found in *La création du service public de l'eau, LONDREA, NEW YORK, PARIS RACONTÉS, AU FIL DE L'EAU* by Christophe Defeuilley, translated by Tang Jun, Beijing: Social Sciences Academic Publishing House, 2019 (the original in French was published in 2017); and *The Sanitary City* by Melosi.

[35] Starting from Michel Foucault, scholars continuously de-constructed the legitimacy of regulating individual bodies during the Progressive Era, such as David Barns' discussion on public health in Paris in 2016: David Barns, *The Great Stink of Paris and the Nineteenth*

Most paradoxically, in the complexity, groups with treason shared certain common moral demands. They all sought social justice, the restraint on laissez-faire capitalist enterprises which were egomaniacal, and the effective, clean, and democratic governments. They framed the standard of justice with their own demands. The key difference in public health among different social classes lies in that in knowledge. As seen by Parton, Pittsburghers (the new immigrants from the countryside, the original urban poor, and the working class) all had their own behavioral patterns and explanations for health, and the life logic behind them stemmed from their inherent knowledge and experience. When the new knowledge featuring universality invaded their lives with strong strength, they felt confused and threatened. However, it didn't mean that other classes remained indifferent and powerless in the process. On the contrary, in the Progressive Era, vulnerable groups in cities on both sides of the North Atlantic found a channel to express their own opinions and appeals through a certain platform or spokesperson. Their acceptance of urban environmental reforms, especially those related to public health, was not a helpless failure or forced to suffer loss, but their own choice of comfort and health that were created by modern technology. For the choice, most of them accept the moral norms required by the city as a public open-space and the scope of freedom granted.

4 What Kind of City It Is?

However, although progressive reformers on both sides of the North Atlantic believed that cities should be planned as public open-spaces and that the maintenance of public health should be the responsibility and function of the government, a basic consensus should be reached on the issue of establishment of good public morality. It did not mean that they had a consensus on what a healthy and good city was. The focus of

Century Struggle against Filth and Germs. Baltimore: Johns Hopkins University Press, 2006. Chinese scholar Zhong Zi also dedicated a chapter to the issue in his doctoral thesis completed in 2020. See Zhong Zi, *Historical Changes of Urban Water Environment Governance in France: A Study Centered on Lyon*, 2020 Excellent Doctoral Dissertations of Peking University. Postcolonial works focus on debates sparked by similar discourse and reform intentions in colonized societies, for instance, the study of the shift of Chinese cities to modernity by medical historians. See Ruth Rogaski, *Hygienic Modernity: Meanings of Health and Disease in Treaty-Port China*, translated by Xiang Lei, Nanjing: Jiangsu People's Publishing House, 2007.

the "municipal housekeeping movement" is to advance public health and morality by purifying and improving dirty environments. At that time, the "playground movement" appealed for the preservation and establishment of amusement parks of different sizes in the city so as to provide enough space for urban populations to spend their spare time in a beneficial way. The advocates of the movement believed that the various activities carried out in amusement parks could not only make people communicate with each other and learn to cooperate with others but also let them escape the strict norms of industrial society and relax their bodies and minds. In 1906, they established the National Playground Association of America in New York City. The following year, more than 200 "playmates" from over 30 cities across the United States gathered in South Region-Chicago Park to hold their first meeting. In the grand event of amusement, many famous progressive reformers at that time stepped into the center of the playground and delivered several speeches on "Play Spirit" and its social function, including Luther Gulick's *Play and Democracy*, Joseph Lee's Pl*ay as a school of The Citizen*, and Jane Addams' *Public Recreation and Social Morality*. In these speeches, there was familiar vocabulary from progressive reformers of the urban age, such as urbanization, industrialization, cities, public spaces, democracy, public interest, morality, evil, and so on. In an article titled *Health, Morality and the Playground*, Elmer Brown, a member of the United States Commissioner of Education, claimed that "the part of the modern park system which most appeals to him [child], and ought to appeal to him, is the part which gives him a chance to do something. For any real lover of human nature—boy and girl nature—the most beautiful thing of all in our public parks to-day is a lot of children hard at play where there is room to play and nobody cares whether the grass grows or not."[36]

However, another group of progressives saw the vitality of urban civilization from the tall grasses and the nightingales in the air (enchanting scenery). For them, the various problems that arouse during the urbanization process were not just urban environments, especially the artificial improvement of environments, such as the improvement of sewage, garbage systems, and residential housing. The construction of libraries, museums, playgrounds, and others. They were not even a simple economic or political problem. When capital from cities put the nature's

[36] Elmer E. Brown, "Health Morality and the Playground," *Proceedings of the Playground Association of America*. Issue I (Aug. 1907): p. 30.

economy on the brink of disintegration, the wave of human migration toward cities also caused the relationship between man and nature to become increasingly estranged, and human beings gradually lost the values that grew up on the near and dear land in the estrangement. A sound artificial environment, a perfect education system, a prosperous material life, and harmonious labor relations couldn't offset the dependence and longing of civilization on nature. Once the dependence on nature was deprived and the desire for nature couldn't be satisfied, problems might occur in civilization itself. These problems not only lied in the increasing scarcity of natural resources on which civilization relies for development but also in the gradual decline of spirit and morality on the path of civilization. It is the message that Frederick Law Olmsted, the founder of American landscape architecture and the most important reformer and planner of American cities, hoped to convey throughout his life.[37]

In the big cities on the East Coast of the United States, there was a group of landscape designers and other green reformers who were deeply influenced by his ideas. The most important figures among them include Charles Eliot, the designer of the Boston Metropolitan Park System, Sylvester Baxter, a renowned Boston journalist and secretary of the Metropolitan Park Commission, Charles Sargent, the Director of both the Harvard Botanic Garden and the Arnold Arboretum, the founder of *Garden and Forest* magazine, renowned botanist and resource conservationist, Robert U. Johnson, the editor of Century Magazine in New York, William Stiles, the editorial writer for the New York Tribune, and the editor of *Garden and Forest* magazine. In the eyes of the group of progressive reformers, the dilemma faced by the United States in the urbanization process in the late nineteenth century was a moral decline caused by nature deficit. Obviously, the demand for "Back to the Land" and the labor on the land as the main way of survival lost its economic and social foundation in the populist movement. The new urban society should establish a new relationship between man and nature. If the so-called "harmony" had long been lost in the long-term confrontation between man and nature, at least a certain degree of balance and some

[37] Since the twentieth century, there has been a continuous stream of research on Olmsted, the most authoritative work is still the biography of Olmsted by Laura Roper, which was published in 1973. Laura Wood Roper, *FLO: A Biography of Frederick Law Olmsted*. Baltimore: The Johns Hopkins University Press, 1973.

other forms of coexistence should be sought. Therefore, they, with cities as a starting point, launched another level of urban reform nationwide, which I call the "urban naturalization" movement, in order to redefine the relationship between man and nature in the urban age.[38]

The urban naturalization reform is an important component of natural reform in the entire Progressive Era, but compared to other major components, for instance, the Conservation Movement, its definition of the relationship between city and nature goes beyond survival and interests and is full of aesthetic enthusiasm and moral needs. Nature centrists like John Muir, who held high the value of the wilderness itself, advocated the urban naturalization reform movement, based on cities, ultimately also focused on the urban population. Their appeal for nature is a unique yearning of the urban age. Those urban dwellers who don't directly seek basic means of survival on the land no longer need to cooperate closely and struggle persistently with nature. For them, food has become a commodity bought and sold in the market, the land is covered by artificial buildings, and nature as a resource is only the responsibility of policy makers and has no direct connection to their own lives. Although they, as consumers, are constantly closely connected with nature, in their understanding, nature is to a large extent a fictional existence and spiritual sustenance for them. These urban naturalization reformers clearly know the transformation of the relationship between man and nature in the urban age. Therefore, they strive to find ways for urban populations to truly perceive and experience nature, so that they can re-establish their emotional connection with nature in different ways.

For the United States, the Natural Movement began in an "urban waste land" on Manhattan Island. Central Park was constructed in 1857 and officially completed in 1873. It covers a total of 843 acres and is located in the heart of Manhattan Island, New York. It is the first true urban park in the United States and also the blueprint for urban parks with natural style in the United States. In the discussion of whether to build a central park, people always referred to Seneca Village, the predecessor of the park, as "wasteland." It was not an undeveloped urban fringe that preserved its original ecosystem. In the first half of the nineteenth century, thousands of people immigrated here, most of which were free black, Irish, and German immigrants. They established several small

[38] Please refer to my book *The City Natural* for the relationship between Olmsted and urban progressive reformers.

villages in the city, raised livestock, and planted vegetables, which were the tangible and small urban ecosystems of the middle of nineteenth century in the United States. The vegetables grown by individuals were supplied to supplement the necessities of daily life. Pigs foraged in garbage piles, and their excrement and other livestock, especially horses', were collected and transported out of the city in the early morning. The excrement became fertilizer for agricultural production in the surrounding areas. Grain was then transported from outside the city into the city, and pigs were slaughtered and consumed after they became fat. The ecological logic operated smoothly most of the time, but it contradicted the hygiene and aesthetic concepts of the middle class, and more importantly, it also posed a risk of infectious diseases. Like cholera which broke out in New York in 1832, 1849, and 1866, respectively. It perplexed New Yorkers in the entire nineteenth century after it heavily hit the areas where working class and urban poor dwelt.[39]

The construction of Central Park put an end to the ecological logic. With original settlements like Senaca Village knocked down, more than 1,600 people were displaced, vegetable gardens were built into lawns, garbage dumps turned into forests, and livestock and poultry grazing grounds were changed into habitats for over 200 species of wildlife, especially birds. Due to the lack of fertile soil and abundant species in the local area, approximately 14,000 m^3 of topsoil were transported from New Jersey, 4 million trees and plants were transplanted here from various places, and over 10 million vehicles of various materials were used to build Central Park in New York. Central Park was a huge project with unprecedented cost. As the chief designer of Central Park, Olmsted painted a huge landscape painting on the land with an area of 843 acres.

Since the construction of Central Park was placed on the governmental agenda of New York, its legitimacy had met with challenges from all sides. In the nineteenth century, the most frequent and intense challenge came from considerations of economic development. In a city like New York where an inch of land is an inch of gold, should such a large area of land be left idle and not to be developed? In the last 20 years of the nineteenth century, it was constantly proposed to convert some land in Central Park for other purposes. Defenders of Central Park wrote, "There is a passion

[39] Historian Ted Steinberg referred to it as the "Death of the Organic City", see Ted Steinberg, *Down to Earth: Nature's Role in American History*. New York: Oxford University Press, 2002: Chapter X.

for 'improving' vacant land, and the only known way to improve it is to cover it up with some construction. The time has not arrived when the people of the city realize that open spaces are quite as essential to health and comfort as solid blocks of buildings or they would never consent to see their property destroyed in this way." Faced with the popular view at that time that "parks are made to use, and not to look at," their answer was: "The beauty of a park is its highest use, and to destroy that beauty is to abuse, and not to use, it."[40]

However, these challenges actuated by economic interests and practical purposes were only a part of the questions that urban naturalization advocates needed are required to answer, and the more formidable challenges they faced came from the critical voices of social justice and ecological logic which represent the grassroots. Those who were forced to leave home and wander about due to the construction of Central Park were the vulnerable groups in urban society, and the picturesque scenery achieved by Central Park was by no means a gospel of Paradise for them. To a large extent, they became victims of the aesthetic tastes of the middle class—of course, many critics of Central Park, purely from a survival perspective, neglected more job opportunities created for new immigrants during the construction process.

The history before the construction of Central Park disappeared for hundreds of years, not until the 1960s and 1970s was it truly excavated when social history research on marginalized groups emerged.[41] However, as early as the late 1880s, the public denunciation of the aesthetic tendency focusing on nature preservation after the construction Central Park appeared in social discussions. The "playground" faction was one of the representatives. Like advocates of urban parks, advocates of amusement parks also believe that public open-spaces for residents to relax and entertain should be preserved in the middle of the city. In the early 1880s, the two formed an alliance to counter the endless development of urban land by real estate developers. However, as the urban land area gradually shrank, the differences between them arose. The amusement park faction believed that urban dwellers, especially the working class,

[40] Editorial, "The Confiscation of City Parks," *Garden and Forest*, Issue 64(1889): p. 229; Editorial, "Keep off the Grass," *Garden and Forest*, Issue 335(1895): p. 291.

[41] The reflection on Central Park from the perspective of social history began with Roy Rosenzweig and Elizabeth Blackmar, *The Park and the People: A History of Central Park*. Ithaca: Cornell University Press, 1992.

needed entertainment activities more suitable for emotional catharsis and strenuous exercise, especially various sports (such as rugby and baseball) that required large venues. They believed that large urban green spaces should not be just objects of observation and contemplation, but be used as playgrounds. They criticized that advocates of urban parks strove to preserve the beauty of nature in the city, which was an aesthetic concept of the urban middle class but disregarded the wishes of the lower class in the city. Advocates of urban parks argued that the demand for nature was not only spiritual but also physiological, the ability to appreciate beauty was not limited to those who had received higher education and had a certain economic foundation, the lower class in cities also had the same ability and needs. The bone of contention between them was whether cities needed nature and whether the so-called general public needed the beauty of nature.

The biggest difference between them and other progressive reformers is their call to nature. In the social reform plan of the latter, humanity is their only concern and the only power in their eyes that determines the development of civilization. Nature, even if not completely eradicated, is at least marginalized. But for these green reformers, a truly civilized urban society had to be built on the harmony between man and nature. If nature isn't in existence, the external environment of human society will be extremely impoverished, and the spiritual world of human beings will also be barren. Their understanding of advancement is built on their admiration for the beauty of nature and respect for the power of nature and also built on their belief that there is a universal and instinctive need for nature in the process of urbanization. They attempted to make people realize that preservation of the beauty of nature in the city was not only based on a certain middle-class aesthetic sentiment but also on the foundation of "the most essential and practical truth," namely, the beauty of nature, along with the clear sky, clean water, green trees, fragrant flowers, and fresh air, are basic needs for people's physical health. In the *Garden and Forest* magazine, Sargent wrote, "It is asserted over and over again, not only by poets and philosophers who give expression to the profoundest truths in our nature, but the curative value of natural scenery is distinctly recognized by the medical profession. All of us have felt the soothing and restful influence of natural beauty, acting in a subtle way through the very highest functions of our being, and tending to establish sound minds in sound bodies." They firmly believed that the integration of cities and nature would let everyone breathe the same air and see the same scenery, thus confusing the boundaries between different social classes, after all, nature would have the same physiological impact on everyone.[42]

Therefore, the reformers of urban naturalization lamented more than once that those who "fight to reduce working hours" didn't realize that they were the biggest beneficiaries and rightful protectors of urban parks. As Charles Eliot, the designer of the Boston Metropolitan Park System and the apprentice who didn't take lessons directly under Olmsted himself, pointed out that the rich could flee from town at certain seasons to enjoy the quiet and peculiar refreshment which came from contemplation of scenery, but the poor were busy running about day and night in the city, and those who had the same need for natural scenery, didn't have the enough money and time. Therefore, they were met only by urban parks.[43] It is precisely in the urban parks that these reformers with urban naturalization strive to realize their democratic ideals and public interests. The fundamental reason lies in their belief that everyone has a need and the ability to appreciate nature. The experience of nature and exchange with beauty should not be monopolized by any class. Their pursuit of social and environmental justice was not only to shorten working hours, increase workers' welfare, improve living conditions, and popularize educational facilities but also to recognize and help realize the ability and rights of everyone to enjoy nature.

5 Nature Purchasing

What a city should be like is an endless topic, it always sparks debate among different interest groups and intellectual groups. At the very least, however, it should be acknowledged that the assertion made by advocates of urban naturalization in the Progressive Era that everyone has the need and ability to enjoy the beauty of nature is not a groundless imagination or an excuse sought by middle class elites for the realization of their own aesthetic purport. After the end of World War II, a significant change brought about by the prosperity of the whole American society was that ordinary people began to ask how to make their lives better. Getting close to nature became an important aspect of the answer to the question. The

42 Editorial, "Natural Beauty in Urban Parks," *Garden and Forest*, issue 488(1897): p. 251.

43 Charles Eliot, "Parks and Squares of United States Cities," *Garden and Forest*, issue 35 (1888): pp. 412–3.

kind of question and answer constituted a new paradox about nature in the urban age, which was similar to the paradox of "loving you to the point of killing you." Suddenly, the soar of industry after World War II and the explosive development of the automobile industry brought people various options to get closer to nature. They could embark on a spur-of-the-moment trip, driving to the national park and seeing magnificent mountains and rivers under the protection of the windshield.[44] They could plant lush lawns in front of and behind their houses. The weeders powered by fossil fuel always kept the lawns flat, and modern insecticides ensured that there was no breeding of weeds and pests they didn't want to see, while modern hydraulic engineering allowed people living in the desert city of Las Vegas to smell the fragrance of grass at any time.[45] Even mass-produced commodities could show, in high-tech forms, the nature experienced with them as media, in order that the distance between urban people and nature can be shortened. The most modern large shopping malls often tend to be the best places for urbanites to come into contact with the nature they desired.[46] "Live with pure cotton to create a breathing space," "Sun-Maid raisins are grown in the sunshine state of California," and "Pabst Blue Ribbon is made of the clearest water." The handsome men and beautiful women in automobile advertisements must be running on the bright and warm seaside, in picturesque deep mountains, and in desolate and peaceful deserts. When buying Apple phones, you will always see delicate and colorful insects and flowers, as well as vast and magnificent mornings and sunsets.[47]

[44] For the discussion on the relationship between cars and wilderness preservation, see Paul Sutter, *Driven Wild: How the Fight against Automobiles Launched the Modern Wilderness Movement*. Seattle: University of Washington Press, 2005.

[45] For the discussion on the relationship between lawns and American society and the environment, see Theodore Steinberg, *American Green: The Obsessive Quest for the Perfect Lawn*. New York: W. W. Norton, 2007.

[46] See Jennifer Price, "Looking for Nature at the Mall," *Flight Maps: Adventures with Nature in Modern America*. New York: Basic Books, 1999.

[47] For the discussion of selling nature in mass production of food and beverages, see the following books: John Soluri, *Banana Cultures: Agriculture, Consumption, and Environmental Change in Honduras and the United States*. Austin: University of Texas Press, 2005; Kendra Smith-Howard, *Pure and Modern Milk: An Environmental History Since 1900*. New York: Oxford University Press, 2013; and Erica Hannickel, *Empire of Vines, Wine Culture in America*. Philadelphia: University of Pennsylvania Press, 2013.

If all of the said things are just accompaniments of life, the suburban sprawl and mass residential construction after World War II are the core of the American lifestyle. Urban historian Kenneth T. Jackson keenly pointed out in his classic work *Crabgrass Frontier: The Suburbanization of the United States* that the persistent love for single-family residence was an important feature that distinguished the United States—as well as Australia, New Zealand, and Canada, which were equally sparsely populated and largely inherited England's anti-urban culture—from the housing culture of the "Old World" countries. Jackson's analysis of the formation of the feature focused on the accumulation of economic wealth and the cultural emphasis on family life, privacy, and private property. He also mentioned the shift in people's attitudes toward nature from utilitarianism to aesthetics after the closure of the border in the late nineteenth century.[48] However, what Jackson did not analyze was that behind the single-family residence was largely the defective reality compromised by millions of immigrants after their dreams of the fertile land of the "Second Earth" were shattered. In other words, in the middle of twentieth century, scaled-down Skibo (the castle of tycoon Carnegie in Scotland), big and small, appeared around American cities. They bought a modernistic single-family residence, with a perfect and smooth water supply and drainage system, constant room temperature throughout the year, as well as a refrigerator, washing machine, computer, oven, and sweeping robot operating inside day and night, Ford sedans filled up with gasoline, which could be driven away at any time, in the double garage of the building, and a woods for leisure nearby the building surrounded by a lawn. What they purchased was not a single-family residence, or not just a single-family residence, but an American life that the middle class already obtained while the working class was striving for. They also purchased nature at the time of purchasing this kind of life.

However, it is in this suburban area representing the American lifestyle that environmental historian Adam Rome told people that the country "has experienced one of the most profound environmental changes in American history. The land that is bulldozed and used for urban development every year is roughly equivalent to the area of Rhode Island. Forests, wetlands, streams, hills, farmland and orchards have all been destroyed to

[48] Kenneth T. Jackson: *Crabgrass Frontier: The Suburbanization of the United States*, p. 13 and Chapter III.

build residential quarters. ... The nearby open spaces have been repeatedly destroyed, taking away children's beloved playgrounds time and time again—the environmental damage around the home is even a matter of life and death, far exceeding the threat faced by distant places like Echo Canyon State Park."[49]

The burgeoning suburban development after World War II was once a gospel of realizing the American dream. Bulldozers, like various machines that had appeared frequently in history, were a symbol of development and prosperity. The people who suffered from the embarrassment caused by war demanded the freedom and prosperity they defended, and a piece of land and a house were the most direct, practical, stable, and reliable response to the demand. In the meantime, the post-war economic recovery required strong support from consumers, and purchasing new houses was the greatest manifestation of the social purchasing power of consumers. Producers fit in readily with the demands of consumers. The huge investment of the American government in suburban development in the 1940s and 1950s was regarded as a fulfillment of its social commitment. Therefore, the tripartite cooperation between the government, economic groups, and the public made the early development of bulldozers smooth, thus each party was satisfied.

But the good times didn't last long. If, due to the faraway distance from the lives of suburban dwellers, the energy consumption problem of the people living in post-war suburban housing didn't raise the alarm of energy conservation for the majority of people, the sewage coming from septic tanks and overflowing in the backyard of suburban residential areas, "detergent cocktails" flowing out of the tap water pipes, the rapid disappearance of mountain forests near the community where children played and adults strolled, and the property losses resulted from natural disasters such as floods and landslides for people living in "ecologically sensitive regions" made the environmental disasters caused by bulldozers directly entered the lives of the middle class. The general public and various professionals were extremely worried about these obvious problems and launched movements to resist them. The progressive image of bulldozers also begun to be questioned. In the 1960s, it was deemed as "an arrogant and conceited tool, a symbol that the country disregarded for nature and

[49] Adam Rome: *The Bulldozer in the Countryside*, p. 8.

destroyed everything." The so-called "progress" relying on various developments and disregarding the limits of nature, which was once regarded as the infallible law in American society, was also marked in quotation marks during the period, and environmentalists often associated it with barbarous and foolish behaviors such as "devastation," "predation," and "murder."[50]

The paradox lies here. For more and more ordinary people, getting away from the city and moving to the suburbs was also the ultimate way to escape pollution. In American society, criticism of cities built with chemical industries and fossil fuels became even more severe. Modern life surrounded by various detergents and insecticides once promised a healthy life, but Rachel Carson's stunning work published in 1962 shattered the peaceful illusion of modern spring. People began to realize that public health was not just about human health. In the absence of guaranteed environmental health, their health was also at risk. As the genius singer and mathematician Tom Lehrer brilliantly satirizes in his black humorous song *Pollution*:

If you visit American city,
You will find it very pretty.
Just two things of which you must beware:
Don't drink the water and don't breathe the air....
See the halibuts and the sturgeons being wiped out by detergents.
Fish gotta swim and birds gotta fly,
But they don't last long if they try.
Pollution, pollution,
You can use the latest toothpaste,
And then rinse your mouth with industrial waste.
Just go out for a breath of air,
And you'll be ready for medicare.
The city streets are really quite a thrill.
If the hoods don't get you, the monoxide will....
Lots of things there that you can drink,
But stay away from the kitchen sink.
The breakfast garbage that you throw in to the bay,
They drink at lunch in san jose.

[50] The description and analysis of the questioning of progress in American society in the 1960s and 1970s mainly focus on *Open space: the First Protests against the Bulldozed Landscape*, the Chapter IV of *The Bulldozer in the Countryside*.

So go to the city,
See the crazy people there.
Like lambs to the slaughter, they're drinking the water and breathing the air.[51]

In an era where consumption became a basic way of life, the middle class, the mainstream American consumers, came to the suburbs in pursuit of a healthier, better, and more eternal quality of life, only to find that their pursuit of comfort in modern life caused them to once again fall into the pollution cycle they hope to escape. Therefore, they formed a new protest group.

Historian Samuel Hays argued in his book *Beauty, Health, and Permanence: Environmental Politics in the United States, 1955–1985* that after the end of World War II, American society shifted from the mode of production to the modes of consumption, and as a result, the environmental movement in the United States shifted from an early resource conservation movement that paid stress on efficiency to an environmental protection movement that attached importance to quality of life. He believed that the contradiction between the consumption of environmental costs by the mode of production and the demand for environmental quality by the modes of consumption was an important power in advancing the development of the environmental protection movement in the 1960s and 1970s.[52] Hays's research on American environmental politics is always thought-provoking, but his division is no doubt too arbitrary. He not only ignored the pursuit of beauty and health in the urban naturalization reformers led by Olmsted before World War II but also lacked understanding of the limitations of consumers themselves. Whether consumers felt scared of the white foam in tap water or protested against the annexation of public open space in the community, whether they showed anger over the destruction of homes by floods or felt regret over the disappearance of birds outside the window, they had a bottom line in their pursuit of environmental quality: environmental quality couldn't

[51] Tom Lehrer, "Pollution," 1965. Tom Lehrer graduated from the Harvard Mathematics Department and taught mathematics at elite schools on the east and west coasts. At the same time, he created a large number of hot songs. The album including *Pollution* was recorded in San Francisco in 1965. After leaving the entertainment circle in the 1970s, Lehrer taught both mathematics and musical history at the University of California Santa Cruz.

[52] Samuel Hays, *Beauty, Health, and Permanence: Environmental Politics in the United States, 1955–1985*. New York: Cambridge University Press, 1987.

conflict with their immediate economic interests. As Rome said, "Consumers' vision is short-sighted: property owners only care about certain parts of the environment."[53]

As individuals, consumers' demands for the environment were limited by their own economic capabilities and knowledge structure. They might not be able to bear higher environmental costs or understood grander environmental issues. Due to short-term benefits, consumers often couldn't take into account longer-term benefits. When they purchased the residences built in the flood plain at a low price or used the residences with septic tanks, they often neglected the current choices at a cheap price that were likely to cost them several times or even tens of times their wealth in the future, and even compromised their health and the quality of life they had been pursuing. However, the contradiction was largely not out of the choices of consumers themselves but resulted from the guidance of the market and economic groups. Developers always make consumers pay the bill of their environmental costs, especially potential environmental costs, so as to achieve the goal of minimizing costs and maximizing benefits.

Taking the use of septic tanks in suburban residences as an example, the cost of purchasing a piece of land that was not using the urban sewer system was much lower than that of purchasing land that had been included in the sewer system. Therefore, for developers, septic tanks had become the simplest and most effective way to solve the basic problems of new residences located on such land.[54] Its cost was next to nothing compared to sewage pipe laying, not only did it require advanced technology, but it was also easy to allocate its cost to households. More

[53] Adam Rome: *The Bulldozer in the Countryside*, p. 13.

[54] Before the book *The Bulldozer in the Countryside*, no historian had analyzed the connection between septic tanks, suburban environments and environmental movements. In his book, Rome pointed out that the problems of suburban septic tanks had undergone three stages of change as a public environmental issue. From the late 1940s to the 1950s, the criticism of the damage to septic tanks mainly focused on public health and community landscape. In the 1960s, people began to pay attention to the issue of water pollution. In the early 1970s, discussions on the issue were no longer limited to human health, but also involved the survival and extinction of other life forms. Rome believed that the discussion on septic tank issues was directly related to the development of environmentalism in the 1960s and 1970s. It not only reflected consumers' pursuit of quality of life, but also highlighted the importance and participation of professionals and government agencies in larger and long-term environmental issues, which had a positive impact on the environmental movement.

importantly, developers didn't need to bear the responsibility of the problems of repairing septic tanks that arose after several years or decades. Therefore, septic tanks became a symbol of suburbs, and the sewage overflowing from septic tanks was also a common sight in the suburbs.

There was no doubt that suburban dwellers were concerned about septic tank issues. On the one hand, they were anxious about the economic losses caused by investment failures, and on the other hand, they were worried about the sanitation issues caused by the pollutants exposed in the air. In the 1960s, when technology began to be questioned by society, the problem of groundwater pollution which caused by septic tanks also attracted the attention of owners. After all, no one thought that the turbulent detergent foam flowing out of the water pipe was a pleasant sight. However, for homeowners, the performance of sewage treatment systems was essentially a consumption issue, and "even in the worst-case scenario, it is typical that homeowners will not worry about the long-term environmental costs due to the use of septic tanks. As a result, national legislators could find ways to alleviate certain issues that homeowners are deeply concerned about, but not address all environmental issues caused by septic tanks. It has been the truth for many years." The reason was very simple. If septic tanks operated well, their economic superiority was extremely significant. Owners were not aware of the complex relationship between septic tanks and ecosystems, and most people were not interested in learning about it. After the immediate environmental issues that threatened the health of their family members were resolved, the coexistence of non-human life forms on the same land was not insignificant for them, but at least cannot compete with their investment ability and returns. Therefore, even if government agencies and professionals had a deeper understanding and longer-term goals of the environmental issues caused by septic tanks, "they couldn't go too far ahead of their neighbors" as well.[55]

More importantly, as a group, the dreams of mainstream consumers had a direct bearing on the American dream. The information behind single-family residence was not only a defense of the sacred status of private property but also a thirst for and realization of American democracy and freedom. The monotonous aesthetic landscape of ranch bungalows located in the suburbs of the United States couldn't deny their

[55] Adam Rome: The Bulldozer in the Countryside: Suburban Sprawl and the Rise of American Environmentalism, pp. 116–117.

rich social and cultural connotations. When the government introduced various restrictions and acts to address environmental issues, economic groups inevitably shifted the increased costs onto consumers. As a result, many impecunious families were unable to realize their American dreams, and the government was also unable to fulfill its commitments to American democracy. When the government tried to restrict the use of land, many laws were introduced to regulate the behavior of homeowners on their private territories, which faced the strongest resistance from homeowners and public opinion, as it, in their eyes, was a fundamental betrayal and infringement of the founding of the United States. When the environmental movement advanced to this point, the call for environmental protection was no longer a popular "motherhood and apple pie" slogan which is harmless and soft. It showed a fighting posture, and its targets were inveterate American values in a large part. The values, firmly supported by the mainstream consumer group in the United States, became the theoretical basis for various economic groups to make profits through environmental devourment. Once the government's environmental policies substantially touched upon the fundamental interests of consumers and economic groups, both the groups who always contradicted each other often formed alliances to jointly resist systems and laws that attempted to fundamentally change Americans' attitudes toward nature. Therefore, the concept of environmental protection was always in straitened circumstances in the battle.

Nevertheless, it should be noted that the new generation of environmental protection movements fermented from the paradox of urbanites being close to nature are indeed different from the urban naturalization reforms of the nineteenth century. The difference was not limited to the shift from the understanding of biological pollution to that of chemical pollution. The struggle which expanded from the middle class as the reformer to grassroots environmental organizations for their own "environmental justice," in terms of its fundamental ideology, redefined the connotations of "freedom" and "health" in cities. The constraint on freedom in the nineteenth century originated from the fact that individual freedom should not be based on the obstruction of others or the interests of the social community. However, in the environmental movement of the twentieth century, we heard an increasingly powerful voice: human freedom should not be based on the deprivation of the right of existence of other life forms and the obstruction of the interests of the land community. For health, the ecological warning brought by *Silent Spring* tells us in the most beautiful yet cruel words that in the optimistic era of swift and vigorous modern medical advancement, our porous and permeable bodies always share a kind of health and fragility with our living environment.

CHAPTER 4

A City upon a Hill

1 Prologue: "New England Vision"

In the summer of 1630, John Winthrop, a Puritan leader from England, stood on the Albella Fleet slowly sailing into Massachusetts Bay, overlooking an unfamiliar continent from a distance. At some point, he delivered his famous sermon, *A Model of Christian Charity*, also known as *A City upon a Hill*. In the sermon, Winthrop eagerly anticipates the realization of a interdependent and autonomous Puritan society that has a special covenant with God. Borrowing the metaphor from the *Matthew · Sermon on the Mount*, he compared the new land and the colonizers who would settle in it to the "City upon a Hill" that would attract world attention. In his view, the continent far from his homeland was a gift of God, and Puritans in England were true Elects of God. Here, they would practice their covenant with God, and establish a New England, a more prosperous, joyful, and morally perfect world than England on the other side of the ocean.

For hundreds of years, devout Christians, passionate patriots, and rational historians have all attempted to interpret Winthrop's sermon. Although the above-mentioned persons have quite different attitudes toward the metaphor of "A City upon a Hill," commending, criticizing, or reflecting, they generally examine the important historical document from the relationship between the metaphor and the emergence of American exceptionalism. Few people discuss or even realize the inspiration from the landscape of the New World for Winthrop's sermon, as well

S. Hou, *Wallless Cities*,
https://doi.org/10.1007/978-981-97-7827-0_4

as the ecological consequences arising from the sermon. If the continent extending beyond the bay that Alberta is about to sail into is bleak and barren, can Winthrop from verdant and lush England have such confidence in claiming that he is an Elect of God, and the continent they arrived on across the ocean is a wonderland bestowed upon them by God? Or perhaps when he looked up at the plain stretching for thousands of miles, would he have the idea that it was the ideal "City upon a Hill"?

However, the land presented by nature to Winthrop can appropriately trigger the inspiration for the establishment of "A City upon a Hill." It is a peninsula almost entirely surrounded by the waters of Massachusetts Bay and Back Bay, with an area of approximately 789 acres (smaller than the Central Park in New York which was built later). The Charles River flows into the sea at the western end of the island, and the link between the island and the mainland is a slender isthmus. A wetland lies to the west of the island, but three hills stand on the island: Trimountain to the northwest, Copp's Hill to the north, and Fort Hill to the south. The island was then called Shawmut, an Indian word with an unclear meaning to date. Some people may interpret it as "a peninsula surrounded by sea," thus its extended meaning is "a place for canoes to anchor" and so on. Some argue that Shawmut shall be regarded as the "land of sweet springs," it is the abundant freshwater resources here that are precisely the reason why Winthrop et al. migrated to Charlestown when they arrived in North America in the early spring of that year.

In *New England Prospect,* a pamphlet published in 1634, William Wood left behind descriptions of the unfamiliar island by early Europeans:

Boston is two miles North-east from Roxberry: His situation is very pleasant, being a Peninsula … Their greatest wants be Wood, and Medow-ground, which never were in that place; … It being a necke and bare of wood: [therefore] they are not troubled with three great annoyances, of Woolves, Rattle-snakes, and Musketoes.[1] the place being too small to containe many, and fittest for such as can Trade into *England*, for such commodities as the Countrey wants, being the chiefe place for shipping and Merchandize. … This Necke of land is not above foure miles in compasse, in forme almost square, having on the South-side at one

[1] After the arrival of the Puritans, the Shawmut Peninsula was renamed Boston to commemorate the homeland in England of many of them, including Winthrop—Boston, Lincoln Lincolnshire. Therefore, by the time Wood's book was published, the place had already been known as Boston.

corner, a great broad hill, whereon is planted a Fort, which can command any ship as shee sayles into any Harbour within the still Bay. On the North-side is another Hill, equall in bignesse, whereon stands a Winde-mill. To the North-west is a high Mountaine with three little rising Hils on the top of it, wherefore it is called the Tramount. From the top of this Mountaine a man may over-looke all the Ilands which lie before the Bay, and discry such ships as are upon the Sea-coast. ... This place hath very good land, affording rich Corne-fields, and fruitefull Gardens; having likewise sweete and pleasant Springs. The dwellers of this place for their enlargement, have taken to themselves Farme-houses, in a place called Muddy-river, two miles from their Towne; where is good ground, large timber, and store of Marsh-land, and Medow.[2] In this place they keep their Swine and other Cattle in the Summer, whilst the Corne is on the ground at Boston, and bring them to the Towne in Winter.[3]

Wood's description is the most detailed record of the Boston landscape when whites first arrived, perhaps with some exaggeration, but it is generally objective and credible. Seen from that, thousands of years before whites actually settled here, Indians greatly changed the ecology of the peninsula with the "slash-and-burn" farming method. Although the continent that Winthrop looked at westwards was still densely covered with virgin forests, the mountains in the peninsula were already open spaces created by Indians. The peninsula itself does not have abundant forest resources, but for these ambitious pioneers, it is not difficult to drive or take a boat to the wooded continent and other small islands two or three miles away. Therefore, when the Puritans led by Winthrop came here, they were able to avoid the great hardships in pioneer work and the worry about the scarcity of resources. In the eyes of these well-informed Puritans, most of whom came from well-to-do merchant families of UK middle class, Boston's unique harbor location as well as water and land transportation connecting to inland plain allowed the Puritans to engage in trade and expand their wealth while distancing themselves from the corrupt society and religion of their motherland. In his sermon, Winthrop

[2] The Muddy River is actually small lakes, ponds, and streams which are composed of dozens of tributaries of the Charles River, located to the west of today's Boston, at the boundary between downtown Boston and Brookline, the suburbs of Boston.

[3] William Wood, *New England Prospect: A True, Lively, and Experimental Description of That Part of America Commonly Called New England*.... London: Printed by Tho. Cotes for John Bellamie and are to be sold at his shop..., 1634: Chapter 10.

made it very clear that wealth accumulation is not a sign of moral decay, but an important proof of the love of a Christian and one of the basic missions bestowed upon his followers by God.[4]

However, where should wealth come from? Winthrop clearly realized that the source of wealth was the land they were about to set foot on and "possess." At the end of his sermon, Winthrop passionately declared that if they love the Lord, obey his will, and fulfill their contract with him, they will survive and multiply: "the Lord our God may bless us in the land whither we go to possess it. But if our hearts shall turn away, so that we will not obey, but shall be seduced, and worship other Gods, … we shall surely perish out of the good land whither we pass over this vast sea to possess it."[5] In the sermon, Winthrop uses the term "possess" twice regarding their relationship with the land. "Possess" is directly related to wealth, and it will be a long process from acquisition of wealth to its use, or may eventually be lost. Even more importantly, it not only reflects a social relationship, but also has profound environmental connotations. It indicates that all nature, organic and inorganic beings on this land, will be completely changed by the process after these Puritans from the Old World walked into the New World.

Therefore, the natural environment of the peninsula offers a real and reliable foundation for the ideal "City upon a Hill" of Winthrop and his followers. Here, the "City upon a Hill" will not only be a symbol, but also a place where people interact with nature. On the peninsula, recently named Boston, a city will emerge, a city where Puritans attempt to elevate their moral beliefs, a city that connects the material and spiritual exchanges of the Old and New Worlds, a city that is fermenting for an independent revolution, a city that is about to become the cultural center of the United States, but also a city that ceaselessly expands through plunder and encroachment on nature, and a city that ultimately makes hills disappear. During the process, the evolution of the socio-economic system led to the subversion and reorganization of nature's economy. In other words, under the principles of the human economy, a land community that originally had run under its inherent laws and had been full of

[4] Chinese scholars are not unfamiliar with the thinking on the issue. In *the Protestant Ethic and the Spirit of Capitalism*"by Max Weber, there is an extremely brilliant exposition, so it is unnecessary to go into details here.

[5] John Winthrop, *A Model of Christian Charity,* Collections of the Massachusetts Historical Society. Boston: 1838, 3rd series 7: 47–8.

vigor became a resource and various types of commodities exclusively for the former to operate. However, it does not mean that nature's economy no longer exists, operates, and resists. While it is being changed, it will also have a complex and profound influence on the society, economy, and different kinds of individuals living in the "City upon a Hill."

The fundamental force driving the process is a unique social system, values, and economic order, in a word, capitalism. American environmental historian Donald Worster summarizes the relationship between nature and capitalism in the following three points: Firstly, nature shall be regarded as capital. Secondly, human beings have the right and even the obligation to use capital for the accumulation of their own wealth. Thirdly, the social order established by capitalism shall allow and encourage the relentless pursuit of individual wealth.[6] Under the specific relationship between human beings and nature, the peninsula with a radius of only four square miles evolved into a part of the changes that occurred in continental North America and even the entire world. In its unwavering and enduring pursuit of wealth, Boston actively integrated into the world market, and the land tens, hundreds or even thousands of miles away from Boston was also incorporated into the construction process of the "City upon a Hill," thus bringing fundamental changes to the soil, water and various forms of life, including human beings.[7]

[6] Donald Worster, *Dust Bowl:* "Preface" of *The Southern Plains in the 1930s*, translated by Hou Wenhui: Nanjing: Jiangsu People's Publishing House, 2020.

[7] For any historian, to rewrite Boston, the earliest city and cultural center in the United States, is undoubtedly a huge challenge. Since the beginning of American history in the nineteenth century, the city has become a research focus. During the evolution of more than a century, the correlational studies, including many masterpieces, are too numerous to be counted. They include *Planning the City on a Hill* by Lawrence Kennedy: *Boston since 1630*. Amherst: University of Massachusetts Press, 1992; Mona Domosh, *Invented Cities: The Creation of Landscape in nineteenth Century New York & Boston*. New Haven. Yale University Press, 1997; Karl Haglund, *Inventing the Charles River*. Cambridge: Harvard University Press, 2003; William Newman & Wilfred Holton, *Boston's Back Bay: The Story of America's Greatest Nineteenth Century Landfill Project*. Boston: Northeastern University Press, 2006; Anthony N. Penna and Conrad Edick Wright, *Remaking Boston: An Environmental History of the City and Its Surroundings*. Pittsburgh: University of Pittsburgh Press, 2009. Worth mentioning among the numerous works concerning the environmental history of Boston is *Eden on the Charles: The Making of Boston. Cambridge* by Michael Rawson: Harvard University Press (2010): The book emerged and won five book awards, including the final nomination for the Pulitzer Prizes in the following year. There was no doubt that Rawson, Cronon's best pupil, delved deeply into his master's masterpiece when he wrote his works about urban history. On the surface, the theme he selected is

2 "America's windfall!"

What on earth is the most far-reaching event in modern and contemporary world history? No historian looked forward to getting a universally recognized answer to such a question. But at least some of them would answer in this way: the Great Geographical Discoveries at the end of the fifteenth century was the most important event in modern history. It not only makes human beings fundamentally re-understand the Earth, the world they live in, but also lets them start re-dividing the different races in the world, drawing world maps, compiling a World Gazetteer and a world encyclopedia of flora and fauna, and launching a scientific revolution. Furthermore, it has pushed Europe, which has gradually entered the Malthusian crisis, out of its bottleneck and provided the material foundation for capitalism, a revolutionary force in its embryonic form, to sweep every corner of the world in the future and completely change nature's economy, human economy, and the relationship between nature and human beings.

American windfall, as Europeans call it, is the greatest discovery in human history. This windfall from the sky was completely unexpected, yet it was so abundant that it seemed to be inexhaustible. Christians saw it as the gospel of God and the reappearance of the Garden of Eden on earth.

the construction of parks familiar to urban environmental history. However, in the stories he told about lake diversion, landfill for urban expansion, bay dredging, making of parks and suburban development in Boston, the internal connection between the city and its hinterland environment can be clearly seen which is strengthened in *Nature's Metropolis: Chicago and the Great West*. Furthermore, *Eden on the Charles: The Making of Boston* introduces the expectations and attitudes of various ethnic groups in Boston towards nature, as well as the conflicts and compromises that arise from them, into the discussion. Through cultural deconstruction of nature, Cronon's contemplation of the "second nature" concept was strengthened yet not analyzed. Fortunately, Rawson's deconstruction is cautious and quite moderate. On one hand, he noticed that people with different cultural, religious, ethnic and class backgrounds have different understandings of nature, so there are differences in the urban landscapes they yearn for. However, on the other hand, Rawson did not completely view nature as a cultural construction or even a state of mind, thus he completely denied the difference between nature and human landscapes. Just as the actual urban planner of Boston's landscape, Rawson, the writer of Boston's urban history, also saw the various limitations that nature placed on Bostonians. Therefore, his works portrays not only the divergences and controversies between people, but also the confrontation, coordination, and dependence between human beings as a common group and nature. Just because of this, compared with previous relevant works, Rawson's new work gives a more comprehensive and profound description of the story of Boston, the old city.

However, for them, the emergence of the continent out of the void did not only inspire religious imagination and piety, or aesthetic shock. Their sharp merchant eyes also saw the hidden benefits in the land, which is still largely in the state of wilderness. How to excavate and utilize the land to continuously increase its value is the mission entrusted to them by Christ.

In Chapter XXV of *Matthew*, there is a fable that goes like this: a rich man traveling into a far country, who called his servants and delivered his goods to them. And to one he gave five talents (a currency unit used by ancient Greece and Rome), to another two, and to another one according to their abilities. The first servant used these five talents to make another five talents. And likewise, the second servant also gained another two. But the third servant dug a hole in the earth and hid his lord's money for he is afraid of losing it. After a long time, the lord of those servants returned home and the three servants came together to pay his respects to him. The first servant gave back five talents and brought five other talents he had gained. The lord praised him as a good and faithful servant, prepared to make him rule over more things, and let him share the joy of the lord. The second servant returned two, along with the two other talents he had earned, and received the same praise from the lord. The third servant presented the talent he had hidden in the earth and said, "Lord, I know you are a hard man harvesting where you have not sown and gathering where you have not scattered seed. So I was afraid and went out and hid your thousand silver coin (one thousand silver coin, quoted from the Chinese version of the *Bible*, is used to replace talent) in the earth. See, here is what belongs to you." The Lord was furious and scolded, "You wicked, lazy servant! So you knew that I harvest where I have not sown and gather where I have not scattered seed? Well then, you should have put my money to the exchangers, so that when I returned I would have received it back with interest. So take the thousand silver coin from him and give it to the one who has ten thousand silver coins. Or whoever has will be given more, and they will have abundance. Whoever does not have, even what they have will be taken from them. And throw that worthless servant outside, into the darkness, where there will be weeping and gnashing of teeth."[8]

In the eyes of sociologists, the fable is interpreted as the famous "Matthew Effect," but if examined from the perspective of environmental

[8] The *Bible* · *New Testament* · *Matthew* Chapter XXV: 14–30, Version from the *Bible* (Chinese Union Version).

history, it conveys fresh information. For a long time, Christians have regarded land as the property entrusted by the lord to them for management. What they need to do is not to hide it to maintain its original state, but to put it into use to continuously increase its value, waiting for the scrutiny of Christ's second advent. Otherwise, they will become lazy and wicked servants, and be thrown into the darkness to weep. Therefore, the fable provides a sacred and reasonable basis for the capitalization of nature. When the Great Geographical Discovery of the late fifteenth century occurred, it was precisely at the time of the emergence of capitalism that, to a large extent, the demand for resources and markets by capitalism stimulated people's exploration of distant countries. The ancient fables in the *Bible* fitted in readily with the new-emerging force and crossed the ocean with the *Bible* in the voyager's backpack. It is capitalism's undisguised, positive, and enduring pursuit of wealth.

When they finally arrived in the New World, the fable also became a high-sounding reason for Protestants seizing the land of the indigenous population of North America. In their eyes, compared with the highly domesticated land of the European continent, the New World was still largely in its state at the beginning of the Creation of the World. Although there were traces of Native American's "slash-and-burn" farming in some areas, including Boston, in the eyes of whites, only permanent residency and cultivation could truly free a piece of land from its wild state, and only in this way could the population people living on the land claim the ownership of the land.

Prior to John Locke's analysis of the relationship between labor and property in the eighteenth century, Winthrop had already expressed similar opinions in a slightly ambiguous and more religious form. The *Bible* once again became the foundation of these Puritans' concept of property. In *Genesis* I: 28, God said to man, "Be fruitful, and multiply, and replenish the earth, and [subdue] it: and have [dominion over] the fish of the sea, and over the fowl of the air, and over every living thing that move on the earth."[9] Therefore, God has entrusted man with the

[9] This is currently the most authoritative translation of the *Bible* in Chinese, the *Bible* (Chinese Union Version), but there are several parts that are not appropriate. The original text is "Be fluid, and multiple, and repeat the earth, and due to it: and have dominion over the fish of the sea, and over the catch of the air, and over every living thing that moves up on the earth. When the *Bible* is translated into Chinese, "subdue" is adapted here to conquer, and so "domination" is to "rule," so that they are closer to the original meaning of the *Bible*. The Chinese translation is based on the King James Version.

mission of conquering the land and ruling over all living beings on it. He demands that they are fruitful on it, rather than setting aside the land. In Winthrop's view, an uncultivated land does not belong to anyone. Only through fences, cultivation, fertilization, and harvesting, that is, through agricultural labor, can land become the private property of someone.

Indians did not completely lack the concept of property rights, but their definition of property rights differed greatly from that of whites. Once whites made a piece of land their private property, they possessed everything on the land as a matter of course, the land and all species on it, as well as complete ownership and use rights over them, which could be inherited, bought, and sold. However, in the eyes of Indians, property rights were neither permanent nor exclusive. Every Indian tribe migrated with seasonal changes and species migration. Their ways of obtaining survival materials were quite diverse, not only relying on agricultural production but also determining food and other means of subsistence for a certain period according to the conditions that the local ecosystem might provide. Even if a piece of land was opened up and left after the harvest time, it might be used by other tribes in the following year. Furthermore, to a large extent, they had no concept of property in land, only the concept of the right to use it. In the territory temporarily inhabited by a tribe, other tribes often had usufruct rights.[10] Environmental historian William Cronon has summarized it in an extremely accurate manner: the property rights of Indians change with the use of ecology.[11]

The concept of property, which is different from that of whites, also determines the attitude of living of Indians on the land. Today, the idea of treating Indians as ecological saints has been proven to be too idealistic, even naive.[12] They not only adopted a wasteful attitude toward nature in a specific period or situation before the start of the colonial era but also were further involved in the wave of land commodification, especially becoming an active participant in the fur trade, not long after the beginning of the colonial era. However, one-sided emphasis on certain

[10] Roman law refers to the right to use the property of others and enjoy benefits without prejudice to that property. Between Indian tribes, a piece of land is used by one tribe for cultivation, another tribe for gathering, a third tribe for hunting, and so on.

[11] William Cronon, *Changes in the Land: Indians*.

[12] In the book *The Ecological Indian: Myth and History*, Shepard Krech made a significant contribution to the deconstruction of the myth of ecological Indians. Shepard Krech: *The Ecological Indian: Myth and History*. New York: W. W. Norton & Co. 1999.

extreme examples between Indians and land does not necessarily indicate that Indians have the same attitude toward nature as later whites. The most crucial difference between them is that whites view land and everything on it as capital that can bring incremental benefits, while Indians primarily hunted and exchanged for their basic physiological need even during their participation in the fur trade. Of course, the consumption of alcohol exceeded the survival needs, but it did not mean an appreciation of individual wealth for Indians to exchange prey for alcohol. However, when whites got fur into their hands, fur became a commodity that continued to circulate in the market, which could bring more profits to fur merchants.

Therefore, with the arrival of whites, continental North America began to experience a long capitalist revolution, which not only affected whites, Indians, or the regime change in a certain country and the extinction of a certain system, but also all non-human organic and inorganic substances such as land, forests, flora, and fauna on the continent. Forced by the capitalist revolution, nature's economy also underwent a long-term transformation known as the Ecological Revolutions by environmental historian Carolyn Merchant, which subsequently changed and rebuilt the relationship between human beings and their environment, the relationship between men and women, and various forms of thought, such as mythology, science, and religion.[13]

3 Hill Land Reclamation

Boston and its surrounding areas were the earliest regions to experience the revolution. As an important commercial port connecting the markets of the Old and New Worlds, various raw materials from the continental North America were transported here from all directions, and loaded onto ships and shipped to the markets of the Old World, especially to the United Kingdom and China, while various daily necessities and luxury goods, such as silk and tea, which were imported from the Old World were also unloaded in Boston, and transported to New England and the southern states. The fur trade that was all the rage for a time in continental North America began earlier than the arrival of Puritans in New England. The initiator of the fur trade was the French, but the changes

[13] Carolyn Merchant, *Ecological Revolutions: Nature, Gender, and Science in New England*. Chapel Hill: University of North Carolina Press, 1989.

in European Fashion led to the pursuit and hunting of Castor canadensis (North American beavers). When Winthrop et al. arrived in Boston, the beavers in the area were on the brink of extinction. In the meantime, due to frequent contact with Europeans in the fur trade, smallpox, cholera, and other pathogens, the most powerful species in European biological travel boxes, were widely spread among Indians, so the Native American tribes near Boston also fell into decline.[14] However, the fur trade did not stop in New England but penetrated the hinterland from north to west, and fur traders in Boston flourished in the continuous decline of the survival line of beavers.

The ecological changes brought about by the extinction of beavers far exceed the scope of a single species. As engineers in the biosphere, beavers often build a small ecosystem in their habitat, where various life forms and inorganic substances are interactive and interdependent. However, when the builders of the ecosystem become extinct, the small ecosystem will also collapse, and all species that rely on it for survival will be affected. Nevertheless, in the eyes of merchants who buy and sell them, they only see the charming fur on the beaver and the great benefits behind its fur, but they cannot even see the existence of the beaver as a life form, let alone the ecosystem it is associated with.

Besides fur, the raw materials exported through Boston at that time mainly included wood, fish products, and ice cakes, among which wood accounted for the largest proportion. The impression of "America's windfall" is largely gotten from the forest resources which seemed like an eternity, and the development history of the New World from the colonial era to the nineteenth century was also a history of conquering forests to a large extent. In the eyes of early pioneers, the forests on the land were not only important resources for civilization development but also enemies for civilization to cut down and annihilate. The sharp axes and iron plows, which always accompanied early pioneers entering the forest, were used to eliminate the chaos and disorder of nature, uproot the forest where heresies and evils hide, and prepare a fertile field for the chosen people to settle and multiply. However, the significance of the forests in New England, despite those, lay not only in providing various building materials or fuels for the "chosen people," but also in becoming the maritime trade center where a healthy profit could be made. In the homeland on

[14] Alfred W. Crosby, *Ecological Imperialism: The Biological Expansion of Europe, 900–1900*, translated by Zhang Suguo: Beijing: The Commercial Press, 2017.

the other side of the ocean, developing capitalism badly needed the raw materials that its own country had already been short of. Therefore, the virgin forests in the New World, which were thousands of miles away, had become a resource for the emerging force to expand. Boston in the midst of a revolution between nature's economy and human economy achieved its own transformation from a peninsula to a city.

However, the part that Boston played in both revolutions was not merely to serve as a link between the Old and New Worlds. Boston also played an important role in the development of industrial capitalism in the United States. It was the revolution in the continental United States that completely changed the landscape and ecosystem of the New World, with a key shift being the commodification of water. Compared to fixed land, it was much more difficult to control, buy and sell flowing rivers. The land trade on the American continent began immediately after the arrival of the Puritans. In 1785, the United States Congress passed the *Land Ordinance of 1785*, which measured and divided American land into a chessboard-type landscape, with each township measuring 6 miles square, and then divided it into regular 36 checks, with each check 1-mile square, or 640 acres. This land plan was launched under the inspiration of and promotion of Thomas Jefferson and became the foundation of subsequent land laws of the United States. The use of geometric patterns to regulate the natural land is seen as a symbol of the Age of Reason since the eighteenth century, it also conveys the ideals of yeoman democracy of the American "Founding Fathers." However, a more practical role of it is that it greatly facilitated the purchase and sale of land and sped up the process of land commercialization.[15]

But the attitude toward water was quite different. For a long time, the people living in the colonies followed British Common Law in controlling rivers, namely, provided that landowners along rivers used the river without changing the natural flow of rivers or affecting the use of the river by other people living riversides, they naturally had the right to use the river. When the land was purchased quantitatively, rivers seemed to be beyond the reach of the commodification. Before the Puritans first

[15] "An Ordinance for Ascertaining the Mode of Disposing of Lands in the Western Territory," *Journals of the Continental Congress*, Vol. 28: pp. 375–81. A Century of Lawmaking for a New Nation: U. S. Congressional Documents and Debates, 1774–1875, Library of Congress, http://memory.loc.gov/cgi-bin/ampage,which was logged in on April 27, 2021.

arrived in New England, they had already built watermills, sawmills, and small canals along the banks of the Charles River. However, over the next two centuries, these workshops ran according to natural law, i.e., only during the harvest season. Besides, they also were influenced by seasonal and annual rainfall.

This situation changed fundamentally as humankind stepped into the nineteenth century. The seeds of industrialization already sprouted in New England. No matter how much Jefferson et al. hoped to keep American factories in Europe and make the New World a pure land of yeomen, industrialization still brought high profits that capitalism could not resist. Meanwhile, some important historical events accelerated the pace of industrialization in North America, especially the Second War for American Independence (the War of 1812), which forced the import and export industries in Boston and other places to temporarily halt. Many supporters of industrial capitalism in the North emphasized the importance of developing industry for the independence of the emerging country. Driven by various reasons, the states in New England became the first regions in North America to enter industrialization processes.

In 1813, the Boston Manufacturing Company obtained its business license through Massachusetts legislation and was officially established in Boston by a group of successful business people later known as the "Boston Associations." It is an event that marks the official commencement of industrialization in North America. At that time, the business people, who accumulated vast amounts of capital in Boston's import and export trade, saw bright prospects for the development of Old England's textile industry in New England. Therefore, in Waltham, a small town on the Charles River just 15 miles away from downtown Boston, they established a Waltham cotton and linen factory where the water power of the Charles River was used to drive textile machines to operate. Soon, the success on the banks of the Charles River and the limitations caused by the flow of the Charles River itself made these capitalists turn to the Merrimack River in Massachusetts, a wider and wilder river, where they established a textile industry town named after Lowell, the founder of Boston Manufacturing Company, thus the textile industry along the Merrimack River was born.

The history of Boston Manufacturing Company attracted the attention of scholars who studied labor-capital relationships, industrial development, and gender history. The Company's emergence brought about changes in social life in New England, marking the beginning of the unique labor management and gender conflicts in the era of industrial capitalism in the United States. Furthermore, it also marks a new conflict between human beings and nature, namely the problem of how to break through the natural characteristics and limitations of rivers, commercialize them, and maximize profits. The process of solving such problems resulted in extremely complex ecological and social consequences.[16]

Steinberg provides a detailed analysis of the process in *Nature Incorporated*. It is first and foremost a problem of quantifying and redistributing natural wealth, and one of the most important innovations adopted by the Company is the emergence of the important concept of "mill power." The water required for the power of a mill power could drive cotton yarn from 3584 spindles. Since the 1830s, textile companies had been able to purchase water solely on the basis of the mill power, but they didn't need to purchase any land along the river. The traditional water rights were broken, and water itself became a commodity. The subsequent process completely broke through the limitations of natural law on the development of capitalism, and water power was allowed to serve the operation of capital in any season of any year. In order to achieve the objective, textile companies went to the source of the Merrimack River where there were several large lakes in New Hampshire, and purchased a total of 103 square miles of lake surface to build dams of all sizes, thus the flow of the Merrimack River was completely controlled, breaking the limits set by nature, making thousands of spindles run day and night along the river, and bringing wealth rolling in from the water.

The process resulted in extremely complex ecological and social consequences, among which the fish and various other life forms in the Merrimack River were fatally impacted. The establishment of large and small dams completely changed the natural flow and rhythm of the water in their habitats, thereby causing countless artificial obstacles for their seasonal migration and disrupting their reproductive cycle. It will take human beings years and months to perceive the subtle changes in the temperature, quality, and quantity of water, but the changes are tantamount to the collapse of heaven and earth for these fish and other life forms that live in such water. The extinction of one or more species, even

[16] For the best research on how industrial capitalism transforms the water flow of New England into a commodity and the ecological and social consequences that arise from such transformation, see Theodore Steinberg's treatise, *Nature Incorporated: Industrialization and the Waters of New England. New Work:* Cambridge University Press, 1991).

tiny microorganisms or duckweed, probably leads to an imbalance in the biological chain in rivers, thereby destroying the entire ecosystem there. Therefore, the natural world, which had been diverse, collapsed after the establishment of a textile kingdom. However, the textile industry drove the rise of other industries in New England and the prosperity of industrial towns along rivers. In the Merrimack River, Charles River and other river basins densely gathered factories of various types, including textiles, papermaking, lumbering, leatherware, furniture, carriages, pianos, cigars, hats, and others, and a large number of immigrants flooded into towns such as Lowell and Wortham. These flowing rivers not only supplied the water power that drives all the machines to operate but also ensured the domestic water supply for coastal and terminal towns. Ultimately, they would also serve as discharge sites for various types of domestic and industrial waste and garbage. Severe pollution destroyed the entire ecosystem in the river basin, and human beings, as a link in the biological chain, were also greatly affected.

But what is Boston's position in the social and ecological changes? Boston could also be deemed as a nature metropolis, which promoted its own development by closely controlling the natural resources of other parts of New England. However, unlike Chicago, Boston itself did not become an industrial city. From an economic perspective, Boston was a commercial center at the bottom from start to finish. Due to tidal effects and the fact that industry in the surrounding areas of Boston relied heavily on water power, the flow was already extremely slow when rivers such as Charles flowed into the urban area of Boston, thus making it impossible to develop industry within the city. Boston's control and transformation of its surrounding areas were based on the abundant capital accumulated by Boston's business people in import and export trade and later industry. What Boston exported to its surrounding areas and even distant hinterland was not the industrial products manufactured within its urban areas, but capital, an intangible yet decisive force. However, as a component of the entire ecosystem, Boston couldn't escape the ecological problems ultimately caused by its capital. The polluted water sources in the city were the most direct issue that was closely related to the population in Boston.

Ge Hong, a native of the Jin dynasty, recorded a story in his *Biographies of Divine Immortals*. Immortal Wang Fangping made an appointment with Immortal Magu, who arrived late. After giving a salute to Wang Fangping, Magu said, "Since I attained immortality until now, I have already seen the Eastern Sea (Donghai) thrice turning into a mulberry field. This time I am visiting Mount Immortal, Penglai, and saw the sea water being much shallower than before, and it is probably about to turn into a hill on dry land again!" Wang Fangping laughed [upon hearing this] and said, "It is no wonder that the Sage said all roads in the sea will become dusty." What Fang Ping and Magu said was originally fabricated by ancients, but the changes from the blue sea to mulberry fields they witnessed is a cock-and-bull story. It records a long ecological history, which is mixed with personal changes, from blue sea to mulberry fields. The change lasts more than ten thousand years, and from mulberry fields to mulberry fields, it will take the time of several generations. However, for Boston, this period of vicissitudes and transformations is completely completed artificially only in just a century.

Many manuals or books about Boston provide the following data: from 1803 to 1863, the West Cove covering an area of 80 acres was reclaimed, and so was Mill Pond with an area of 70 acres from 1804 to 1835, South Cove with an area of 86 acres from 1806 to 1843, East Cove with an area of 112 acres from 1823 to 1874, South Boston waters with an area of 714 acres from 1836 to 1988, South Bay with an area of 138 acres from 1850 to 1988, Back Bay with an area of 580 acres from 1857 to 1894, Charlestown waters with an area of 416 acres from 1860 to 1896 and Fenway with an area of 322 acres from 1878 to 1890.[17] A hill land reclamation movement that occurred in the nineteenth century gave birth to a new Boston. The area of the city was tripled as a result, and the narrow bottleneck-like isthmus seen by Winthrop et al. upon their first arrival became as wide as the peninsula. The distance between both banks of the Charles River decreased by two-thirds, while the hills of the "City upon a Hill" almost disappeared from the map, leaving only a few broken ridges of the signal hill, the highest peak of the Tramount. When the reclaimed soil in Boston itself was not sufficient for use, Bostonians purchased clay, another natural resource, from its surrounding areas to fill it into rivers and seas and help the city expand.[18]

The measure of "hill land reclamation" has complex social, and environmental reasons. Boston's original geographical space was too narrow,

[17] Richard Wilkie & Jack Tager, *Historical Atlas of Massachusetts*. Amherst: University of Massachusetts Press, 1991.

[18] For a classic discourse on Boston's reclamation project, see Walter Muir Whitehill, *Boston: A Topographical History*. 2nd edition, enlarged, Cambridge, Mass.: The Belknap Press of Harvard University Press, 1959.

and by the early nineteenth century, there was no room for development. However, the urban population continued to rise, thereby making reclamation a direct and effective way to expand its territory. Especially at that time, Boston was gradually losing ground in competition with New York. Whether in population or commercial attractiveness, the laurels of the largest city in the United States were won by the former colonial city built by the Dutch. For Boston's political and business elites in hopes of establishing more accessible ports, it was a weighty consideration how the past glory in the competition with New York could be recreated. On the other hand, Boston's population underwent significant adjustments in the nineteenth century. Boston received a large influx of new immigrants, especially Irish immigrants who believed in Catholicism. The arrival of new population means a new workforce for future western cities. However, the new immigrants, for a city like Boston that developed in a strict Puritan tradition, brought not only the power to increase urban wealth but also the danger of eclipsing urban morality. Many middle and upper-middle classes began to migrate to surrounding cities. In order to urge the latter to stay here and resettle new immigrants, Boston's response was to build new communities for different social classes through reclamation.[19]

From an environmental perspective, in the nineteenth century before the emergence of modern means of transportation, new immigrants put a city upon a hill to many inconveniences in real life. After reclamation, the construction of infrastructures like urban roads, transportation, buildings, and drainage systems was more convenient. However, more importantly, the severe pollution of the Charles River became the root cause of the reclamation of the Back Bay area.[20] The Charles River, which flows through Boston, became a natural sewer in this area after the City of Boston was built. Since the eighteenth century, the waters around Boston were nagged by domestic sewage, as well as wastewater

[19] Nancy S. Seasholes, *Gaining Ground: A History of Landmaking in Boston*. Cambridge, Mass.: MIT Press, 2003: pp. 4–7. The book, currently the most comprehensive work on the Boston reclamation project, provides a good overview of the entire reclamation process, especially reclamation technology. However, there are significant shortcomings in analyzing the environmental changes brought about by reclamation.

[20] See William Newman & Wilfred Holton, *Boston's Back Bay: The Story of America's Greatest Nineteenth-Century Landfill Project*.

from the sugar and wine-making industries. In the first half of the nineteenth century, Boston, the most developed and public city with public spirit in the United States, firstly built a substantial underground drainage pipe network. Thus sewage flowed into the Back Bay through the pipe network in hopes of discharging sewage into the ocean through the principle of tides. However, although the use of water supply pipes increased water consumption and ensured the use of flush toilets, the amount of feces discharged from flush toilets increased accordingly. Because a large amount of feces was not washed away by the tidal current, they returned to the coast through the returned tidal current and remained on the shore. By the middle of the nineteenth century, in the report of the City of Boston, the Back Bay was directly described as "a great cesspool covered in decaying green trash, several miles wide, resembles a cauldron with wide-open mouth which emits a foul odor from the explosion of decaying materials at the bottom."[21] Compared with other treatment methods, reclamation seemed to be the most practical and effective way after that area was regarded by people as a septic tank and wasteyard.

Driven by different reasons, Boston changed the original appearance of the City upon a Hill, as well as the ecology and social system here. For a growing and expanding city, the reclamation project that lasted for a long time cannot be simply defined as violent and unbridled ecological damage. These areas accounting for a considerable proportion of the region, for instance, the Back Bay, were a restoration of the damage already caused by industrialization and urbanization to a greater extent. Compared with other parts of the United States in the same era, the New England region represented by Boston had not only a more consistent and clearer sense of the aesthetic function of nature but also a more transcendental expectation for the proper appearance of its city.[22] More and more Bostonians realized that the value of nature to civilization was far beyond the measure of the economic benefits of human society. Nature had its own value, as well as the spiritual and aesthetic values required for the development of civilization. During hill land reclamation, Olmsted was employed to design the Emerald Necklace, one of the earliest urban park systems in the United States. The Back Bay Fens, which was built by reusing natural

[21] City of Boston, *Documents of the City of Boston for the Year 1849*, no. 36. Boston City Printers, 1850, pp. 3–4.

[22] A discussion on transcendentalism and Ralph Waldo Emerson's ideas can be found in Chapter VI of the book.

water flow and local plants on the reclaimed septic tank, became the most brilliant diamond on the necklace.

4 A World in a Wild Flower

In 1888, Edward Bellamy published the novel *Looking Back: 2000–1887*). It is a typical science fiction about utopian. The protagonist Julian West, a Boston native, went to the year 2000 from the year 1888 while sleeping soundly. When he woke up from his dream, he found that Boston in the twenty-first century had become the ideal world of universal harmony: There is a prosperous economy as well as flourishing since technology, each takes what he needs, everyone is wealthy, and happy, and there is no war, class, monopoly, even greed, jealousy, or selfish desires. However, at the time looking back retrospect, West vividly remembers that in Boston in the late nineteenth century, politicians secured personal gain, financial oligarchies competed for profits, the gulf between rich and poor was enormous, the workers were confronted with their employers, the city was dirty and noisy and urban dwellers felt extremely anxious and confused.

At the beginning of its publication, the book "*Looking Back*" was widely read nationwide, its overwhelming popularity even caused a shortage of printing paper. Within three years, over 500,000 copies were sold, thus it was ranked first on the list of best-sellers in the United States for several years. If we explore the sensation that the book caused within its specific social context, it is not difficult to understand why a utopian novel with a story and text that is not the best choice can achieve such great success. As mentioned previously, the late nineteenth century was a period of great change in American history. The rapid development of industrialization and urbanization not only brought unprecedented prosperity and convenience but also brewed social problems at different levels and aspects. Boston, as reviewed by West, is a true portrayal of the restless American urban society. Although the political ideals of Bellamy's socialism are difficult to recognize by mainstream American society, the urban civilization he conceived is a model that people yearn for. Order, stability, and prosperity under equality were the stated objectives of the urbanization process in the United States after the Civil War. The scientific spirit touted in the book came along with high morale, it had been looked up to as the standard at that time since the Victorian era and was paid homage to.

However, in the 400-page (first edition of the original) blueprint of the city, only a few lines are used to depict its natural environment: "At my feet lay a great city. Miles of broad streets, shaded by trees and lined with fine buildings, for the most part not in continuous blocks but set in larger or smaller inclosures, stretched in every direction. Every quarter contained large open squares filled with trees, among which statues glistened and fountains flashed in the late afternoon sun. Public buildings of a colossal size and an architectural grandeur unparalleled in my day raised their stately piles on every side. Surely I had never seen this city nor one comparable to it before. Raising my eyes at last towards the horizon, I looked westward. That blue ribbon winding away to the sunset, was it not the sinuous Charles? I looked east; Boston harbor stretched before me within its headlands, not one of its green islets missing."[23] Compared with the full and vivid description of the urban cultural environment in the book, the narration becomes increasingly pale and fuzzy. Boston in the twenty-first century that Bellamy conceived is rich, clean, healthy, and comfortable, but there is very little space left for people to experience nature. The Charles River and the green island in Boston Harbor are staked over. The street trees constitute the natural environment which embodies one aspect of urban civilization, for the neatly and uniform trees precisely reflect solemn and standard urban civilization.

However, for the actual designers of City of Boston such as Olmsted, Sargent, Eliot, and Baxter, the view of ignoring nature or setting the city against nature is not only a misunderstanding of the relationship between the two, but also a misreading of the evolution of civilization itself. They believed that cities built on natural ruins were unstable and barren, and only cities that sought integration in conflict with nature had vitality and charm. A truly civilized society shall cover and contain everything, in which nature plays a decisive role. In their view, the American civilization in the late nineteenth century was not already in the stage of excessive civilization, as criticized by some intellectual elites at that time, and therefore the civilization was required to retreat to a more simple and natural state. On the contrary, they believed that the current level of civilization was still far from sufficient because in the evolution process of urban civilization, nature was often neglected or at least not taken seriously.

[23] Edward Bellamy, *Looking Back*, translated by Lin Tiandou et al. Beijing: The Commercial Press, 1963.

In other words, civilization encompassed not only industrial development, prosperous material, flourishing science, democracy, art, literature, morality, and other creations and inventions that were made by human beings, but also the abundant natural resources and the protected natural beauty—not merely mountains and forests in the distance and within the boundaries of cities, but also an order and sustainability that was won in the symbiotic cooperation between nature and humanity. As the most magnificent symbol of modern civilization, the city should be a landscape that includes all the aforementioned factors, especially the natural landscape.

This group of reformers was acutely aware of the feelings of "Back to Nature" that permeated the urban middle class. They themselves undoubtedly represented to some extent a specific love of the city, which simple life, pure morality, direct and close contact with nature, and all aspects related to the past rural life were lingering in their memories.[24] However, they not only indulged in such an emotion but also tried to expand it into a more practical field. Most of them were not natural writers[25] like Thoreau or John Burroughs. Their more important social roles were professionals in their own fields; they only actively and directly participated in and led urban planning. For them, nature was not a memory of an agricultural society that faded away in their vague dreams, but a practical existence that could be touched and applied in the real and future era of the urban industry. In their plans for social reform, they inevitably introduced some ethical ideas and behavioral patterns from traditional American agricultural ideals to position the moral coordinates of cities, just like as transcendental believers believed in the purification

[24] Historian Peter Schmidt believes that "Back to Nature" movement emerged in the United States in the late nineteenth century included the construction of urban parks, the rise of natural education and the popularity of natural literature. See Peter Schmitt, *Back to Nature: The Arcadian Myth in Urban America*. Baltimore: Johns Hopkins University Press, 1990,reprint.

[25] John Burroughs, a famous nature writer on the East Coast of the United States in the late nineteenth century, was on par with John Muir on the West Coast, they are known as the "Two Johns." But unlike Muir, Burroughs loved the rural scenery of states in New England and New York, while Muir admired the desolate and rugged ranges in California and Nevada. Burroughs appreciated the beauty of nature from the perspective of civilization, while Muir's conception of nature, which was closer to Thoreau's, considers all things in nature with an equal pantheistic worship.

effect of nature on morality, but it was the vanished agricultural society that they were not trying to restore.

Since Olmsted, Boston's reformers have had a clear urban ethos and spirit. In *Garden and Forest* magazine conducted by Sargent, an editorial states: "Today, It is a very narrow point of view to continue to regard the rapid rise and increasing importance of cities as a pure evil. The development of health science has made urban life much healthier than it was in the first few decades of the twentieth century, and the more complete facilities in the city have also increased its comfort several times. If the social and political importance of many rural communities is clearly being weakened, then what cities win is what countrysides lose." Their goal in the "Back to Nature" movement is not to restore a lost rural paradise, but to establish harmony between civilization and nature in the era of urbanization. They hoped to bring nature back into the daily lives, practices, and landscapes of citizens. On the one hand, Boston should be reshaped from a landscape perspective in hopes of redefining the relationship between citizens and nature. On the other hand, their response was the rise of horticulture among Boston's middle class, which was an important component of the "Back to Nature" movement. Its advocates believed that horticulture was a direct way for urban populations to communicate with the land in the post-agricultural era, which could change the distance between urban dwellers and nature. Its most direct function was to beautify urban environments of different scales in the simplest way possible, but it also had distinct moral connotations: it could not only curb the exaggerated materialism of the middle and upper bourgeoisie, but also alleviated the anger and dissatisfaction of slum dwellers and improved the moral cultivation of the entire urban society.

In 1829, a group of intellectual elites in Boston established the Massachusetts Horticultural Society, the first horticultural society in the United States. Horticulture, a popular leisure activity among gentlemen in Europe, especially in England, since the eighteenth century, was also favored by contemporaneous American squires. George Washington regarded himself first and foremost as an American farmer, he preserved a garden which he planted carefully in his plantation. In Monticello, the mountaintop plantation of Thomas Jefferson, horticultural expos could almost be held. Jefferson's favorite hobby is cultivating new plant varieties. The enthusiasm of both founding fathers for horticulture can well represent the favor of the middle and upper class in the late eighteenth century in the United States. However, during this period, the United

States was basically still an agricultural society, and Jefferson, who lived in Virginia plantations, was still building his ideal agricultural society. However, in the early nineteenth century, Boston in the north had already formed a commercial city with frequent internal and external transportation and trade to plunder the natural resources of its surrounding areas. Therefore, it was a different significance that the first horticultural society was established in Boston, one of the largest commercial cities in the United States at that time.

The Puritan tradition in Boston had a profound impact on the shaping of Boston Brahmins, the elite class in the city. On the one hand, they viewed wealth accumulation as a mission bestowed upon them by God in the New World. On the other hand, they believed that material accumulation and enjoyment should not be the highest pursuit of life and that pure morality and devout faith were the standards for measuring God's chosen people. Therefore, the elite class in the city had always been seeking a balance to prevent materialism from spreading unchecked. Boston became one of the wealthiest cities in the United States through domestic and overseas trade. After the descendants of local Puritans accumulated a vast amount of wealth through various means such as trade and investment, they felt a moral crisis about their survival way. But how could the moral crisis be eliminated? Charities and volunteer work became important forms for them to return their wealth to society. Just because of this, Boston's education and public works both became a model for other cities in the United States. However, the way of investing back into society with wealth was mainly adopted in the public domain but had no direct impact on the individual's spiritual world. At that moment, horticulture began to rise and develop into a movement.

The rise of the movement had physiological, intellectual, and social reasons. Many urban reformers at that time believed that everyone had an instinctive desire to come into contact with the land to some extent. Planting on the land or at least a small piece of soil was directly touching nature in a specific way, thus production was quite different from pure reading, appreciation, and even nature sense. In the late nineteenth century, the rise of cities provided a huge market for commercial horticulture. When many people chose to purchase cut-flowers to bring nature to their homes, advocates of urban naturalization promoted potted flowers they cultivated, because compared with cut-flowers, potted flowers could transmit more or less the vitality of nature. In an article published in *Garden and Forest*, the author explores the importance of rooftop gardens for urban dwellers. After stating the effects of moral education, the author concludes that "the most practicable immediate use of the roof garden is for the families of professional men and others who long for some contact with growing things, a bit of nature, wild or tame, all to themselves."[26]

The instinctive love for nature could also be awakened through the practice of horticulture. This kind of passion was different from the appreciation of natural landscapes. The latter was a more profound emotion mixed with memory and imagination when civilization develops to a certain extent, but the former, especially in its primary stage, was an intuition that even "savages" possess. The appreciation of natural landscapes was often limited to the prostration of a majestic, admirable, and unfamiliar natural beauty, while horticulture could uncover the attachment to the plain, simple, and even familiar nature itself that lay dormant in people's instincts. At that time, almost all advocates of horticulture were enthusiasts of natural landscapes, but they hoped that people would not overlook the ordinary yet real side of nature by emphasizing too much on the prominent part of it. They hoped that people could penetrate those extraordinary landscapes and touch the essence of nature. In the eyes of advocates of horticulture, this was also the charm of horticulture. Through daily, year-long irrigation, fertilization, and cultivation, people could not only perceive the natural beauty among ordinary flowers and plants but also experience an instinctive love for nature.

For those Boston Brahmins who were obsessed with horticulture, horticulture could inspire a certain degree of interest in exploring the mysteries of nature and scientific research. In the eighteenth and nineteenth centuries when natural history grew vigorously, many upper and middle classes on both sides of the North Atlantic Ocean made inquiries about the laws of nature in different ways. However, the plants in the gardens of the people who couldn't travel far to tropical rainforests or Devil's Island revealed some natural secrets to such people. Especially in the second half of the nineteenth century, when the scientific spirit of the Victorian era was high, many people who received modern scientific training longed to participate in natural processes. While understanding the role of natural processes, they hoped to discover the influence of human power on the evolutionary process.

[26] E. P. Powell, "Housetop Gardens," *Garden and Forest*, issue 212 (1892): pp. 125–6.

Francis Parkman, one of the most eminent historians in American history, made a representative statement when he assumed the presidency of the Massachusetts Horticultural Society in 1875: "It (Horticulture, translator's note) is an art based on science, or on several sciences. When pursued in its highest sense and to its best results, it demands the exercise of a great variety of faculties, and gives scope to a high degree of mental activity .... The mind of the true cultivator is always on the alert to detect the working of principles and carry them to their practical application. To read the secrets of nature and aid she in her beneficent functions is his grateful and ennobling task."[27]

Parkman's personal experience can also represent the attitude of the elite class toward horticulture at that time. Although he is a Harvard historian, he greatly enjoys the fun and challenges that horticulture brings to him. Especially when his health restricted him from continuing to venture into the wilderness and perceive the natural beauty and power, he devoted all his spare energy to his garden. Not only did he become the chairman of the oldest horticultural society in the United States in 1875, but he was later appointed as the first horticultural professor at Harvard Bussey Institution.[28] Although there are not many amateurs like him who have achieved such great success in horticulture, there are plenty of people who have similar backgrounds and interests. In the late nineteenth century, horticulture became an important component of the burgeoning nature-study movement at that time. Environmental historian Kevin Armitage pointed out in his research that: Nature-study advocates encouraged children to gain scientific knowledge through rational, objective experimental methods while learning to become morally engaged through intimate contact with nature. They tried to reconcile what had often seemed diametrically opposed values.[29]

[27] Francis Parkman, "The Presidential Address to the Massachusetts Horticultural Society," in 1875, quoted from Charles Haight Farnham, *A Life of Francis Parkman*. Boston: Little, Brown, 1901: 33–34.

[28] Due to health reasons, Parkman's position at Bercy School was soon taken over by Sargent. Parkman was also one of the earliest followers of forest resource conservation in North America, inspiring the forest conservation of the White Mountain in New Hampshire.

[29] Kevin Armitage, *The Nature Study Movement: The Forgotten Popularizer of America's Conservation Ethic*. Lawrence: University of Kansas, 2009.

However, the rise of horticulture in the nineteenth century, by its nature, was still a tradition unique to Boston where Puritanism and transcendentalism are combined. Horticulture attempted to clean up the social evil through nature. Nature-study advocates believed that in the process of cultivating, anticipating and admiring the growth of plants, the natural purity and beauty would care for the beautiful side of human nature. Although the need to rely on the work on the land to obtain daily necessities had come to an end in cities, the moral reverence for agricultural life still survived. Therefore, when many people gave up their farm life and came to the city to seek survival and when the urban barriers and factory regulations no longer allowed their residents to have the opportunity to work on the land, at least horticulture could allow them to have the opportunity to contact the land with their own hands, purify their souls stained by materialism and hedonism in the city.

Historian Tamara Thornton points out that, in Boston before the Civil War, horticulture first appeared as a hobby among gentlemen before emerging as a popular movement, with flourishing organizations and periodicals. However, during this period, due to the connection between horticulture and rural life as well as morality, gardening was deemed to cure the upper class's moral failings, especially the materialism and boorishness created by greed and ambition. At the same time, the appeal of horticulture lay in its ability to calm the "spirit of unrest" and make the nation more stable. After the Civil War, the advocates of horticulture switched their attention to the working class and the new immigrants, who, in the eyes of Puritan descendants in Boston, were often associated with various sins (such as alcoholism, gambling, laziness, adultery, plagiarism, and even various serious criminal cases) in the city due to poverty and the Catholic tradition of new immigrants. Thornton believes that horticulture "was to function as an antidote to the moral failures." She argues that behind this horticultural movement lay the belief "that horticulture taught republican virtues: hard work, thrift and the sacrosanct worth of private property."[30] Therefore, horticulture became a means of social improvement for progressive reformers.

When the monotonous industrial life and gloomy urban environment squeezed nature out of the lives of urban dwellers, nature was also

[30] Tamara Plakins Thornton, "Horticulture and American Character," *Keeping Eden: A History of Gardening in America*, ed. by Walter, T. Punch. Boston: Bulfinch Press, 1992: pp. 200 & 189.

dislodged from the range of their sight. Therefore, the questions that these reformers are constantly seeking answers to are how to reintegrate nature into the urban environment, how to dissolve the indifference of urban people toward nature, and how to make them have a deeper understanding of nature. The formers, convinced of everyone's innate desire for nature, were also convinced that nature had the ability to enhance human physiological and moral health. But through their observation, they realized that the alienation of urban life from nature also caused people to forget their instinctive thirst for nature and natural beauty. They not only attempted to integrate nature into the urban landscape but also strove to infuse nature into the daily lives and minds of urban populations. They established horticulture for urban populations, an important way of the connection between life and nature, but their thinking and practice were not merely limited to the reform of the environment and life of a person, a family, or even a community, but also based on the reconstruction of the entire urban landscape. What they tried to change was people's fundamental definition of cities, and they wanted to integrate nature into the evolution process of urban civilization through the construction of the Boston urban park system.

5 Combination of City and Nature[31]

In May 1892, the Governor of Massachusetts appointed the Boston Board of Park Commissioners, which is obliged to "consider the reasonable establishment of many public open spaces in Boston and its surrounding towns."[32] Ten months later, the Board submitted two reports to the state legislature, written by its secretary Sylvester Baxter and landscape designer Charles Eliot. They pointed out in the reports that the proposal to establish these urban open spaces was not only unquestionably rational but also urgent. Both illustrated and incisive reports undoubtedly convinced members of the Massachusetts Legislature. In May 1893, the official Metropolitan Parks Commission was established in Boston

[31] Some of the content from the old article the *Integration of Nature and City: The Construction and Inspiration of the Boston Metropolitan Park System*, which have been adjusted, are adopted in the fourth section the book. (*World History*, August 2009, No. 4, pp. 73–85).

[32] "An Act: To Establish a Board of Metropolitan Park Commissioners and to Define Its Powers and Duties," *Boston Metropolitan Park Report*. Boston: 1893.

by legislation, that is to say, the Boston Metropolitan Park System, the first metropolitan park system in American history, would be established among the 12 cities and 24 towns surrounding Boston. After the operation of more than a century, this system was preserved, developed, or constructed around 20,000 acres (approximately 80 km^2) of urban public space in various forms, and it became a model followed by other American cities such as suburbs of Washington D.C., Cleveland, Chicago, and Minneapolis.

Before that, Boston had already been constructing the Boston urban park system, named the "Emerald Necklace" by Frederick Law Olmsted. The discussion about building the Boston urban park system first appeared in the 1850s, but it was not until the end of the Civil War and the intensification of urbanization in the late 1860s that it was truly included in the agenda of Boston's municipal construction. Its comprehensive construction coincided with the large-scale wetland reclamation in Boston. From the perspective of geographical boundaries, some parts of the park system were also located on top of the reclaimed wetlands, for instance, the Back Bay Fens, as stated earlier. The physical landscape of the "Emerald Necklace" was entirely the product of the mind of the genius, Frederick Law Olmsted. Compared to the challenges and encroachments during the construction of Central Park in New York and after its completion, Boston's park system basically won the unanimous support of Boston legislators and the public opinion community although it emerged relatively late. Its root cause was still the moral values and aesthetic concepts that were dominant in Boston for the long term.

The "Emerald Necklace" is an L-shaped chain park system that starts from Boston Common in downtown Boston and winds westward to Franklin Park located in Jamaica Plain, Roxbury, and Dorchester in the suburbs of Boston. It is about 7-mile walk but covers an area of approximately 1100 acres. The "Necklace" is strung together by Park Drive and The Riverway, and the "Diamonds" on the chain include Boston Common, Boston Public Garden, Commonwealth Avenue Mall, Back Bay Fens, The Riverway, Olmsted Park, Jamaica Pond, Arnold Arboretum, and Franklin Park.[33]

[33] During the construction of the "Emerald Necklace," Frederick Law Olmsted and Sargent collaborated in multiple ways to incorporate the Arnold Arboretum into the Boston urban park system. The City of Boston Government leased the land belonging to Arnold Arboretum to Harvard University for a term of 10,00 years, which can be renewed,

The design of Boston's urban park system already contained Olmsted's pursuit of the integration of civilization and nature. However, as Boston's urban boundaries further expanded in the late nineteenth century and more surrounding areas were incorporated into its urban system, an urban park system could no longer meet the further development of urbanization. It was in such a context that the concept of a metropolitan park system began to emerge. At that time, Olmsted's health was from bad to worse, and he undertook the landscape design of the first World Expo in the United States—the World's Columbian Exposition. Therefore, the new park system was taken over by his genius disciple, Charles Eliot, a direct descendant of the Boston elite family. Nevertheless, in the reports written by Eliot and Baxter, we can still clearly see the influence of Olmsted on them. In his report submitted to the Boston "Metropolitan Parks Commission," Baxter pointed out that establishing an urban park system is "largely a health issue, but it can enhance the physical and moral health of the community from a broader perspective. A screamingly obvious fact is that for those who are immersed in the hustle and bustle as well as chaos and turmoil that are innate in urban life, in order to maintain a healthy and beautiful human nature, they shall often take the opportunity to break free from the shackles imposed by the state of life. The best way to obtain such an opportunity is to escape into more natural and comfortable environments."[34]

With such ideas in mind, the group of progressive reformers in Boston initiated the construction of the first urban park system in the United States. The fundamental idea of the park system still lay in the traditional moral appeals of Puritans and the natural aesthetics of transcendentalism. However, the special era they lived in made the urban park system have its own unique features which are different from previous urban parks. Firstly, Boston already expanded into a metropolis centered around the

at a price of US $1. The City of Boston Government would provide all the costs for the planning and renovation of Arnold Arboretum, and the staff of Arnold Arboretum would be employees of Harvard University, with research conditions and salaries provided by Harvard University. The collaboration made Arnold Arboretum a natural garden, rather than just a plant exhibition hall and research institute. Afterwards, the two persons collaborated to design Commonwealth Avenue Mall. Sargent is a loyal admirer of Olmsted's art, and Olmsted needs Sargent's professional knowledge in horticulture, dendrology and botany to make up for the deficiency in his knowledge background.

[34] Sylvester Baxter, "Secretary Report," *Boston Metropolitan Park Report*: p. 9.

City of Boston and surrounded by several towns, suburbs, and countryside. Therefore, the park system was initially established with the city as its spatial scale while without its boundaries set by simple administrative divisions, which reflected its distinct urban characteristics. Secondly, the diverse terrain and landforms of Greater Boston provide a vast space for the designer's romantic imagination, and the romantic aesthetic concept made the system aimed at the protection and restoration of natural landscapes. As a result, Boston Metropolitan Park was characterized by its main body with a system based on naturally formed landscapes. Compared to the naturalistic style of other urban parks in the United States, it emphasized local natural geography, vegetation, and animals. Thirdly, one of the themes of progressivism was to professionalize urban society by emphasizing the authority of expertise in the management of social affairs. Thus, the Boston Metropolitan Park System, from boundary determination to design and construction, and then to management improvement, was entirely decided by professional landscape designers. Meanwhile, it was precisely because of the establishment of the leadership of professionals that the urban park system could break the constraints of administrative divisions and was free of any negligence in management. Therefore, another characteristic of the system was specialization. Finally, the social philosophy of these progressive reformers, namely their unwavering belief in American democracy, was consistently implemented in the construction of the park system. They strove to not only resist the tradition of natural beauty which is monopolized by a certain social class but also make it a space shared by all ethnic groups in society. As a consequence, the park system is imbued with the democratic spirit of progressivism.

In 1910, William B.De Las Casas, Chairman of the Boston Metropolitan Parks Commission, wrote in an article that "Boston's natural terrain is not suitable for use by a large population. It is such a peninsula, almost a lonely island, standing loftily on a bay in the middle of a hill composed of three elliptical ice mounds. It is surrounded by islands and peninsulas that have the structures similar to the lonely island, but they are isolated from each other. In the middle are rivers and bays, as well as large stretches of swamps that meander and extend to the glacial slopes of the surrounding hills where are covered with almost entirely rocks."[35] However, the increasing population pressure and the prosperity of industry and commerce forced Boston to look for any way to break through the limitations of nature. Through wetland reclamation and land annexation, the population of downtown Boston in the late nineteenth century increased to nearly 450,000. With a covered area of nearly 90 square miles, downtown Boston was not only one of the largest industrial, commercial, and port cities in the United States but also a cultural center in North America.

However, the pace of Boston's urban sprawl did not stop there. It incorporated surrounding towns and countryside into its development network, thus forming the Boston urban system in the 1880s. In 1896, Eliot wrote, "Surrounding Boston and forming with Boston the so-called metropolitan district lie thirty-seven separate and independent municipalities, comprising twelve 'cities' and twenty-five 'towns' all of which lie either wholly or partly within the sweep of a radius of eleven miles from the State House. The population of this group of towns and cities is about one million of people, and the total of taxed property about one thousand millions of dollars."[36] These towns were connected by convenient roads and transport, and their economic relationships were inseparable, so personnel exchanges were extremely frequent. Similarly, it is noteworthy that they also shared common interests in many other aspects of urban life. The report of the Boston Metropolitan Park System Commission states, "In a metropolitan area composed of a large city and its suburbs - such as London, Paris, New York, Chicago and similar smaller areas like Boston - in addition to common needs and interests in policy, drainage, water supply and transportation, open-space reservations shall also be added."[37] Therefore, as Baxter said, Boston, as a city, "is essentially a community in terms of its social intentions and purposes although it was divided by political lines into many towns."[38]

Compared with the urban society, a community that couldn't be separated by simple administrative divisions was the land itself located

[35] William B. de las Casas, "The Boston Metropolitan Park System," *Annals of the American Academy of Political and Social Science*. Vol. 35 (March 1910): p. 64.

[36] Charles Eliot, "The Boston Metropolitan Reservations," *The New England Magazine*. Vol. 21 (Sept. 1896): p. 117.

[37] "The Report of the Board of Boston Metropolitan Park Commissioners," *Boston Metropolitan Park Report*: p. xxi.

[38] Sylvester Baxter, "Secretary Report," p. 3.

within the city. Under the rigid and straight lines of the Greater Boston administrative map, there were towering mountains, rugged rocks, meandered rivers, and twisty coastlines. Their existence was not constrained by artificial boundaries but often extended across several towns. The Blue Hills Reserve, which was first incorporated into the Boston metropolitan system, covered an area of 7000 acres and crossed four administrative divisions: Quincy, Dedham, Milton, and Randolph. However, the system composed of vegetation, animals, soil and water flow within it was a complete natural economic system closely related to human life within reach, but it was independent of the political and economic order of human society. For example, in Middlesex Fells with a total area of about 3000 acres, rocky mountains, prairies, wetlands, oak and walnut forests, lakes, ponds, and others were scattered into such 5 towns as Medford, Winchester, Stoneham, Melrose, and Malden. Similar to the Blue Hills Reservation, its natural boundaries were completely irrelevant to human boundaries.

Therefore, how to protect and manage these natural zones located between or on the edges of several administrative divisions was a problem that Eliot, Baxter, and others needed to solve. Baxter pointed out in his report that the distribution of public open spaces in Greater Boston was extremely uneven, with parks and botanical gardens scattered throughout the south, while without such spaces in the north. The reason for such a situation was simple yet intriguing. Their wealth and power were concentrated in the south, so a single administrative area was formed where they could pay for the construction of urban parks on their own account. However, the north was divided into many small communities by political rather than natural boundaries although the population in the north accounted for half of the total population and was on a more rapid upward trend, thus it was impossible to establish such public spaces there. Baxter simultaneously saw that conflicts of economic interest between towns and local protectionism in each town made such towns unwilling or unable to undertake public affairs that crossed the boundaries of the city. The power of the first level administrative unit in Massachusetts County declined, so it didn't have funds or manpower to establish various public open spaces between several towns in the county.

In that situation, Eliot established the Trusteeship of Public Reservations in 1891, the first state-level nature conservation agency in the United States. With its semi-private nature, it protected Massachusetts through donations, purchases, and other means, especially historical sites and natural scenic areas along the Boston coast were available to the public by resisting private and corporate development.[39] However, Eliot quickly realized that such institutions still lacked the ability to protect and manage thousands or even thousands of acres of natural scenic areas although they were highly efficient in establishing small public open spaces. Therefore, in his letter to Charles F. Adams, Chairman of the interim Metropolitan Parks Commission, written in 1892, Eliot pointed out that the establishment of such large public open spaces "strongly calls for the efforts of the entire urban community."[40]

The many reasons of society, nature, politics, and economy made it inevitable the establishment of a park system based on the urban scale, and the new technologies generated by urbanization and industrialization made it possible the emergence from such a system. In between them, the most significant development was the one of transportation technology in the late nineteenth century. Located in the northeast region where the economy was most developed in the United States, trains became a widely used means of transportation connecting various towns in Greater Boston in the late nineteenth century. The emergence of trams was another symbol of fast and affordable transportation, and their routes extended to every corner of the city during the Progressive Era. For urban dwellers of Greater Boston, wild areas such as the Blue Hills Reservation, Middlesex Fells, and Lynn Forests which were located around Boston were no longer unreachable.

Although various convenient transport facilities shortened the time for urban dwellers to travel to the natural scenic areas located on the outskirts of the city, they also accelerated the speed at which the behemoth of the city erodes natural space, especially the erosion of nature by real estate development industry was the most significant. The iconic industry of urbanization process not only transformed many natural landscapes that were originally on public land into private industries but also completely destroyed countless beautiful mountains and clear water in the so-called development. Baxter warned in his report that Boston is "likely to become

[39] It is the world's first organization of its kind, and the Land Trust Committee in the UK was directly inspired by it and was established in London four years later. The details can be found in Gordon Abbott, Jr., *Saving Special Places: A Centennial History of the Trustees of Reservations: Pioneer of the Land Trust Movement*, Ipswich, Mass, 1993.

[40] Charles W. Eliot, *Charles Eliot: Landscape Architect*. Boston; New York, Houghton, Mifflin, 1912: p. 381.

a vast desert composed of houses, factories and shops, the natural landscape on the land will be buried in the vast desert, just as a sand dune line moving along the coast ruins and wipes out forests and fields. The desert results primarily from human activities, which is difficult to save by an oasis, is threatening the beautiful natural place on a larger scale."[41]

Buildings were seen by contemporary people as a basic indication of urban progress, while Baxter's voice precisely represented the question posed by his colleagues to the prevailing concept of urban progress. The ultimate goal of the Boston Park system was still to improve the urban environment, but the improvement was to preserve and restore the various styles of open spaces in Greater Boston. In 1897, Baxter wrote an article titled *How Boston Has Systematized Its Parks, A Lesson for All American Cities*, which was published in *The Century* magazine. He wrote in the article, "Intelligent growth, instead of the haphazard expansion of the former periods, should be the aim in the municipal activities of the present age Modern science has placed in our hands the means essential to such growth, together with knowledge of what the circumstances demand. Of cardinal importance are the adaptation of every scheme for municipal development to topographical requirements, and a wisely considered use of the natural features of a site in determining the permanent character of landscape surroundings." Baxter continued to point out that Boston was different from other large cities on the East Coast. It was surrounded by mountains and water but it was difficult to expand. Therefore, despite the development over three centuries, it still preserved large areas of natural scenery within reach of the city: "(It) is a maritime city lying upon an island-studded bay, with estuaries penetrating a strikingly diversified landscape; a region not modeled upon a grand scale, such as distinguishes the country about New York, but with a broad and liberal graciousness in the charmingly picturesque commingling of hills and valleys, woodlands and marsh levels, with many ponds and various clear streams, and margined by the sea in an irregular coastline of bluffs and rocky headlands alternating with shinning beaches - the whole country toned and softened by nearly three centuries of human occupancy."[42]

[41] Sylvester Baxter, "Secretary Report," p. 3.

[42] Sylvester Baxter, "How Boston Has Systematized Its Parks, A Lesson for All American Cities," *The Century*. Vol. 54 (Oct. 1897): p. 952.

Such a geographical environment gave the Boston Metropolitan Park System unique features that distinguished it from previous urban parks. Although it also had more artificial landscapes, such as Boston Common and Commonwealth Avenue, it encompassed much more than that. At that time, most of the urban parks in the United States were reconstructed into artificial natural spaces between the cracks of urban reinforced cement. For example, Central Park in New York, whether its original soil structure or vegetation condition was a willy-nilly result left with nothing better than the second choice and was not suitable for the establishment of urban parks. However, the foundation of the main works of the Boston Metropolitan Park System was an ecological system composed of local natural elements such as air, soil, terrain as well as flora and fauna. Although Eliot, Baxter, et al. were well aware that these sceneries were inseparable from human activities and were "toned and softened" scenery, nature remained the basic driving force of shaping these scenes, and the Boston Metropolitan Park System was to assist nature in reserving these beautiful scenes through manual labors.

Therefore, in Eliot's report submitted to the interim commission, he pointed out that the metropolitan park system based on Boston's geographical environment should include five basic components. Firstly, open spaces in coastal areas; secondly, the islands and beaches on the bay; thirdly, the line of tidal flats at the estuaries of several rivers; fourthly, two to three wide wild forests located on the outskirts of human settlements, and fifthly, countless small squares in densely populated areas.

In 1893, the report by Eliot and Baxter prompted Massachusetts to establish the Boston Metropolitan Park System by legislation, which was funded by the state government and implemented by the Metropolitan Parks Commission. On the one hand, as analyzed earlier, a metropolitan park system went beyond the scope of a single administrative division at the city and town levels, and as an urban area, Greater Boston was not a true administrative concept, so it couldn't be managed by one city government or several local governments. On the other hand, in the view of progressive reformers advocating the system, the park system was a long process which takes decades, or even hundreds of years for its consummation. However, politics which was full of variables and transients couldn't guarantee the long-term stability required by the system. Therefore, a metropolitan commission that was detached from local politics, unaffected by political uncertainty factors, yet supported by legislation, officially emerged. As Baxter pointed out in his article, "What is needed is to connect the various scattered public holdings, make them convenient of access, ... bring all under a uniform system of administration. This can best be done under the proposed metropolitan park administration."[43]

The commission was very different from traditional ones. Firstly, it went beyond the limits of local politics; Secondly, fundamentally, this new metropolitan park commission emphasized the authority of experts, namely landscape designers. The composition of traditional commissions often consisted of local political, commercial, and intellectual elites who often had no knowledge of the things they were in charge of, while experts were restricted by them. Taking the New York City Parks Conservancy appointed in 1897 as an example, among the four commissioners, two were local politicians who invested in real estate, one was a builder and the other was a banker. They not only lacked professional knowledge of park administration, but in many cases, they might disregard the integrity of the park due to commercial or political interests.

However, the Boston Metropolitan Parks Commission tried to avoid such misconception since its establishment. Eliot, a young landscape designer, and Baxter, an urban design expert, were appointed as chief designer and committee secretary respectively, and given the greatest freedom in funding budgets, geographic surveys, boundary setting, road construction, and even the selection, planting, and trimming of plants. As the authoritative publication of the environmental protection movement at that time, *Garden and Forest*, said: "An individual park ought to be an organized work of art. If a park system with various outdoor activity facilities can reasonably meet the different needs of all classes and ages, more research is needed. Just here, the example set by Boston should be a model for every city. Professional advisers were appointed before an acre was bought. They were consulted not only in a general way as to the selection of park sites, but, after making a complete study of the whole question, they selected the sites and worked out their boundary lines."[44] In several subsequent articles, *Garden and Forest* repeatedly pointed out that the Boston urban park system could serve as a blueprint for future

[43] Sylvester Baxter, "The Boston Metropolitan Park Movement," *Garden and Forest*, issue 207 (1892): p. 62.

[44] Editorial, "Park Lands and Their Boundaries," *Garden and Forest*, issue 452 (1896): p. 421.

urban parks in many aspects, but it was most praiseworthy that it unreservedly trusted and applied the professional knowledge of landscape design.

The ethos shows the temperament of the specialized era in the late nineteenth century. With the increasingly refined classification of disciplines, a large number of emerging professions began to establish their conduct and discipline and seek social recognition. A major theme of progressive reform was to recognize the authority of the professions, such as engineers, doctors, and lawyers, in the government and society, thereby establishing a social order in the urban age, a rigorous system that emphasizes efficiency and rationality. Landscape architecture was one of the many emerging professions. In the late nineteenth century, American landscape designers began to clarify their professional scope and principles. In his letter to a friend in 1897, Eliot argued that "landscape architecture includes and covers landscape engineering, landscape gardening and landscape forestry," which "means the designing of all things and arrangements necessary or desirable for the use and convenience of human beings occupying the surface of the earth."[45]

As loyal progressive reformers, the enthusiasm of Eliot et al. for achieving the public interest in the design process of the Boston Metropolitan Park System was another layer of persistence beyond specialization. They hoped that in their park, people could breathe the same breath and enjoy the same scenery, and the boundaries between class, religion, and education level would temporarily disappear. The concept of democracy and equality also became the driving force behind the design of the Boston Metropolitan Park System by Eliot et al. in the future. They believed that the emergence of such a system prevented the natural landscapes that should be shared by society from being occupied by a few people, and its service targets should be various social classes, especially the working class and the urban poor. Eliot pointed out that the establishment of the Boston Metropolitan Parks Commission "proved that a great and complex American democracy was vibrant in achieving the effectiveness of beauty and the value of public spaces. Similarly, the

[45] Eliot to Mary C. Robbins, Dec. 2, 1896, Charles Eliot Collection, Frances Loeb Library, Harvard Graduate School of Design.

democracy also had the ability to cooperate and foresight, the Commission made all-out effort for the purposes it believed in ...".[46] As Baxter said, "In a democratic community, nothing should be held too good for the common people."[47] The Boston Metropolitan Park System not only integrated natural beauty with democratic ideas but also introduced the former into the public domain. In terms of material, it was open to the public completely free of charge, and in terms of ideology, it broke the prejudice that appreciation of natural beauty was exclusive to social elites. During the process of shaping the urban environment of Greater Boston, the understanding and perception of nature and beauty among various social classes was also reshaped.

Its designers and advocates didn't want their parks to be one or several lonely spaces that were ill-adapted to the urban civilization they were in, and even diametrically antagonistic to it. On the contrary, although they believed that the existence of urban parks would provide urban populations with a natural environment that they were accustomed to, as well as the tranquility, simplicity, and freedom contained within such an environment, they still regarded parks as an indispensable component of urban civilization for they thought that parks embody the essence of openness, magnanimousness, and civilization. As stated in an article in *Garden and Forest* "public parks are quite as essential to the health and comfort and morals of the city as a pure water-supply or a good system of sewage, ... a civilized community can no more flourish without them than without hospitals, libraries, museums, colleges and churches."[48] Eliot et al. attempted to emancipate nature from the boundaries of parks and expand its scope and imagination to the entire urban civilization.

In order to achieve the ideal, they rushed to call for the preservation and development of public open spaces with natural scenery in and outside the city. Such open spaces might not only be swallowed up by the current economic development but also be built into factories, shops, and parking lots. It also possibly became private property of the upper class in society. At that time, the sign of "No Admittance" would have been set up at the gate although the scenery is still attractive. In the second half

[46] Charles Eliot, "The Boston Metropolitan Reservations," p. 122.

[47] Sylvester Baxter, "A City's Small Pleasure Grounds," *The Century*, Vol. 55(Dec. 1897): p. 315.

[48] Editorial, "Small Parks for New York," *Garden and Forest*, issue 380(1895): p. 222.

of the nineteenth century, these urban naturalization reformers unremittingly fought against threats from both sides by lobbying governments, collaborating with other media, and stimulating public opinion. In the meantime, they repeatedly emphasized the openness and public nature of such open spaces, so they should be open to the public completely free of charge and equipped with various cheap and convenient transportation facilities around them. Only in this way, "everyone, irrespective of high or low birth, will breathe the same breath of nature and art, enjoy the same scenery, and there is no hate or conflict between them."

Naive? Or so it might be. But all social reformers are naive to varying degrees. It is the naive pursuit that has driven the construction of thousands of parks in cities and the preservation of tens of thousands of acres of natural scenery. Many social activists are still criticizing the natural beauty sought assiduously by these urban naturalization reformers as the aesthetic imagination of the middle class themselves. Some passionate wilderness-huggers believe that the nature embraced in the city is a product of culture but a false illusion. However, even these critics cannot deny that the thoughts and actions of these early urban reformers with environmental awareness are not abstract theories or artistic fantasies, but are closely related to the daily life experience of everyone in the city. The ecological logic they carefully constructed can be found in the graceful and green trees on both sides of the street to the urban parks covered with grasses and pines. Meanwhile, the social justice they strove to achieve can be seen in the thousands of visitors of different races and social classes who come to urban parks.

PART III

Expanding Cities

CHAPTER 5

City Without Borders

1 Prologue: Viva, Las Vegas!

Viva, Las Vegas is undoubtedly the sexiest musical in the history of American film. Dices are catching everyone's eyes, wheels are quickly whirling, neon lights with various colors are flashing on and off, towering hotels seem to be kissing the cloud, as well as Elvis Presley is playing the guitar loudly, and Ann Margret is twisting her hips with abandon, the city is immersed in love, joy, singing, and dancing. In the young and gorgeous city, all dreams will probably come true, and all adventures will pay off. The city is Las Vegas, a city that makes the night last quite a long time. But the night cannot stay at last. In the morning everything recovers from the revelry at night. When the first faint rays of dawn brighten the brown horizon in the distance, the city that shone brightly at night suddenly became glum, forlorn, and serene. The sky is crystal clear, and a stretch of the Sierra Nevada Mountains is like the curtain of Las Vegas. The silhouette of the city sets off hard rufous rocks, on which there is no tree or grass. The rough texture tells the visitors of the city that they are in the desert.

In the movie, Ann and Elvis Presley (Hillbilly Cat) fly a helicopter over the Hoover Dam. Ann excitedly and proudly tells Hillbilly Cat, a young man from a central town, who is mad at racing, "Look, Hoover Dam, this is a new wonder of the world!" The graceful girl dressed in red was born in the city. Her father, recruited by the Bureau of Reclamation, came to the southwestern desert. He, together with a labor force that increased by

S. Hou, *Wallless Cities*,
https://doi.org/10.1007/978-981-97-7827-0_5

5000 people per week, poured 3.4 million cubic yards of cement into the running Colorado River to embed a huge arch dam into the stone walls of the Andesite Canyon, thus creating a modern engineering miracle and opening up the modern urban miracle of Las Vegas.[1]

The so-called miracle refers to the fact impossible things are created in impossible places. Las Vegas is such a miracle. The city was officially founded in 1905. Although most American cities are young compared to cities in the Old World, it is especially true for Las Vegas, which was born in the early twentieth century, and its true development began in the second half of the century. Its biographer, Hal Rothman, sees the city as a weird paradigm of the postindustrial age: "It didn't have fertile land or rich mineral veins; railroads didn't meet, highways didn't cross there. Banks didn't seek out Las Vegas, developers didn't fashion it into the next paradise, corporations didn't come to the desert to establish new headquarters, and people certainly didn't come looking for the little oasis to put down roots."[2] It is not even like Southern California, which has a pleasant Mediterranean climate due to the regulation of the Pacific although it is dry. Thus, it can sell its leisurely vacation destinations to the wealthy and idle middle class.[3]The original destination of the city was more likely to be an abandoned transit town after the gold rush subsided. It is located in the Mojave Desert, one of the hottest and driest places in the world. If entire America were the second Earth that provides endless abundance, the Mojave Desert could dampen all the ardor for the inexhaustible myth and make it completely disillusioned. Therefore, Las Vegas people had ample reasons to take pride in, after all, they boasted sharp minds and quick actions. Recognizing the enormous changes in consumer behavior and cultural psychology in the United States after World War II, they perfectly combined their nostalgia for the West, their passion for technology, their escape from the mundane reality of life, their desire to end long and boring marriages, and the luxury that the middle class

[1] Donald Worster, *River of Empires: Water, Aridity, and the Growth of the American West*, translated by Hou Shen, Nanjing: Yilin Publishing House, 2018, pp. 242–244.

[2] Hal Rothman, *Neon Metropolis: How Las Vegas Started the Twenty-first Century*. New York: Routledge, 2002: Preface and p. 3.

[3] For research on how the metropolitan areas in southern California can utilize their natural resources, break the magic spell of drought, shape modern American lifestyles and redefine urban functions, see Lawrence Culver, *The Frontier of Leisure: Southern California and the Shaping of Modern America*. New York: Oxford University Press, 2012.

yearned for and could afford. They made full use of air conditioners and automobiles and turned the forgotten desert town into a vacation destination. In Rothman's view, the city that can "start the 21th century" seemed to declare the victory of the city, one of the greatest inventions of humanity, to the whole world.

Throughout the entire movie, Ann and Hillbilly Cat enjoy all the excitement brought by the miraculous city, release the dopamine of their youth. In the end, Ann tells Hillbilly Cat that she wants a real tree. When the loafer woke up to reality and married an attractive woman, they built up to the American desire for a family, a stable community, and a house of their own in front and at the back of which there are lush lawns and several shady trees. Las Vegas in the imagination of people was mostly a surreal world similar to quadratic element, where there were beautiful and magical casinos, ballrooms, fountains, imitation shows, convenient channels for marriage (and divorce) in the south of Las Vegas Strip. On the surface under the neon lights, Las Vegas seemed to be a place quite different from other cities that were disciplined by industrial order, but a real world like any other city after its dazzling halo was pierced through. Compared with other big cities in the United States, Las Vegas was even a world that was required to strive for the possibility of survival in the harshest natural environment. Tourists, who came from all over the world because of its reputation, enjoyed the seafood feast in the gambling town, and felt excited and frustrated before slot machines and roulette, but few were not able to see or care about such a reality. Meanwhile, most of 2.2 million people living in that metropolitan area, like every regular urban dweller, ate, and dressed, worked a nine-to-five job, and ran the residence and life that Ann yearned for in the movie. In the desert, where such a life came from and how they maintained it were the problems that they had to confront directly.

The most direct answer to such questions was the entertainment industry centered around casinos. Indeed, unlike most cities, Las Vegas was a city that did nothing productive. Its rise and rapid development didn't stem from trade, manufacturing, or a hub for agricultural products like Chicago and Kansas City. It was a city that vividly manifested the spirit of "amusing ourselves to death," which stimulated the primitive sensual impulses, sexual desire and greed of human beings. Ironically, while being extremely faithful to the human body, the spirit was also a rejection and avoidance of the real world on which the body depended. Of course, economic historians of the city would not be satisfied with such answers. They would strive to reveal the essence of capital operation in the city, and its complicated relationships with the capital in the Los Angeles metropolitan area, the United States, and even the world. Social historians would see the laborers who made the dice of the city roll, their ethnic composition, experiences, and systems of the city as well as the social injustice and discrimination beneath the glossy exterior of the city. However, as an environmental historian, the first thing he might think about was that other historians might turn a blind eye to water, the crucial element for the survival of the city. Whether the city was a land of paradise or a city of sins, whether a person was a hasty passing traveler or a permanent resident, whether he was a wealthy Arab tycoon who spent gold on one throw or a penniless and homeless person who was addicted to drugs, water, for them, was the most basic need, while for Las Vegas, its existence wasn't natural. At the buffet and poolside in the casino, it created the illusion of abundance, but for those who actually lived there, scarcity was grave and stern, a reality they had to confront and adapt to every day.[4] In other words, Las Vegas has never escaped the curse of the desert on it.

At first, Las Vegas was a place that provided water for survivors and travelers in the desert. "Las Vegas" means "fertile grassland" in Spanish. It is a small oasis in the Mojave Desert, and a home to argali, coyotes, foxes, bats, vultures, hagworms, and various insects. Over the past 2000 years, Paiutes have drank their fill of spring water in its desert countless times, chased large lizards, and picked wild onions amidst its low, thorny weeds

[4] The problem of water shortage has long existed in the southwestern United States, but before the 1990s, apart from a few visionaries, there was a general consensus among the public and government that technology could overcome the difficulty in scarcity of natural resources to provide sufficient water sources for cities and countries in the Southwest. After the 1990s, the obsession with technology gradually receded, but the water problem became increasingly severe. The Southern Nevada Water Authority (SNWA), founded in 1991, was the result of the response to such issue. However, there was always controversy over the water consumer that consumed water most in the area. SNWA believed that the water consumed by local residents, especially for lawn irrigation, was much more than that by tourism industry supported by hotels and casinos. However, many voices from the public believed that it was a pretext for SNWA to pursue economic growth and evade responsibility. See: Alex Prud' homme, *The Ripple Effect: The Fate of Fresh Water in the twenty-first Century*. New York: Simon & Schuster, 2011. See also: "Fear and Water Us in Las Vegas," https://blogs.lt.vt.edu/adameen1/tag/las-vegas/, which was logged in on August 21, 2020. This chapter does not intend to discuss the debate, but aims to reflect on the connection between cities and vast ecosystems by looking back upon the history of Las Vegas.

and shrubs. Like most western towns, the first to arrive here after the Catholic Spaniards were devout Mormons. The place became a resting place connecting Salt Lake City, their stronghold, and Los Angeles. What attracted them to stay here was the spring water.[5] Later, the Mormons left. People who made the journey to Los Angeles still rested here because there was a reliable water supply. After the railroads between Salt Lake City and Los Angeles were open to the public, it still became a small transit station due to its geographical location as a desert oasis.

The turning point that converted Las Vegas from a water supplier to a water grabber occurred in 1935 when the Hoover Dam was completed, in fact, it should have been earlier than the completion of the Hoover Dam. The fate of Las Vegas or another similar place was changed from the time when irrigation agriculture was developed on a large scale and the desert was made to blossom like a rose. Rothman said that Las Vegas had no hinterland, but in fact, Las Vegas made the farmland, ranches and orchards in the whole West its hinterland, and also made the Colorado River its water source. Its neon lights, slot machines, and fountain shows relied entirely on the hydroelectric power from the Hoover Dam. Its dazzling array of pastries, fruits and steaks also relied on irrigation agriculture supported by water diversion channels built in arid areas. More importantly, 2.2 million people living there on a daily basis and millions of tourists every year (a total of 42 million in 2019) need water, a large amount of water, to drink, wash, irrigate lawns, keep pets, and fill swimming pools. Before 1970, the local water sources in Las Vegas were still enough to meet the need of daily consumption and tourisms. However, with the advent of the city's booming period, 90% of its daily water

[5] After the death of Joseph Smith, the founder of Mormon Church, Mormons began to migrate westward and established a series of western towns as they advanced westward along the Mormon Trail. They were the whites who first implemented irrigation agriculture in the southwest, but they were ultimately unable to control the Colorado River as a result of capital and technology. Mormons had a significant impact on the natural world of the western United States, but research on its environmental history is relatively scattered. In 2019, a collection of papers on the environmental history of Mormon was published. It is currently the most comprehensive work on the study of Mormon's environmental experiences from various perspectives such as faith, practice and experience. Jedediah S. Rogers and Matthew C. Godfrey, *The Earth Will Appear As the Garden of Eden: Essays on Mormon Environmental History*. Salt Lake City: University of Utah Press, 2019. For the irrigation agriculture of the Mormons, see "The Lord's Beavers" in the Chapter III of *River of Empires: Water, Aridity, and the Growth of the American West* by Donald Worster.

supply came from Lake Mead, the artificial reservoir behind the Hoover Dam, which made the Colorado River its water hinterland. Ironically, as the ecological and economic influence of the city grows stronger and stronger, its ecological independence gradually disappeared.

All existing studies indicate that the Colorado River is constantly shrinking, and its river system is "experiencing the worst drought in the history of the basin. In January 2000, the water level of Lake Mead, one of the most important reservoirs of the River, dropped approximately 130 feet [about 40 meters]".[6] The wildest and freest river on continental North America suffered one defeat after another in the tussle with the alliance of capital, technology and development of twentieth century. It seemed that the United States had finally completed its "Manifest Destiny" on the river and its conquest of nature. However, the river alerted humanity who reveled in its triumphs in a state of scarcity. Then, how will Las Vegas continue its development?

German ecologist Wolfgang Haber believed that humankind didn't realize that they had led themselves into several "ecological traps" in the progress of civilization and technology. The setting and excavation of these traps were based on humans' superiority over other living and non-living nature, which rests on five main foundations: "command of fire requiring fuel; controlled production of food and other biotic substances; utilization of metals and other non-living materials for construction and appliances; technically determined, urban-oriented living standard; economically and culturally regulated societal organization."[7]In the paper, Hubble points out an important meaning about modern cities, namely, cities are not just entities that operate within the administrative boundaries of cities, but also a way of life and consumption concept which form an ecological relationship different from agricultural society. Driven by an unwavering philosophy of growth and empowered by technology, new ecological relationships break the rhythm and limits of nature, transform impossible things into resources, and build cities in places it was considered impossible to build cities. However, the existence

[6] All data regarding water use in Las Vegas and drought in the Colorado River are sourced from the website of the Southern Nevada Water Authority (SNWA). https://www.snwa.com/importance-of-conservation/responding-to-drought/index.html, which was logged in on August 23, 2020.

[7] Wolfgang Haber, "Energy, Food, and Land-The Ecological Traps of Humankind," *Environmental Science and Pollution Research-International*, Vol. 14 (2007): pp. 359–365.

of Las Vegas not only declares the victory of technology and capitalist consciousness, but also constantly reminds people of a fundamental fact that even the most surreal cities in the most developed countries require water, food, and energy, and have far more material needs than those supporting a village or a traditional city. All of those also make the city fall into a deeper "ecological trap," in a more fragile state.

If we say that on continental North America, nature once showcased its magnificent existence as a healthy, diverse and bountiful "Second Earth," nature, which seems to be domesticated, uses natural disasters (such as floods, hurricanes, earthquakes and wildfires) as well as slow but more ordinary crises (such as virus spread, soil degradation, loss of species diversity and resource depletion) to tell humankind that it always possesses spontaneous and unrestrained power after a series of confrontations with newly-arrived human beings and their ideologies. The focus of the chapter is not to repeat the familiar plunder and destruction of nature that are made by cities, but to capture the ecological paradox that has occurred on the land as a result of urban sprawl, reflect on the myth of inexhaustible resources that inspires Americans to pursue what cities should be, and examine the efforts and transformation that are driven by the reality of scarcity and disasters. The fact that all cities have a certain emotional need from nature is just as important as the material needs of cities from nature although the ways to meet such need may vary greatly in different eras and regions. In the United States, it has a special expression, namely, wilderness, which is also the most interesting aspect in the loop of the ecological paradox.

2 Immigrants Flocking in Droves

There are many ways and standards to measure the rise and fall of a city. After the birth of the modern concept of Gross National Product (GNP), GDP and a series of precise figures accompanying it have become the main indicators that are commonly used. However, an increasing number of urbanists, sociologists, needless to say, historians and cultural analysts, have found that figures are dry and powerless. They believe that the figures actually obscure the process of individual life, and even to some extent distort or at least narrow the understanding of history and reality although such figures may seem objective. When scholars determine whether a city has vitality with the city as a unit, many of them call for the need to break free from the constraints of figures and observe the lives of ordinary people, the use of public facilities such as hospitals, libraries, museums, and parks, the popularity of various cultural activities, and the accessibility of education at different levels to various groups in the city. However, whether the state of a city is calculated through figures or the rise and fall of a city is described with words, people are the core of urban research, including the quantity, ethnicity, and class as well as their livelihoods, activities, and desires of people. A city cannot exist without people. It requires people, many people, including producers, service providers, and consumers. A city cannot be divorced from people who can eat, drink and poop, as well as can indulge in love and reproduction, and contemplate poetry and distance while looking at the starry sky and the sea. Only when quantity and density of persons reach a certain value can cities survive, prosper, and create wealth. Although the significance of urban-environmental history is the emphasis on the fact that human beings cannot create cities alone and maintain the effective operation of cities if they are divorced from nature. It is also required to examine these human beings, the environmental history of the city, and the ecological bonds that connect them to the city.

The United States is a nation of immigrants, and American cities are cities of immigrants. Some immigrants came first, some later. Some settled permanently, others were just passing travelers. Some traveled far away across the ocean, others migrated internally. Some were Anglo-Saxon, others were Irishmen, Southeastern Europeans, or Jews. Some were whites, others were non-whites. Immigration historians often care more about the efforts and struggles that immigrants made to survive and grow on a new land, the slaughter and destruction they carried out, and the enslavement and discrimination they suffered after arriving in the United States (or North American colonies). The reasons why they migrated were generally described with just a few lines of text, nothing but long-term wars, political oppression, economic depression, disasters, and famines. Of course, there is no lack of detailed analysis of individual reasons, such as religion, freedom, wealth, love, and escape, in the migration research on the micro level.

Ecological reasons are intertwined with various motivations of the age and individuals. They are often downplayed even in the writings of environmental historians. If we retrieve every large-scale autonomous migration wave again since the colonial era, we will find that there are often profound ecological driving forces behind it. Migration is the response of these immigrants, who ultimately decided to leave their hometowns, to the tense ecological relationship in their hometowns, and also a reflection of their desire for nature's wealth in the areas where they were about to arrive. The starting point of immigration is often the countryside. When immigrants left their homeland, due to their self-awareness and survival skills, they hoped to find new land on the new continent and continued their familiar life as farmers in the unfamiliar land. However, their destination is mostly the city. The migrating population forms a huge ecological loop: under the pressure of the ecological pressure of their hometown, immigrants move to cities. When they flooded into the city, their demand exceeded the production capacity of the suburbs, and they began to seek new opportunities in more open areas, new farms, ranches, plantations, and orchards require stricter land development to meet the growing demand of urban markets. Subsequently, new environmental issues arose, so people once again turned to cities to seek livelihoods. Since 1800, such an ecological cycle has become one of the basic motive forces for the transformation of American society and culture. Population pressure led to an ecological crisis, which triggered migration and intensified urbanization. Urbanization brought new ecological crises and population pressure. However, it is worth noting that it was an open cycle, and the ways to reduce pressure and transform crises changed with the change in environment, culture, technology, and power.

The city was deeply embedded in the ecological loop and always played a multi-directional role. It served as not only a receiver for the immigrants who flooded in, or a container, as Mumford put it, to alleviate the enormous population pressure on rural areas through job opportunities and mass production, but also a hotbed for brewing environmental problems. Such case was manifested in two aspects: on the one hand, a large amount of resources from its hinterland, including food, water, fuel, and other raw materials, were grabbed from its hinterlands to maintain the operation of the city, supply necessities for the survival of growing population and meet the increasing consumption desire, thus fundamentally changing the original ecosystem of those areas. On the other hand, the hinterlands had to receive a large amount of urban waste, to meet the consumption needs and quality of life of the city, thereby aggravating the environmental pressure on areas outside the city, such as countries, forests, deserts, oceans, and atmosphere. Therefore, only by knowing well the operation of the ecological cycle can the various layers of connections affected by immigration and the operation and evolution of American cities be understood.

The immigration wave set off by Irish immigrants in the mid-nineteenth century clearly presented the loop of the ecological cycle and the change in American society and culture. Behind these ragged immigrants was the Irish Potato Famine (1845–1849) that was well known by people, or the Great Famine in English.[8] During the famine, one million people starved to death or died from the post-Famine plague, and over one million people chose to immigrate. The continuous flow of subsequent immigrants changed Ireland's population composition at last. Compared with famines in Chinese history, for instance, the "Ding-Wu Great Famine" that occurred between 1877 and 1878, the absolute death toll was not high. However, at that time, the total population of Ireland was only 8.4 million. The Irish Potato Famine was the greatest disaster that occurred in modern European history in peacetime.

Ireland has never been a wealthy country in history, and the daily life of ordinary people is a matter of "what can and can't be eaten." In fact, before the Enlightenment, hunger was also common in Europe. The imagination of fairy tales and fairyland always unfolded around a roasted goose with thick grease. Scarcity was the discovery of the "Second Earth" and the full utilization of the basic ecological reality of the Old World.[9] However, the "Irish Potato Famine" was not simply a repetition of the old incubus of scarcity. It had the familiar themes of the disasters that occurred all over the world throughout the ages: ruthless nature, incompetent governments, greedy landlords, and suffering

[8] There is an immense number of related historical research. Scholars have discussed the different dimensions of land famine in Ireland. Some scholars have focused on the analysis of the economic, institutional, political, and ecological reasons behind its occurrence and its impact on Ireland. Another group of scholars focus on the study of Irish immigration, including how they arrived in the United States, personal experiences, the living conditions of Irish laborers in the United States, and the changes in American political culture, society, and religion that they made. There is still little research on the relationship between Irish immigration and urban environment. In general, research on the environmental history of overseas immigrants always focuses on the change in the agricultural and fishery ecology of the United States, while research on the urban-environmental history of immigrants, due to the background and interests of environmental historians in the field, lays particular emphasis on Italian Americans and Latinos and focuses on the relationship between social equity and environmental justice.

[9] In *The Great Cat Massacre and Other Episodes in French Cultural History*, Darnton provides a fascinating analysis of the state of hunger in Europe, with fairy tales from France and Germany as examples. Robert Darnton, Chapter I of *The Great Cat Massacre and Other Episodes in French Cultural History*, translated by Lv Jianzhong, Beijing: The Commercial Press, 2014.

farmers. However, it was largely a product of the affluent modern world, and could even be said to have opened up "the late Victorian Holocast," as Mike Davis called.[10] Of course, the famine, which Davis referred to as the "Secret History of the 19th Century," refers to several catastrophic disasters that occurred in China, India, Egypt, Brazil, and other countries from 1870 to 1914. The natural cause of these famines was the severe drought induced by the El Niño phenomenon over the past few decades, and the natural driving force behind the "Irish Potato Famine" was the Phytophthora infestans that caused potato late blight. However, they shared deeper ecological and economic logic in terms of natural triggering factors beneath the surface.

Davis wrote, "Millions died, not outside the 'modern world system', but in the very process of being forcibly incorporated into its economic and political structures. They died in the golden age of Liberal Capitalism; indeed, many were murdered ... by the theological application of the sacred principles of Smith, Bentham and Mill."[11] In his view, the expansion of laisser-faire capitalism incorporated the periphery of the modern world-system into its commodity economy to make it a colony of the former, disrupt their own agrarian economic order, and ultimately give rise to groans and suffering under the shadow of the glorious Victorian era.[12] There is no doubt that the discussion incorporating the famine, which occurred almost simultaneously in different regions of the eastern and western hemispheres, into the context of global economy and ecology has a strong penetrating power, and is also in line with the harsh critique of capitalism and colonialism in contemporary thought. It is not intended to defend capitalism or fully agrees with Davis's reflection on it. However, it must be pointed out that although Davis saw El Niño and acknowledged its wild power, he was not interested in exploring deeper ecological factors but only focused on the logic of the capitalist economy. He evaded the other side of the famine from the "Irish Potato Famine" to the "late

[10] Mike Davis, *Late Victoria Holocausts: El Nino Famines and the Making of the Third World*. London, New York: Verso: 2001.

[11] Davis: *Late Victorian Holocausts*, p. 9.

[12] Chinese scholars such as Hao Ping and Xia Mingfang also paid attention to the relationship between the commercialization of agriculture in North China and the Great Famine when discussing the "Ding-Wu Great Famine." See Hao Ping: *Ding-Wu Great Famine: A Study on Shanxi's Disaster and Relief in the First Year of the Reign of Emperor Guangxu of the Qing Dynasty*, Beijing: Peking University Press, 2012.

Victorian Famine," the thirst and ability for reproduction, the biological instinct of human beings, and the enormous pressure on the land which was caused by the population explosion in these areas. However, it is at the level that human beings established the deepest connection with the rest of nature, which cannot be separated in any era.

However, even if the connection has existed from ancient times, it does not mean that it is immutable. In fact, all connections will change with the changes in the subjects connecting the two ends, and so will the increase in population size and the pressure borne by the rest of nature. Compared with the ancient ghosts hovering in the Roman Empire or Ming Dynasty, the "Ghost of Malthus" that appeared in the nineteenth century had a new aspect. The discovery of the "Second Earth" was not only about land and the resources on it across the ocean but also about minerals and species that could be transported to the Old World. The good harvest of grain which was brought by new species to the Old World supported the possibility of the reproductive instincts of human beings, thereby maintaining sustainable population growth. When these species, such as potatoes, corn, sweet potatoes, and pumpkins, were widely cultivated in the Old World, they not only guaranteed population growth, but also fundamentally changed the original ecology of the Old World. The diversity of ecosystems that were once preserved due to their unsuitability for cultivation were reclaimed and domesticated due to the strong adaptability of new crops, thus resulting in further simplification of the ecosystems in those areas and changing the relationships between human beings and other species together with their habitats.

The tremendous changes that occur in the ecosystem were synchronized with the transformation of the world economic system, namely, the process of commodifying nature by the capital that has been observed, pondered, criticized, or praised by innumerable scholars. The subject of domesticating land is no longer ordinary farmers seeking ample food and clothing, nor is it just a country hoping to naturalize a region, transfer population pressure, and obtain more taxes by reclaiming land. Intangible capital from distant cities joined the domestication of land, and even surpassed farmers and the country to a large extent, and became the decisive force in what the land should produce, how it should be produced, and how many it should be produced. In the writing of environmental historians over the past few decades, it has become the process that people are familiar with. However, what is often neglected in existing historical writings is that farmers living on land and suffering from heavy exploitation or nature being plundered and exploited is not passive victims, they have their own ways of resistance. James C. Scott told the story of the rebellion art of an ordinary peasant with low profile in *Weapons of the Weak*. The peasant challenged the injustice imposed on him by loafing on the job, deserting his post, feigning ignorance, and other humble means which can erode the new economic and legal order.[13] However, what Scott did not point out was that reproduction itself was the most effective weapon for the weak. Driven by conscious continuity of a clan and unconscious reproductive instincts, they ensured that they were not engulfed and could continue to exist. But the problem was that such resistance, together with distant capital and direct or indirect political power, increased the pressure on the land and ultimately led to suffering. Similarly, nature also possessed its own weapons to resist the oppression of capital and state apparatus. In other words, the logic of capital could not arbitrarily take whatever it desires from nature. Droughts, floods, hurricanes, earthquakes, the disappearance of soil fertility, and viruses affecting various organisms including humans, crops, and livestock are all natural confrontations against the new order and authority. They often forced institutions and systems that established the order to make adjustments, and in extreme cases, they, along with other weak individuals became disruptors of the order.

The failure of the potato crop was a typical modern disaster. Potatoes were the most successful species in the Old World after the Columbian Exchange. However, the reproduction of domesticated crops was different from species that were stigmatized as weeds, which could quickly spread in suitable natural environments and even replace native species as invasive species. As a result of artificial selection, even the crops which were difficult to domesticate still require human care and promotion. Under the co-action of population growth, urbanization, and colonization, potatoes evolved from a small-scale garden crop to a basic crop which was widely monocultured in Ireland. The population of Ireland tripled in the century before the outbreak of the Irish Potato Famine. Although the rate of population growth began to slow down in the half century before the outbreak of Irish Potato Famine, the contradiction between such rapid population growth and the sudden pressure

[13] James C. Scott, *Weapons of the Weak*, translated by Zheng Guanghuai et al., Nanjing: Yilin Publishing House, 2011.

on the land also became apparent in the nineteenth century.[14] However, at that time, the crisis in Ireland was no longer a local crisis in Ireland. If population growth stemmed from the joint efforts of plant reproduction in the New World and local Celtic reproduction, the outbreak of the crisis was closely related to London, capital and the empire. To a large extent, the crisis was a product of both urbanization and industrialization in the British Empire, known as the "Victorian Calamity" by Davis. In other words, population growth and potato late blight, or any single factor, did not inevitably lead to such massive deaths and immigration. The colonization of Ireland was inextricably linked to the outbreak of the Great Famine.

After the enactment of the Act of Union 1800, Ireland completely became the political, economical and ecological hinterland of England. In order to meet the increasing needs of urban consumption of the latter, the tract of land in Ireland was converted into pastures, but the products from them were transported to British cities such as London, Liverpool, and Birmingham. The high-yield and hardy potatoes that could grow on marginal lands were not only a panacea for the relief of domestic food pressure, but also a source of energy for the expansion and prosperity of British cities. If human beings seemed to have gained cooperation from nature in terms of successful domestication and cultivation of potatoes, and even conquest of other species, which had evolved on the same land for millions of years, with such alien species, such cooperation was not definite or firm. Single cropping violated the fundamental principles of ecosystem diversity, so when natural backlash arrived, it also had the destructive power enough to sweep everything. The entire Europe was in turmoil as a result of the potato late blight, which largely became an ecological factor in the European Revolution of 1848. Communism was not the only ghost hovering over Europe. The Famine confirmed the Malthusian theory of population. However, compared with the famine, plague, and war in the approach of "Active Suppression" put forward by Malthus, a more effective suppression method for Europeans in the Malthusian Trap since the sixteenth century was the immigration to the "Second Earth." The undeveloped nature of the Americas became the endpoint of the long ecological chain and the starting point of a new set of chains.

[14] Cormac O Grada, *The Great Irish Famine*. Cambridge: Cambridge University Press, 1989, pp. 5–7.

The great Famine quickly triggered immigration. In only five years from 1846 to 1851, approximately one million Irish immigrants came to the United States. Throughout the 1840s, more than half of the immigrants were Irishmen. From 1850 to 1920, Irish immigrants continuously flooded into the United States. A total of 3.3 million to 3.7 million persons came to the United States. These immigrants were not the poorest Irishmen. The persons at the other end of the scale could afford the boat fare, or died of hunger and illness in times of famine. However, most Irishmen didn't have the economic capability of their migration after landing in New York, so most of them chose to settle in cities. At the same time, cities on the East Coast of the United States such as New York, Boston and Philadelphia were in a period of rapid development, they needed a great many of labors and consumers. The boosters of any city clearly knew that the operation of capital called for resources and labors, which was the most basic material reality of urban existence.

The failure of the potato crop was only the most extreme example of the urban–rural ecological cycle in both the Old and New Worlds throughout the nineteenth century. In fact, going to the New World was not the dream initiated by the Famine. At the beginning of the eighteenth century, Irishmen already resided in continental North America. However, the truly large-scale immigration began after the Napoleonic Wars. With the rapid growth of the European population, the land and resources of continental North America naturally played the role of a safety valve for Europe. In the 30 years before the outbreak of Irish Potato Famine, over one million Irish immigrants had already arrived in the United States. Meanwhile, millions of Germans also synchronously arrived in the United States on a large scale. Apart from the outbreak of the Irish Potato Famine in the 1840s, the number of German immigrants was the largest for most of the nineteenth century. Even before Hitler started World War II to scramble for "Lebensaum" (living space), ordinary Germans had already individually gone abroad in search of new living spaces. Unlike Irish immigrants, these German immigrants, excluding a small number of political refugees of 1848, had a clear goal; they immigrated for the land, and for cultivating the land to settle. As Irish immigrants filled American cities, they headed toward the Great Lakes and the Great Plains and filled America's breadbasket with their farm products. From the beginning, their agricultural actions were already in the midst of capitalist machinery, and their production was not to meet the needs of their families or neighboring towns, but to support the operation of cities located hundreds of, thousands of miles away from their settlements, and even across oceans in the twentieth century. Cities prospered as a result of the farm products they provided.[15]

Along with or slightly later than the Germans, immigrants from southern Europe, Eastern Europe, vast regions under Tsarist Russia, East Asia, and South Asia arrived. The political and economic reasons that drove them away might vary greatly, but behind them, they all left an overcrowded homeland where soil was eroded and species were withered. They also came to the "Second Earth" in search of their "living spaces." Thus, the migration driven by famine in the Old World happily combined with the abundant natural resources of the New World, under the lubrication of capital, thus ushering in the urban age of America.

3 Expanding Bubbles

These persons who constantly migrated between land and sea had initiated waves of global ecological circulation that have not yet declined so far and turned ecological exchange and circulation between urban and rural areas into a global movement. American urban theorist Neil Brenner reinterpreted the construction and deconstruction of contemporary cities on the basis of "planetary urbanization" put forward by Henri Lefebvre to attempt to give new definitions to "city" and "non-city." Its core lies in completely breaking down the boundaries of contemporary cities in cultural and geographical terms and in ecological terms, blurring the differences between cities and hinterland, and even wilderness, and viewing urbanization as a dispersive, dynamic, and uneven force and phenomenon on a planetary scale. Compared with Lefebvre, Brenner was more concerned with the ecological impact of "planetary urbanization," and it is precisely at such level that he constantly emphasized the necessity of thinking and planning cities on the "planetary" scale. He wrote, "Capitalist-dominated urbanization has always been anchored in the widespread extraction, production and consumption of fossil fuels far from centers of major cities. The model also directly involves global ecological plunder that may permanently change the Earth's climate, and

[15] The data on the number of immigrants comes from the Library of Congress, https://www.loc.gov/classroom-materials/immigration/, which was logged in on September 21, 2020.

potentially cause unprecedented amounts of pollutants and toxic waste to infiltrate into the soil, oceans, rivers and atmosphere in the Earth."[16]

Brenner's observation of contemporary cities is undoubtedly insightful, but as historians, we need to ask how the hinterland has transformed from the place surrounding urban area into the dubious free flowing space, how the world has entered the process of "planetary urbanization," and what role the United States plays in it. In a broad sense, the "Second Earth" itself was developed as a colony, a vast hinterland of Europe. Whether cities with a strong religious tradition like Boston or cities like New Amsterdam that had always focused on commerce were closely related to the cities and trade networks on the other side of the Atlantic. As the hinterland of Europe, plenty of raw materials from North American, such as fur and wood as well as subsequent cotton and wheat, were continuously transported to Europe. More importantly, it continued to receive an endless stream of surplus population from Europe. The rapid growth of cities like Boston and New York also relied on their plunder of cod and whales, the ocean's hinterland, and exploration of overseas markets. To some extent, since the discovery of the "Second Earth," the entire Earth, driven by immigration and trade, has been in a process of "planetary urbanization," and the relationship between cities and their hinterlands has taken on a planetary scale. Of course, the problem goes beyond that. Brenner's definition of "planetary urbanization" shall be rethought if cities are put back into history. Has the discussion about the hinterland, wilderness, and other places that are deeply influenced by cities but still largely retain their own operating mechanisms really become meaningless? Or are they still, implicitly or explicitly, constraining the spread of cities, challenging the confidence and optimism of cities?

To a large extent, we can indeed construe American history as the creation history of the continental hinterland, but it has created more than one hinterland, just as we cannot assume that Chicago is the only inland center. On the contrary, the process of environmental change has created hundreds or thousands of hinterlands, which are connected to

[16] Neil Brenner, *City, Territory, Planet: Studies in Critical Urban Theory*, Translated by Li Zhigang, et al., Beijing: The Commercial Press, 2019, p. 215. For the discussion on Lefebvre's "planetary urbanization", see Henri Lefebvre, "Dissolving City, Planetary Metamorphosis", *Environment and Planning D: Society and Space*, Vol. 32 (April, 2014). See also Neil Brenner ed., *Implosions/Explosions: Towards a Study of Planetary Urbanization*. Berlin: Jovis, 2014.

cities to form economies of all sizes, like active and shining bubbles. Every city, even every small town, has its own bubbles. The bubbles vary in size and form, often short-lived and fragile, and their rise and decline change according to the times. So many bubbles are wavering in the tide of capital, and there is no unified rule to regulate them into a harmonious whole. Instead, they collide and conflict with each other to scramble for larger bubble space. They can also merge with other bubbles to dissolve into larger bubbles. As their volume increases, they are more prone to breakage. They also continuously generate new bubbles in the process of breakage and recombination. These bubbles aren't perfectly spherical in shape. They are often irregular and can extend inland, even expand overseas. Urban boasters were the first to blow bubbles. They, as Cronon said, were loyal supporters or practitioners of capitalism. However, to maintain the bubble and keep it growing required labor and a large number of people who constantly migrated to cities. It was in a world of rapid population growth and high mobility that bubbles of all sizes in the United States were able to expand rapidly and last for one or two centuries, or even longer. Chicago and the Great West in Cronon's works were like that. As Chicago grew, its bubbles continued to expand. However, it is worth noting that Chicago's bubble was never closed and was unique to it. In the nineteenth century, many towns in the west of the Appalachian Mountains were creating their bubbles. The hinterland generated by all the cities like Minneapolis, St. Louis, Denver, and Kansas City often overlapped with each other to a large extent. No city had absolute control over the direction in which these bubbles extend.

The process that these bubbles constantly expanded was a history composed of cities and their hinterlands. In the discussion of the history of the western United States and cities, there was an enduring question: was it Turner's Frontier Theory that explained the development of the western United States, or was the theory of the "urban frontier" greater in explanatory power? In his thought-provoking paper published in 1893, Turner depicted a rapid yet still gradual evolution of American history, which is to some extent a pony edition of human history. He wrote, "The United States lies like a huge page in the history of society. Line by line as we read from west to east we find the record of social evolution. It begins with the Indian and the hunter, it goes on to tell of the disintegration of savagery by the entrance of the trader, the pathfinder of civilization; we read the annals of the pastoral stage in ranch life; the exploitation of the soil by the rising of unrotated crops of com and wheat in sparsely settled farming communities; the intensive culture of the denser farm settlement, and finally the manufacturing organization with city and factory system."[17] Since the 1930s, the criticism of Turner has been incessant. Like all historians of the nineteenth century, Turner's definition of civilization and barbarism, his imagination of the westward movement of idyllic pastoral, his praise of conquering the wilderness, and his support for the legitimacy of "Manifest Destiny" have all been criticized by historians from different fields, especially historians of New West. The book does not intend to evaluate Turner's significant defects on the issues of race, class, and gender, but only focuses on Turner's discussion of the development order of the West, as it directly relates to the relationship between cities and their hinterland.

On the issue, Turner's critics not only came from the United States but also from numerous Canadian scholars. The former is represented by Richard Wade's classic work *The Urban Frontier: The Rise of Western Cities, 1790–1830*, while the latter points out the differences between the history of Western Canada and American history. In Wade's view, the development of the western United States was led by cities: "The towns were the spearhead of the American frontier," and commerce and manufacturing opened the door to the western Ohio Basin, the West in Wade's view. Then, the expansion of agriculture started. However, as Cronon pointed out, the thinking of Wade and his successors mainly focused on the interior of administrative boundaries of cities, but seldom touched upon urban–rural relations and the relationship between city and

[17] Frederick Jackson Turner, "The Significance of the Frontier in American History," which was originally presented at the Chicago Annual Meeting of the American Historical Association in 1893 and was published in the *Annual Report of the American Historical Association*, 1893, pp. 197–227. The current data sourced from the archives of the American Historical Association: https://www.historians.org/about-aha-and-membership/aha-history-and-archives/historical-archives/the-significance-of-the-frontier-in-american-history, which was logged in on September 22, 2020.

nature.[18] Cronon's inspiration came more from the discussion of Canadian scholars about the relationship between city and country, but his acceptance of the latter was cautious. He did not explicitly state that a significant difference between the United States and Canada in the issues of cities and the West lay in the role played by the state, but he didn't believe that city and country were a simple relationship of controlling and being controlled. In Cronon's view, Turner's Frontier Theory was not just a declaration of victory at the end of the nineteenth century. On the contrary, he always believed that the theory still had its value and enlightening significance even now. He wrote, "... the center of metropolitan empire—and of Turner's frontier—was the marketplace of modem capitalism. When Turner spoke of the frontier as 'the outer edge of the wave', what he unintentionally described was not some implicitly racist 'meeting point between savagery and civilization' but the ongoing extension of market relations into the ways human beings used land—and each other—in the Great West." So in Cronon's view, if there was a leading force, it was not the independent and fearless ordinary pioneers in Turner's writing, nor the Western cities that Wade or Canadian scholars believe, but rather the ubiquitous power of capital and the market relationships that it established and spread across frontier lands. At the end of the chapter, Cronon wrote, "As village became metropolis, so frontier became hinterland. The history of the Great West is a long dialogue between the place we call city and the place we call country. ... Viewed from the banks of the Chicago River, the Great West is both an urban empire and a countryside transformed."[19] In other words, cities created hinterlands, but the rural world on the frontier was not necessarily a product of cities.

[18] Richard Wade, "Urban Life in Western America, 1790–1830", *American Historical Review*, Vol. 64 (October 1958): p. 14. For a more detailed discussion on such issue by Wade, see Richard Wade, *The Urban Frontier: The Rise of Western Cities*, 1790–1830 Cambridge: Harvard University Press, 1959. Urban historian Wang Xu had a wonderful discussion on Wade's research. Mr. Wang Xu believes that the leading role the city played in the Ohio Basin, as Wade discussed, is only a case, and is more common in the West Coast and mountainous cities. See Wang Xu: Chapter III of *The Pattern of American Urban Development from Urbanization to Metropolitanization*. For the discussion on the issue by William Cronon, see *Nature's Metropolis: Chicago and the Great West*, pp. 70–72, including notes.

[19] William Cronon: *Nature's Metropolis: Chicago and the Great West*, pp. 72–75.

Compared with a country, or "domesticated ecology" like the country or pasture, the ecosystem in a state of nonhuman control, which is known as wilderness, does not inevitably become hinterland.[20] For aborigines in the North American continent that made a living by gathering and hunting, the land was the place where they directly obtained food and means of subsistence. Although a few goods enter the exchange channel, they didn't constitute a relationship between the land as a market and urban production or energy supply. In a self-sufficient agricultural economy, although the country formed a hinterland relationship with the city in the form of taxation, their production logic was not the logic of capital or commodity economy, but the logic of meeting their own needs and adapting to local ecology.

If we temporarily shift our sight away from the banks of the Chicago River in the nineteenth century and extend the time scale back to the colonial history of the East Coast, we can discover another relationship between city and country in the North American continent, which is different from the one generally imagined. The widely accepted cognition of urban–rural relations originated from Cronon's first work, *Changes in the Land: Indians, Colonists, and the Ecology of New England*. From a certain perspective, the interpretation of urban–rural relations in *Nature's Metropolis: Chicago and the Great West* was a continuation of Cronon's thinking on the commercialized land of New England in the *Changes in the Land: Indians, Colonists, and the Ecology of New England*. Although the city was only marginalized in the discussion in the *Changes in the Land: Indians, Colonists, and the Ecology of New England*, the commodity economy remained the protagonist of land change, or more directly, land degradation, in New England.[21] In the book, fur hunters, timber dealers and farmers all directly participated in the process of breaking down the integrity and divinity of land and commercializing land. Thus, the rural

[20] Wilderness is one of the most complex and controversial term in American environmental thought, and there is numerous discussions about it. The book does not intend to intervene in the discussion of the term. Those who are interested in it can consult Roderick Nash, *Wilderness and the American Mind*, translated by Hou Wenhui et al., Beijing: China Environmental Science Press, 2015. William Cronon ed., *Uncommon Ground*, particularly his own article "The Trouble with Wilderness; or, Getting Back to the Wrong Nature"; when reviewing the refection on wilderness from the perspective of non-cultural analysis, see Donald Worster, "The Higher Altruism." *Environmental History*, Vol. 19 (Nov. 2014): pp. 716–720.

[21] For the discussion on the book, see the previous chapter "Emerald City".

image of New England was solidified, with a basic interpretive model of capital-commodification-agriculture-environmental degradation formed.

However, in 2004, Brian Donahue published *The Great Meadow: Farmers and the Land in Colonial Concord*. With a focus on the agriculture of Concord, Massachusetts, during the colonial era, the book breaks down such stereotypes and provides another paradigm of the urban–rural relationship that once existed in continental North America. Donahue believed that the Concord lamented and evaded by Thoreau was a basic self-sufficient, sustainable agricultural ecosystem combining farming and animal husbandry, which the farmers living in Concord established in the colonial era, formed after the War of Independence due to the expansion of the external commodity economy and the pressure of growing population internally. Such form was more similar to the country of early modern England where the rhythm of nature was followed, with the land carefully cared for and diverse crops planted but they left. In terms of social and economic organization, it was a collective, cooperative system basically independent of the city.[22] It is worth noting that Donahue himself, apart from being a historian and professor, is a steadfast observant of organic agriculture. He has a special sense of closeness to traditional small-scale farming methods, which made it difficult for him to examine the domestication and even domestication of nature in the agricultural form from a more critical perspective than Cronon or Worster had. He refused to see the reality of environmental degradation and farmers' hardship under the cover of its picturesque rural landscape even when he returned to England which the Concord farmers had emulated. Raymond Williams' attempt in *The Country and the City in the Modern Novel* is precisely a deconstruction of the rural illusion of a "beautiful past."[23] However, Williams, from an ecological perspective, did not truly ponder on the impact of traditional agriculture on nature, the natural hatred of farmers toward the wilderness and the wild species living on it, especially the ecological reality where the growing population had pushed the carrying capacity of land to its limit. Even in Concord as Donahue described, the expansion of less than 200 years had led to the disintegration of the natural economy here,

[22] Brian Donahue, *The Great Meadow: Farmers and the Land in Colonial Concord*. New Haven: Yale University Press, 2004.

[23] Raymond Williams, *The Country and the City*. London: Chatto and Windus, 1973.

and the so-called sustainability was simply a conclusion drawn from the observation at what time scale.

Donahue's analysis of the natural economy of Concord during the colonial era may not be fair and just, but in his analysis, we can indeed discover a city-hinterland model that is different from the one between Chicago and the Great West, a looser, neighboring, and ecologically closer model, which was more in line with the city-hinterland relationship in agricultural society. In the model, the development scale of a city is largely determined by the ecological carrying capacity of its surrounding hinterlands, and many cities in New England and their hinterlands follow such basic principle to some extent in their growth process. Boston didn't develop into a super large city like New York, Chicago, or Los Angeles. In that city, although traders also dominated, its urban elite class, constrained by a deep Puritan tradition, held a more cautious attitude toward excessive expansion at the retrieval of its history. Compared with other larger cities in the United States, the city and other cities in New England had a sincere belief in community interests and public morality since its establishment. The recklessness of laissez-faire capitalism was always criticized by mainstream public opinion, and even subject to legal restrictions there.[24] In other words, the expansion of Boston bubbles has two directions and two dominant modes. A large part of its wealth came from the ocean and overseas trade, and in that direction, it was as greedy as any laissez-faire capitalist in the world, it pursued maximum profits and relentlessly squeezed nature until cod absconded (an interesting fact is that Boston's business elites often retire early and go home to grow flowers).[25] However, a relatively mild and moderate relationship was established between it and the hinterland that supplied the most basic but important resources, such as food, water, and energy. Of course, this did not mean that its capital had not extended further to the West.

New York represents the fusion of these two types of bubbles, or rather; the first large bubble continuously expands and swallows the second smaller bubble. On the path of expansion, New York carried through the spirit of laissez-faire capitalism. It didn't stop plundering its oceans and overseas resources, and when its surrounding hinterlands

[24] For the unique public spirit that Boston has developed since its establishment, see Sam Bass Warner, *Streetcar Suburbs: The Process of Growth in Boston, 1870–1900*. Cambridge: Harvard University Press, 1962, and the Chapter III of the book.

[25] See the discussion on Boston horticulture in Chapter II of the book.

couldn't continue to meet its expansion, it began to search for a broader hinterland further away, the Great Lakes. The rapid rise and development of Chicago was closely related to the expansion of New York. In the early stages of the development history of the former, it fundamentally existed as the hinterland of New York, it processed the resources needed by New York and transported them to New York market through the Erie Canal. The expansion of New York gave Chicago its vitality, but as Chicago gradually grew and matured, it separated from the bubbles of New York and formed its own bubbles, even formed a competitive relationship with the bubbles of New York to a large extent.

At that time, free capital in the United States had crossed the Appalachian Mountains. From the national level to the social moral level, it seemed that there was no longer any restraint on its operation, and only nature itself could limit the inflation of bubbles. From New York to Pittsburgh, St. Louis, Chicago, Kansas City, and others, all of these large cities that emerged in the first half of the nineteenth century still adhered to old principles of site selection and strove to leverage their waterway, geographical advantages. However, as Cronon showed, the reason for truly liberating Chicago from the constraints of nature and achieving significant growth is not by waterways, but by railways. The territory, politics, economy, and ecology of the United States changed greatly due to the octopus-like expansion of railroads.[26] With the disappearance of transportation restrictions imposed by the winter and summer seasons, the hinterland of a city could be located thousands of miles away. In fact, it had also to expand in this way, because, at that time, the urban demand for resources was no longer just food, water, building materials and fuel. The use of fossil fuels and the industrial revolution that it brought about had completely changed the urban demand for resources, thus forcing all cities to rely to some extent on a distant or even exotic hinterland, and also forcing these distant places to change greatly in ecology and

[26] So far, Richard White's masterpiece *Railroaded: The Transcontinentals and the Making of Modern America* is the most comprehensive and in-depth description of the relationship among railroads and American politics, economy and society. However, interestingly, White had little discussion on environmental issues related to railroads although he actively participated in the research on environmental history in the 1980s and 1990s. White was the earliest scholar to propose the "cultural turn of environmental history," but his turn was so thorough that he completely abandoned it. Richard White, *Railroaded: The Transcontinentals and the Making of Modern America*. New York: W. W. Norton, 2011.

society as a result of the needs of distant cities. The existence of rivers was to provide not only drinking water and irrigation water but also water for power generation. In the early expansion of bubbles in New York, Boston, Chicago, and Kansas City, people could easily distinguish their edges without much effort, and their hinterlands had clear directionality and certainty. But after the transportation revolution brought about by fossil fuels, the clear hinterland disappeared, bubbles were broken and the hinterlands became unsteady. When people came to Las Vegas, they didn't see a single giant bubble, but innumerable floating and moving bubbles of all sizes. Paradoxically, it was there that new constraints on the expansion of laissez-faire capitalism emerged which stemmed from a scarcity of critical resources that forced capitalism to make peace with the state. While technological and capital support of the state was accepted, capital also forced the state to relinquish its own development freedom and accept the capital's control.

The urbanization process in the west of the Rocky Mountains was basically synchronized with that on the Great Plains, and the driving force behind their growth was also consistent, the unlimited expansion of laissez-faire capitalism, but the premise of expansion was materiality, which was a continuous source of resources to supplement its energy. On the path of civilization development, human beings have been exploring new resources. When a resource is depleted, people have confidence in finding alternative resources. For example, the forests of the continental North America were the resource that was most easily accessible for hundreds of years, and until the 1880s, except for a few persons with breadth of vision, people still believed that the forests of North America were endless. During the period, although steam engines and railroads had emerged in succession, coal began to be widely used. In most cities in the United States, materials directly from forests were still the basic building materials and fuel. However, soon the fear of "deforestation" emerged, and people on the east coast began to urgently discuss how to contain the expansion of deforestation. From George Perkins Marsh to Charles Sargent, to Gifford Pinchot, they called for a resource conservation movement to ensure that these forest resources were "wisely utilized and effectively managed." However, in the eyes of economists, engineers, and urban boasters, a more effective approach was to find alternative resources. Coal, oil, and natural gas replaced wood as fuel, while reinforced concrete and cement became the most important building materials in cities. Although the burning has made the Earth warmer and warmer, and billions of tons of nondegradable substances have been added to the Earth's surface in buildings, at least they have helped Americans out of the crisis brought about by deforestation and go into the civilized urban and industrial era.

However, if forests as resources can be replaced by new fossil fuels and building materials, forests certainly have value beyond resources. Some resources as important ecosystems on Earth, for example, water, cannot be replaced by other things. Compared with the humid East, the western United States precisely lacks such crucial basic resources. Without water, there would be no agriculture. Without agriculture, there would be no country that supplies urban demand. Without country, cities formed by the population attracted by the mining industry cannot grow. Even if resources could be allocated through railroads to places further away, the West would never be able to break through the bottleneck of development and end the situation of being constrained by the capital of the East. However, more than one large city was formed in the West at last, but the "Metropolitanization" of several large cities, as discussed by urban historian Wang Xu, was established due to the conquest of water.[27]

In the *River of Empires: Water, Aridity, and the Growth of the American West*, Donald Worster referred to the West as the "ecological variant of the modern world-circling culture of capitalism." Its parent was a complete set of cultures nurtured by the humid climate of the eastern United States. It formed expectations for nature's products in its own natural environment, framed social values, and explained systems and order. But when transplanted to the West with an arid climate, it, on the one hand, attempted to practice the parent's existing concepts and behaviors in a new natural environment, on the one hand, in the context of the scarcity of water, a crucial natural element, went against the expectations of the parent's culture to establish an equal, free and democratic society in the new region. A new power structure emerged in the process of striving to create abundant resources, but ceaselessly encountering natural constraints. In the end, a new empire, which could be compared in wealth and ideology with its parent, was created. In the cycle of scarcity-abundance-scarcity, the creators of the new water empire felt an unprecedented need for strong organizational power in American history. As Walter said, "Making the once scarcity abundance and placing

[27] Wang Xu: *The Pattern of American Urban Development from Urbanization to Metropolitanization*.

the previously inaccessible, elusive and stingy nature within the reach of private individuals is the most important and fundamental ecological role of capitalist countries. In the West, the extent to which the role plays a part is unmatched by any other part of the United States." On the one hand, the state had the strongest capital; on the other hand, it possessed the most advanced technology. The combination of the two formed a "technocracy" over the nature and water resources in the West. The cities in the West were not born under technocracy, but their expansion and growth had to rely on the maintenance of technocracy.[28]

4 The Presence of the State

The cities and hinterland of the United States are indeed primarily created by new economies and deeply influenced by modern capitalism. Cronon is undoubtedly correct in this regard. From early fur traders to supply stations on the westward road, to chambers of commerce which were established in sequence, to interstate or even multinational monopoly groups, a continuous flow of private capital flew into the West. Cronon also noticed the collusion between technology and capital. It was those who possessed capital and mastered technology that created the market and controlled the production process of bubbles. Both the federal and state governments made minimal contributions in the initial process of urban development. In addition to Washington, D.C., other American cities were born in the capital bubbles. During the process of westward movement, Mormons established several small towns, but most of them were engulfed by other bubbles before they formed. It was precisely because the leaders of Salt Lake City had transformed into owners of capital and technology themselves that Salt Lake City could expand into a massive bubble. However, Cronon and many scholars who wrote the history of American urban creation overlooked an important fact: the State was always present. Actually, compared with Europe, Asia, and even Canada, the State in the early history of American cities had a blurred and even imperceptible appearance. But when we link up the history of American cities into a whole, rather than the biography of individual cities to examine the State, the State emerges from the veil of capital. In the nineteenth century, perhaps it was not yet a massive and omnipotent machine,

[28] Donald Worster: *River of Empires: Water, Aridity, and the Growth of the American West*: p. 334.

but it remained an indispensable organizational force in the process of urban development.

Since its founding, the United States has been a state with capitalism as its basic economic and cultural belief. The function of the State is not only to maintain security but also to encourage individuals to pursue the maximization of wealth through the establishment of laws and systems. However, it is not proper to rely solely on beliefs and ambitions or capital and technology to maximize wealth. A true material foundation such as soil, water, energy, minerals, plants, and animals is really required to achieve the goal. Therefore, from the founding of the State to the end of the nineteenth century, the measures at the core position of functions and powers of the United States were to acquire more land and the resources on it for the United States through plunder, purchase, war, and annexation. At the same time, the State was also a sponsor of transportation and communication. From the construction of the Erie Canal, to the establishment of courier stations from east to west, to the extension of telegraph lines representing civilization and entering the wilderness, to the large-scale land grants received by railroad companies, the State and cities were never absent from the rise process on the land where aborigines once galloped.

The entire process of acquiring land was a familiar one. In 1803, Thomas Jefferson eagerly signed a treaty with France, known as the "Louisiana Purchase," before obtaining approval from Congress. It is the largest land transaction in the world history. A vast area of 2.14 million km^2 in the west of the Mississippi River, through the transaction, was incorporated into the territory of the United States of America, thereby doubling the territory area of the United States. From 1818 to 1853, the territory of the United States continued to expand westward, whether through purchases, treaties, wars, or mergers, all the processes of land acquisition were distinct state actions. At this point, the United States transformed from the earliest 13 eastern states to a continental country spanning North America and being adjacent to two oceans. In 1867, the United States purchased Alaska from Tsarist Russia in internal and external troubles, adding nearly 600,000 km^2 to its territory. In 1898, Hawaii was merged. They both became the 49th and 50th states of the United States in 1959, respectively. The map of the United States finally took its current form.

Faced with such a vast territory, the young state adhered to the spirit of the *Land Ordinance of 1785* and divided its territory into grids for distribution. Just this time, the hand of the State became clearer. It no longer only appeared as a land dealer in the process of land distribution, but more as a spokesperson and defender of the public interest. During the stalemate of the Civil War in 1862, the Homestead Act, which played a decisive role in the fate of the American West and the lives of thousands of men and women, was enacted. Afterwards, with the signing of a series of related acts, nearly 1/10 of the territory of the United States was eventually granted free of charge to nearly 1.6 million homestead farmers, thus ensuring their pursuit of freedom and wealth. A large amount of land was gifted to railroad companies to encourage private enterprises to join the country's camp and work together to achieve the so-called "Manifest Destiny" of the State, a continental empire that comprehensively developed nature. The land belonging to "you and me" was marked with the symbol of "private property, no trespassing."

The symbol symbolizes the enormous paradox that emerged during the process: from the founding of the state to the late nineteenth century, the basis for the implementation of America's laissez-faire policy was the expulsion, massacre, confinement, and enslavement of others on the land. These otherized groups include other ethnic groups within the human species, Indians and slaves, and non-human species, native plants and animals. The process of removing others was achieved through active guidance or even direct participation by the State. If blacks and slavery are an issue that was always debated by the government and the public in the United States, they are unanimous in the expulsion of Indians together. The US federal government demonstrated the power of the State in the destruction of blacks' culture or the extinction of their means of subsistence, rather than a laissez-faire style. It wrote far more than a social and cultural history, but also an ecological history. As early as 1831, when Tocqueville wandered the American continent, he had a keen observation of the close relationship between Indians and the changes in their natural environment: "By making their game flee, it is as if you made the fields of our farmers sterile. Soon they almost entirely lack the means of existence."[29] That year was the second year after the implementation of the Indian Removal Act, Tocqueville witnessed the migration on the "Trail of Tears," and recognized the role played by the federal government in

[29] Alexis-Charles-Henri Clérel de Tocqueville: *De la démocratie en Amérique* (*Democracy in America*), translated by Dong Guoliang, Beijing: The Commercial Press, 1996, p. 377.

it. He wrote, "The dispossession of the Indians often takes place today in a regular and, so to speak, entirely legal manner. ... Therefore, the states, by their tyranny, force the savages to flee; the Union, by its promises and with the aid of its resources, makes the flight easy. These are different measures that aim at the same end."[30]

Of course, Tocqueville was born early and did not witness the collective environmental violence that was completely driven by the federal government on the Great Plains. The federal army was called out to slaughter bison. General Sherman, who led a northward march toward the seaside, ultimately defeated the Confederate States of America and turned the American Civil War into an ecological "total war," became a vanguard in clearing the bison barrier after the end of the American Civil War.[31] During the process, the most representative remarks were made by Lieutenant General John M. Schofield. When he recalled in his autobiography his hunting of bison by commanding the main forces around 1870, he wrote, "I wanted no other occupation in life than to ward off the savage and kill off his food until there should no longer be an Indian frontier in our beautiful country."[32] The Indians, bison, and the State were clearly presented in the declarative sentence. Although some scholars believe that the US federal government has never directly issued a policy to hunt bison, the disappearance of bison should not be solely the responsibility of the federal military, but the deployment of the military is a state action.[33] They joined the hunting of bison market, made scattered individual behavior a legalized violence of the State, made it patriotic behavior similar to America Civil War, and ultimately led to the near extinction of a dominant species that once occupied the entire Great Plains.

Throughout the process of land acquisition and distribution, the official and civilian discourse basically reached a consensus. The State acquired land in order to achieve maximum individual freedom and created a democratic and moral country different from those in the Old

[30] Ibid, pp. 378 and 392.

[31] In this sense, the State's re-integration of resources and transformation of land landscapes in the form of a total war reflects another aspect of the State's environmental governance. Brady Lisa. *War Upon the Land: Military Strategy and Southern Landscapes during the American Civil War*. Athens: University of Georgia Press, 2012.

[32] Schofield John. *Forty-Six Years in the Army*. New York: Century Publisher, 1897.

[33] Isenberg Andrew. *The Destruction of the Bison: An Environmental History, 1750–1920*. Cambridge, U. K.: University of Cambridge Press, 2000.

World. In the process of creation, the State should and should only provide protection for its citizens seeking wealth and freedom, but be free of other interference. The strongest support behind the laissez-faire policy was the "America's windfall," a land that seemed to have inexhaustible resources. In other words, the nature of North America allowed the United States to promise freedom to its people, and the ideological core of the State's laissez-faire policy toward its citizens was the conquest of nature.

By the early twentieth century, as mentioned previously, when the development of the hinterland encountered the most steadfast obstacles of nature, laissez-faire capitalism finally cooperated with the State and transferred the ultimate power to the latter in order to obtain the capital and technology to conquer harsh nature. From this point on, the State declared its unquestionable role in determining urban affairs and officially transitioned from the backstage to the stage. The role would only become expansive due to the expansion of bubbles, as the urban demand for hinterland became more complex and was no longer satisfied by resources within the boundary line. When World War II ended and the United States together with the industrial world it led grew at an unprecedented pace, the development of its hinterland was no longer an urban affair, but an important issue in international politics. From fossil fuels to timber and to minerals required for other industrial civilization, the State was urged to be more active and powerful in building its international hinterland, and the driving force behind it was precisely the ecological paradox formed between the growing urban demand and the shrinking hinterland resources.

However, at a time when the State started to go into action and directly sought broader and more abundant hinterland for the greater development of new cities, it also began to play a role in preventing cities from expanding unboundedly their own bubbles under the drive of laissez-faire capitalism. The link between the city and nature was not just the exchange of materials and energy between them. As natural resources from distant regions endlessly flew into large cities in the East to support their growth and expansion, a reverse trend centered around cities such as Boston and New York slowly spread into the United States and even various corners of the entire continental North America. It was a trend of thought. Although its flow had a much slower impact on the landscape compared to the capital flowing out of cities, it still left its impress on the natural system of the New World, or more enduring and profound impress.

In the trend of thought, its mainstream could be referred to as nature conservation, environmental awareness, and reform that emphasizes not only utilitarian resource conservation but also aesthetic and ecological significance. The reform was another joint effort between the State and cities. People living in cities called on the State to preserve the nature lost in the process of urban sprawl through legislation. The State, as a protector rather than a developer, began to emphasize another important role in nature management. In Brenner's view, building national parks and setting up wilderness reserves were also important components of "planetary urbanization." These areas were also urbanized, and the only difference was that their function was to serve as the hinterland of urban leisure and entertainment. However, if we delve deeper into this period of history, from the original intention of its construction to the current reflection, what we see is more complex ecological thinking. Ultimately, their existence would transcend human self-contemplation and have their own way of operation and value when these regions were established under the driving force of urbanites' desire to bridge their "emotional rift."

5 Value of Mount Desert Island[34]

There is a small island on the coastline of Maine in the northeast corner of the United States. For a long time, for urbanites living in New England and New York which is more than 200 miles away from the south of the island, the island was just a term on the map, "Mount Desert Island," an unattractive term. The term "desert" here does not refer to an arid and rainless desert, but to desolate and useless land. Therefore, if labeled in Chinese, the small island is called a "Huangliangshandao (desolate mountain island)." It was named by Samuel de Champlain, a French explorer in early seventeenth century. In 1604, when Champlain arrived at the small island, he saw seven or eight steep peaks standing on the winding coastline. The slopes were covered with virgin forests made up of birch,

[34] The Sections IV and V of the chapter adopt some of the content from the old text "Places far from Cities: America's Wilderness and Urban Green Reformers in the late 19th Century", which has been adjusted after being included in the book. Hou Shen: "Places far from Cities: America's Wilderness and Urban Green Reformers in the late 19th Century", editor-in-chief Xia Mingfang: "New History" Vol. VI, Beijing: Zhonghua Book Company, 2012.

fir, and pine, which extended all the way to the exposed rocks above the treeless line, giant granite under glacial action. Although Cadillac Mountain, the highest peak of the island, is only 470 meters above sea level, it is widely regarded as the place in the United States where sunrise can be seen earliest. The coastline is covered with jagged reefs. Just as Su Shi, a famous poet, scholar, and artist of the Song Dynasty in Chinese History, was described in *Memories of the Past at Red Cliff*, "Jagged rocks tower in the air, Swashing waves beat on the shore, Rolling up a thousand heaps of snow." However, in the eyes of Champlain, who still had the lingering fear of wilderness in the Dark Ages (medieval Europe), the small island was so dark and desolate, and the Abaniki of the Aboriginals on the island were so savage and horrible. Therefore, "Mount Desert Island" the name of the small island even though there were lush forests that nourished countless wild animals and nurtured vibrant natural vitality.

During the period between the seventeenth century and the 1850s, Mount Desert Island experienced several tides in the affairs of men, from the earliest French fishing boats to the tireless footsteps of fur hunters, to the unsuccessful preaching of the French Jesuits, to several wars between British and French colonizers, to the incessant struggle between the French and aboriginals, to the War of Independence and the Second War of Independence in 1812. In the end, the island was included in the territory of Maine, USA, and the Abnaki completely disappeared. The island was sparsely inhabited by fishermen and farmers, the former caught and sold fish, while the latter mainly made a living by felling trees in the local virgin forests and sold wood. It was also included in the capitalist market system; however, due to its location in the northeast corner, inconvenient sea and land transportation, and successive years of war, the lives of the island residents were still quite simple.

However, in the 1860s, like many small islands along the coast of Maine, the peaceful landscape and lifestyle of Mount Desert Island were disrupted by urbanites from New York, Boston, and other places. The expansion and pressure of the city forced its residents to seek natural comfort and a summer resort in summer. Soon, private villas of all sizes called cottages were built on Mount Desert Island. Real estate developers and the tourism industry saw the hidden "business opportunities" here, so various hotels were built on it. Land jobbers, like the shadow following developers, also sneaked into every corner of the Maine coastline. In order to facilitate the buying and selling of land, real estate speculators, who completely disregarded natural lines and landforms of the island, followed the tradition of the United States since its founding to divide coastal land into a network of small square blocks and sold them to private individuals. And these new houses for the poor were crude and shabby, those for the rich were flamboyant and ostentatious, and the roads and paths that were temporarily paved often ran straight through the forest, but were too much filthy. Urbanites, based on their habits toward townhouse, cut down the woods before and behind their houses, transported soil from other places, planted wide lawns that were neatly trimmed, and decorated rose beds between cliffs, green pines, and gorge banks, thus creating a landscape that was ill-adapted to the rough and wild mountains, seas, wood and stones around them.

In 1890, young Charles Eliot, the designer of the future Boston Metropolitan Park System, saw such a scene when he accepted the advice of his spiritual mentor Frederic Law Olmsted to explore the natural landscapes of New England and seek inspiration and sources for creation. The precipitous rocks, huge waves, ancient trees, and strange rocks here fully conform to his aesthetic imagination immersed in the trend of thought in transcendentalism. The diverse species as well as complex and ever-changing landscape here enlightened him, a landscape designer, and even the ups and downs of human history here also afforded much food for thought. After all, Eliot was the first person to propose the protection of famous historical sites in the United States.[35] However, the new things here made him angry and even fearful. In an article titled "The Coast of Maine" published in the magazine *Garden and Forest*, Eliot wrote, "The lamentable feature of the situation is the small amount of thought and attention given to considerations of appropriateness and beauty by the builders and inhabitants of the summer colonies of the coast. Indifference in these matters works ill results everywhere, but nowhere is lack of taste quite as conspicuous as on the sea-shore. Both corporations and individuals are guilty on this head." He further pointed out, "The real danger of the present situation is that this annual flood of humanity, with its permanent structures for shelter, may so completely overflow and occupy the limited stretch of coast which it invades, as to rob it of that flavor of wildness and remoteness which hitherto has hung about it, and which in great measure constitutes its refreshing charm." Eliot also observed that a few people already realized the existence of such beauty, so they bought

[35] Namely, the Trusteeship of Reservations (1891) of Massachusetts, as mentioned in Chapter III.

considerable tracts of land, surrounded them with walls, and turned them into their private property. He warned that if such a situation was not cut at its root, very little public space was left for ordinary people to appreciate the natural beauty.[36]

However, Eliot did not indulge in such laments. He urgently demanded that citizens with a "public spirit" should take action to change the existing situation. He believed that human beings had been a consensus that "the importance to human happiness of that refreshing antidote to city life which fine natural scenery supplies, nor is it necessary to remind them that love of beauty and of art must surely die if it be cut at its roots by destroying or vulgarizing the beauty of nature." He asked, "Can nothing be done to preserve for the use and enjoyment of the great unorganized body of the common people some fine parts, at least, of this sea-side wilderness of Maine?" Therefore, Eliot suggested that local residents and governments, and even the federal government, should preserve the beauty of the wilderness through legislation, purchases, donations, and other means, as well as the right of ordinary people to appreciate and enjoy such beauty.[37]

The voice given by Eliot on Mount Desert Island was a distant response to his passionate call in the Boston Metropolitan Park System. By their nature, they were completely consistent. In the view of Eliot and early environmental reformers of his time, civilization and nature were inevitably complementary; the vanishment of nature meant the death of art, the disappearance of nature would make cities desolate, and once nature was destroyed, man would no longer be complete. However, at that moment, Eliot's sight and imagination completely penetrated the boundaries of cities and even metropolises and reached the wilderness, what seemed to be the farthest region from civilization, where he discovered the tonics and barks of civilization, as Henry David Thoreau had said.

Today, when rolling out the map of the United States, we can see multicolored green of all sizes covering the continent between the Pacific and Atlantic oceans, they represent national parks, national forests, and state parks of different sizes. These greens mostly are located in places far away from large cities such as Boston, New York, and Chicago. They

[36] Charles Eliot, "The Coast of Maine", *Garden and Forest*, issue 104 (1890): p. 87.

[37] Ditto.

create a landscape that is completely contradictory to the urban landscape, and the cultural significance that it expresses is quite different from urban civilization. However, these areas which seem to be remote and wild have an inseparable connection with cities. Their emergence not only reflects the psychological needs of urban dwellers in the industrial era and the material needs for the continuous development of urban civilization, but also to a large extent, they were founded by the appeal of urbanites who live in cities and build urban landscapes.[38] Ultimately, they will also transcend the values defined by urbanites on the base of the aesthetic interests and needs of their own, and become the final preserve of the natural world which gradually die away during urban sprawl.

6 Transcendence of the City

In 1864, Olmsted accepted a job as the supervisor of the Mariposa Estate, which took him to California, the west of continental North America. There, he found himself quite close to some of the most magnificent natural landscapes in the world, but at the same time, he was truly immersed in the real frontier life. Compared with the romantic imagination in the "Leatherstocking Tales," the real world was far more cruel and gloomy. Pioneering Christians realized that it was not a frontier where they could fulfill their dreams of freedom and democracy on the land, but a diverse yet singular miners' society composed of people of different skin colors, races, nations, and social strata. There, Olmsted saw expelled Indians, fugitive or newly liberated blacks, Chinese people who flooded into the Golden Mountain during the gold rush, ubiquitous Mexicans, Italians, and Germans newcomers who just came to the New World, and "gentlemen" like him from the East Coast. The society was far more

[38] For research on the status and significance of national parks in the history of environmental movements, see Stephen Fox, *The American Conservation Movement: John Muir and His Legacy*. Madison: University Wisconsin Press, 1985; The representative works concerning the management history of national parks include Richard Sellers, *Preserving Nature in the National Parks: A History*. New Haven: Yale University Press, 1997. The representative works in the field of biographies include Donald Worster, *A Passion for Nature: The Life of John Muir New York:* Oxford University Press, 2009. The representative works reflecting on the elite awareness and the costs paid by marginalized groups in the process of establishing national parks include Karl Jacoby, *Crime against Nature: Squatters, Poachers, thieves, and the Hidden History of American Conservation*. Berkeley, California: University of California Press, 2001.

diverse in terms of race and language than the urban world on the East Coast where Olmsted lived, but it was almost barren and primitive in terms of culture. People repeated the same monotonous work every day, and struggled and jostled against one another to obtain the most basic means of subsistence. It differed markedly from the well-organized and sound society of New England. In it there was a unique system of disorder and moral decay in the borderlands.

In the rough frontier life, Olmsted increasingly affirmed his observations several years ago in the southern part of the slave plantation, namely, that the antonym of "civilization" was not "nature," but "savage," and it was full of violence and corruption. Nature before his eyes could excite all the good imaginations of humanity, but the society around him often made him slip into a brooding depression and frustration. Therefore, Olmsted threw himself into two main tasks during his three years when he lived in California. Firstly, he attempted to end the state of primitive life in the Mariposa Estate. Secondly, he called for the protection of the pristine natural beauty of Yosemite Valley and the nearby redwood forests. For Olmsted, the ruthlessness toward nature was as barbaric as the cruel exploitation of their kind by humans. Therefore, these two tasks which seemed to be contradictory were consistent in nature, even the same. Their fundamental objective was to make the society more civilized. In his report on protecting Yosemite, Olmsted pointed out that "The power of scenery to affect men is, in a large way, proportionate to the degree of their civilization and to the degree in which their taste has been cultivated. Among a thousand savaged savages there will be a much smaller number who will show the least sign of being so affected than among a thousand persons taken from a civilized community. This is only one of the many channels in which a similar distinction between civilized and savage men is to be generally observed."[39]

Compared with his first task, Olmsted's second task seemed to be more successful. At least in 1864, President Lincoln granted the land of the Yosemite Valley and redwood forests to the California government with the premises that they should be held for the construction of a state park for public use, which was the first park authorized by the government to be established solely for the protection of natural beauty in the United

[39] Frederick Law Olmsted, "Typed Transcription of Draft of Preliminary Report upon the Yosemite and Big Tree Grove," The Evolution of American Conservation Movement, 1850–1920, the digital archive of American Memory, Library of Congress.

States and the world.[40] Olmsted was designated as one of the board members to manage the authorization, and in 1865, he wrote a report for the board. The report systematically discussed for the first time the reasons why the government had the responsibility to protect the beauty of wild nature, and was also the first report on how to manage the natural landscape. In the report, Olmsted pointed out that there were two reasons why the government was required to take responsibility for protecting the region. The first reason was that by developing the tourism industry, "the direct and obvious pecuniary advantage which comes to a commonwealth." But the second reason was what Olmsted believed to be "more important." He wrote, "It is the main duty of government, if it is not the sole duty of government, to provide means of protection all citizens in the pursuit of happiness against the obstacles, otherwise insurmountable, which the selfishness of individuals or combinations of individuals is liable to interpose to that pursuit." Appreciation of the beauty of nature is a component of the pursuit of happiness.[41]

On the issue of how to manage natural landscapes, Olmsted's exposition of his views was very clear, which became the fundamental creed of landscape management in United States National Parks for over a hundred years, that is, its wild state and natural style shall be maintained as far as possible. In these places far from the city, Olmsted's landscape design philosophy is an adaptation to local conditions. A sense of domestication and order in their natural style was still required for Parks in cities as a result of their close proximity to people's lives. Therefore, nature should be rescued from its own chaotic state, and its unstable factors should be removed, then it was rationalized and humanize in certain aspects to meet the daily physiological and aesthetic needs of urban dwellers. However, what was preserved in national parks was the system, wildness, and unrestraint of nature. Therefore, Olmsted suggested that people should make any so-called renovations or improvements on the original landscape of Yosemite as little as possible, except for extremely

[40] The California government's management of Yosemite was extremely poor. From the 1860s to the 1880s, Yosemite was in a state of overgrazing for a long time, and some miners and farmers also entered the region for profit. Therefore, at the call of John Muir, together with Robert U. Johnson, the editor-in-chief of Century Magazine, Olmsted, Sargent and others, Yosemite became a national park in the 1890s.

[41] Olmsted, "Typed Transcription of Draft of Preliminary Report upon the Yosemite and Big Tree Grove".

necessary amenities, such as a very small number of places for board and lodging as well as roads. ####In his report, he wrote, "The first point to be kept in mind then is the preservation and maintenance as exactly as is possible of the natural scenery; the restriction, that is to say, within the narrowest limits consistent with the necessary accommodation of visitors, of all artificial constructions and the prevention of all constructions markedly inharmonious with the scenery or which would unnecessarily obscure, distort or detract from the dignity of the scenery." All artificial constructions, especially roads within national parks, were required to be designed by professional landscape designers, as only their professional vision and aesthetic taste could know how to integrate artificial elements into nature.[42]

The establishment of Yosemite State Park ushered in a new era of aesthetics in the relationship between Americans and nature. In 1872, the US federal government established Yellowstone National Park, the first national park in the world, among the giant rock, springs, black pine, and gray bears at the source of the Yellowstone River in the northwest of the United States. The purpose of establishing the national park was described in the Act passed by the US Congress as follows: "[the land] is hereby reserved and withdrawn from settlement, occupancy, or sale under the laws of the United States, and dedicated and set apart as a public park or pleasuring-ground for the benefit and enjoyment of the people; ... Such regulations shall provide for the preservation, from injury or spoliation, of all timber, mineral deposits, natural curiosities, or wonders within said park, and their retention in their natural condition. ... He [the Secretary of the Interior] shall provide against the wanton destruction of the fish and game found within said park, and against their capture or destruction for the purposes of merchandise or profit."[43] Although at that time, the social function and deeper ecological value of national parks still needed to be interpreted and developed over the next hundred years, the act clearly defined two aspects: firstly, the national park was open to all people and serve them. Secondly, the national park shall protect the mountains, rivers,

[42] Olmsted, "Typed Transcription of Draft of Preliminary Report upon the Yosemite and Big Tree Grove".

[43] "An Act to Set Apart a Certain Tract of Land Lying near the Head-waters of the Yellowstone River as a Public Park, Mar. 1, 1872, The Evolution of American Conservation Movement, 1850–1920, the digital archive of American Memory, Library of Congress.

soil, vegetation and animals within it from the erosion of commercial interests. The first point expresses distinct characteristics of democracy, while the second point emphasizes its non-utilitarian orientation.

Olmsted's two tasks in California were not completed, and soon he accepted the call of the East and returned to the world he grew up and was familiar with. However, when he arrived there, he found that the society in the East, which had already ended its frontier era, also required his efforts in California's mining areas. While his eyes were ceaselessly focused on the fate of Yosemite and the redwood forests, he saw the natural beauty of the East Coast that was also thrilling groan and disappear under the impact of industrialization and urbanization. Shortly after returning to the East Coast, he, along with a group of intellectual elites from Boston and New York, began appealing for the protection of the northernmost Niagara Falls in New York State.

As people ride on the "Maid of the Mist" to cross the splashing waves of Niagara Falls today, let the boat drift freely as it continued its motion across the vast expanse of the river, shake off the dust on their clothes on the lofty mountains, and wash away the dirt on their feet in the vast river and stand aloof from the world, it seems that the place is just an inexhaustible treasure supplied by the Creator, and its landscape has remained unchanged since ancient times. It is not Central Park in New York, but a green patch built within the city's grid, it's not Fenway Park in Boston, either, it's a product of the city's hill land reclamation. Except for those who have a deep understanding of its history, few people associate the giant cascading waterfall with landscape designers like Olmsted. However, the fact is that it is precisely the years of hard work and calls of a group of urban naturalization reformers, represented by Olmsted, which preserved the grandeur of the natural landscape and made it seen by the public. And it is precisely the design of Olmsted and others that perfectly integrates manpower into the landscape and makes people forget its existence.

Niagara Falls was one of the earliest tourist destinations in New England states and New York in the East. As early as the eighteenth century, Europeans had come from afar to worship it. However, compared with its tourism development, another profit generated from the giant waterfall was obviously more attractive to industrialists. Since the mid-eighteenth century, like all rivers in the eastern United States, the giant waterfalls and its source, the Niagara River, had been the main suppliers of energy for sawmills, mills, and tanneries along the river. By the mid-nineteenth century, it had been largely monopolized by large private water & electricity companies, and became the source of electricity, one of the greatest good news of the Industrial Revolution. The land around the giant waterfalls was almost entirely owned by the power companies, and the natural miracle had been completely incorporated into the capitalist market system by that time. In the 1870s, tourists who came here could only catch a quick glance and were immediately forced to leave after gasping in admiration. However, it was also at that time that the giant waterfalls began to attract not only intellectual elites seeking literary and artistic inspiration in nature, but also a large number of ordinary people from the urban middle class.

In order to preserve the beauty of nature and to make more people appreciate it, a movement called "Free Niagara" was launched on the initiative of the Olmsted and others. Apart from Olmsted, the backbone of the movement also included Frederick Church, a representative figure of the Hudson River School from New York, Henry Hobson Richardson,[44] Olmsted's long-term collaborator and one of the most famous architects in America, and Charles Eliot Norton, the famous literary critic, progressive reformer and professor at Harvard University.[45]

In fact, in its subsequent development, the movement almost included the major intellectual elites of Boston and New York. In 1880, Olmsted in collaboration with James T. Gardiner, the director of (New York) State Survey, and others wrote a survey report on Niagara Falls submitted to the Governor of New York, which included a petition signed by more than 100 elites of all circles from New York, Boston and Washington

[44] Richardson came from the South but received education at Harvard University before studying architecture in Paris. After returning to the United States, he settled in Brookline, Boston and created the "Richardsonian Romanesque" style in architectural history. His representative works included the Trinity Church and other buildings in Boston. He collaborated with Olmsted for a long time. In Olmsted's landscape design, his architecture could be often found as both of them fitted well. After Richardson passed away in 1887, Olmsted felt a deep sense of loneliness as if they couldn't talk about their experiences together any more after the person with the same heart and mind as him departed this life.

[45] Norton is a famous literary critic and an important disseminator of European Romantic literature and aesthetics in the United States. Many people consider him the most elegant gentleman in the United States. He is the cousin of Charles W. Eliot, President of Harvard University, and also the uncle of landscape designer Charles Eliot. Norton was extremely critical of the development direction of American civilization, and attempted to reform progressivism through education.

as well as the United Kingdom, Canada and others. These undersigners included not only the then Vice President of the United States, Chief Justice, Senator, Congressman, Governor, Archbishop, presidents of Harvard University, Yale University, and other universities, as well as intellectual elites such as Ralph Waldo Emerson, James R. Lowell, Romantic poet, Charles Dana, renowned New York journalist, Asa Gray, Harvard botanist, and Francis Parkman, Harvard historian, but also Canadian figures and several renowned British scholars, such as John Ruskin, romantic aesthetic critic, and Thomas Carlyle, Scottish writer. In the grand petition from both sides of the Atlantic, they wrote: "Objects of great natural beauty and grandeur are among the most valuable gifts which Providence has bestowed upon our race. The contemplation of them elevates and informs the human understanding. They are instruments of education. They conduce to the order of society. They address sentiments which are universal. They draw together men of all races, and thus contribute to the union and the peace of nations."[46]

Therefore, Olmsted and others referred to the protection of the natural miracle as the "sacred obligations to mankind."[47] Fundamentally speaking, the report written for the giant waterfalls in New York in the East is no different from the report written by Olmsted for Yosemite on the Western Frontier a few years ago. It still emphasized that the government should take measures to take responsibility for protecting the giant waterfalls and making it for the enjoyment of the people. However, in the report, Olmsted laid more stress on the desire and need for nature and wildness, which were generated under the pressure of urbanization, as well as the necessity of protecting them therefrom.

The "Free Niagara" movement had distinct city characteristics, not only in the sense that it is a preservation movement driven by urbanization and industrialization, but also in the form in which it operates. It actively borrowed the role of mass media in the city, drew public attention through media dissemination, and formed public opinion to urge politicians to carry out reforms through legislation In 1881 and 1882, after fierce opposition from Governor of New York Alonzo B. Cornell,

[46] Olmsted, "Special Report of New York State Survey on the Preservation of the Scenery of Niagara Falls," *Documents of the Assembly of the State of New York*, Vol. 6. The Evolution of American Conservation Movement, 1850–1920, the digital archive of American Memory, Library of Congress.

[47] Ditto.

Olmsted, Norton and others began supporting the publication of a series of articles about the giant waterfalls in major newspapers in New York and Boston, thus the movement emerged from the intellectual elite and entered the public opinion. At last, State of New York was forced to establish the Niagara Falls Reserve through legislation in 1885, and the state government began gradually buying back land from private enterprises to construct the first New York State Park.

In 1887, two years after the establishment of Niagara Reservation State Park, Olmsted and Vaux, his former co-designer of Central Park in New York, were once again invited back to the giant waterfalls to design the roads there and the Goat Island located in a corner. Olmsted was once more caught up in the dilemma he faced in the national park, namely, the way to coordinate the conflict between an increasing number of tourists and the preservation of the primal and wild beauty of nature, which will be the challenge that every landscape designer and national park manager will face in the future. Its emergence represented a great success in the efforts of these green reformers to awaken people to love for nature, and was also an important manifestation of the democracy they hoped to achieve. However, the inevitable destruction of nature itself that came with such phenomenon was contrary to their desire to preserve the wilderness. Olmsted did not really solve the dilemma, or the only thing he could do was what he consistently insisted on doing. "We came here to restore and protect the natural environment of Niagara Falls, not to add anything to it."[48]

Since the 1890s, Olmsted himself had gradually faded from the cause of reform, but urban naturalization reformers didn't stop there. When Eliot arrived at Mount Desert Island, he discovered the wilderness value that Olmsted, Norton, and others sought. However, Eliot's premature death prevented him from seeing the natural landscapes of the Maine coast preserved during his lifetime. In 1914, when war broke out in Europe on the other side of the ocean and the cloud of gunpowder floated over the battlefield, Charles W. Eliot, Eliot's father, the most important president in the history of Harvard University and the famous educator who truly transformed Harvard from a former modern humanistic divinity

[48] Olmsted, "Special Report of New York State Survey on the Preservation of the Scenery of Niagara Falls," *Documents of the Assembly of the State of New York*, Vol. 6. The Evolution of American Conservation Movement, 1850–1920, the digital archive of American Memory, Library of Congress.

school into a comprehensive university in the modern sense of the word, wrote an article titled *The Need of Conserving the Beauty and Freedom of Nature in Modern Life* for The National Geographic Magazine. Throughout his life, Charles William Eliot turned Harvard University into a sacred place of modern science. He devoted his extremely limited leisure time to the study of the issue of labor relations, which was a concern for more people in his time. He was not indifferent to urban naturalization reform with his son as a leading figure, on the contrary, he supported his son to the best of his ability. But it was not until his beloved son passed away in 1897, when he was in extreme pain, that he began to truly throw himself into the reforms facilitated by his son. In 1900, with his support, Harvard University established the first Department of Landscape Architecture in the United States, where Frederick Law Olmsted Jr., son of Olmsted, first offered the course of landscape architecture in American. In an article published by Charles William Eliot in 1914, we once again read about the sense of nature and civilization American progressive urban naturalization reformers.

In the article, Charles William Eliot keenly captured the characteristics of the progressive reform forces that had emerged since the 1880s. He wrote, "Over the past century, it has been thoroughly proven to civilized people that the various evils accompanying the growth of modern cities and factory systems have developed too far beyond the limit of human tolerance. However, these evils were the result of individualism, a form of freedom that emerged in nineteenth century civilization, especially in the first half of the nineteenth century. In the past 40 years, another form of freedom, the freedom of association and collective action, has begun to suppress some of the evils nurtured by individualism, thus improving the human environment."[49] He discussed at great length the negative impact of harmful urban life and factory systems on people although he did not clearly point out how urban life is harmful. He seemed to assume that everyone knew exactly the urban evils he was talking about, so there was no need to state them one by one. After reading the entire article, we obviously feel that the evil mentioned by Charles William Eliot is a moral degeneration although he also lamented how urbanites living in artificial light kept themselves away from fresh air and beautiful scenery. But most importantly, he believed something had gone wrong

[49] Charles W. Eliot, "The Need of Conserving the Beauty and Freedom of Nature in Modern Life," *The National Geographic Magazine*. Vol. 26 (July 1914): pp. 67–74.

with American morality in the process of urbanization. He complained that "pernicious excitement and evil pleasure" exclude "beneficial intellectual interests." Therefore, he would like to see more "completely healthy spiritual pleasure and noble enjoyment" among urban populations, and such enjoyment was found in nature.

At the end of the article, Charles William Eliot came to the conclusion that the establishment of a new national park would cure existing moral illnesses. Hence, he suggested turning "Mount Desert Island" into a national park. He wrote that the power of the federal government to enforce preservation of nature was more active in the distant West, which was not densely populated; hence "the evils of urban life and factory system were not well developed." Thus, he concluded by asking whether it was not the right time to establish a national park in the East, where the "detrimental effects of urban life" were most pronounced.

Therefore, in 1919, the National Park Service of the United States, which was set up three years ago, established the first national park on the Mount Desert Island of Maine in New England. It was initially named "the Lafayette National Park" in honor of one of the founding fathers. In 1929, it was renamed the "Acadia National Park," which is the ancient Greek legend of harmonious coexistence between nature and man. Frederick Law Olmsted Jr. and Beatrix Farrand, one of the most celebrated female landscape architects in the United States who grew up under the influence of Olmsted, designed the roads for the new national park. Driving all the way along the road between the Cadillac Mountain, one side is a billowy sea, and the other side is a forest inflamed with autumnal tints. The river gurgles between cliffs that rise a thousand feet high on either side, and the wind generates high waves. There people discover the value of the Mount Desert Island.

More than 20 years after the establishment of Acadia National Park, Aldo Leopold rethought the relationship between wilderness and civilization in the sand country of Wisconsin. He wrote this passage: "in the history of the human species, two changes are now impending. One is the exhaustion of wilderness in the more habitable portions of the globe. The other is the world-wide hybridization of cultures through modem transport and industrialization. Neither can be prevented, and perhaps should not be, but the question arises whether, by some slight amelioration of the impending changes, certain values can be preserved that would otherwise be lost." In the next paragraph, Leopold gave his answer, hoping that civilization should do what it can to preserve the wilderness which was struggling on the brink of extinction, and thus to preserve a certain values that were also facing an impasse. He wrote, “This is a plea for the preservation of some tag-ends of wilderness, as museum pieces, for the edification of those who may one day wish to see, feel, or study the origins of their cultural inheritance.”[50]

The wilderness and its value in Leopold’s mind are not entirely the same as those spoken of by urban green reformers in the late nineteenth century, for instance, Olmsted. There were no roads, inns, and even the most basic artificial facilities in Leopold’s wilderness, which also rejected the form of entertainment that relies on modern technology. In the understanding wilderness, Leopold went further than Olmsted and others did. What he saw was an ecological panorama and the subtlest change and rhythm in every part, rather than a mere magnificent landscape. And the values in preservation of wilderness, which he called for, imply respect for and even humility toward nature and the organic and inorganic existence in it.

Compared with Leopold’s wilderness, which preserved its pristine state as much as possible, the wilderness of Olmsted and others was still mixed with artificial traces and rational constraints. Their values was fundamentally grounded physiological and spiritual needs of human beings, rather than in land ethics centered on ecological center, which Leopold sought for. However, we cannot deny that in the national parks they call for, the dominant force is the economic system of nature, not that of man, which is a difference that deserves to be taken seriously between national parks and urban parks or purely artificial buildings. The values they found in nature also broke free from the utilitarianism that had long dominated American civilization, thus the idealistic vision in aesthetic level was realized in the landscape of the United States and the aftermath of the romanticism that had haunted the elites was transformed into the voice of the public of the urban age. They left a rich legacy for the continent, not only ecological heritage like national parks, but also cultural heritage that stirred up the imagination of wilderness aesthetics. These became important theoretical origins and ecological foundations that shape Leopold’s ecological thought of transcendent wilderness and the ecological eras that rose afterwards. They did not follow the frenzy of industrialization in the second half of the nineteenth century, but questioned the development

[50] Aldo Leopold, *A Sand County Almanac*, translated by Hou Wenhui, Beijing: The Commercial Press, 2016, pp. 187–188.

direction of civilization with reflection and action. They also realized that a truly civilized society would retain its rightful place for the existence of the wilderness, otherwise civilization would become a truly desolate island.

CHAPTER 6

City of Sea of Grass

1 Prologue: Knowledge of a Child in Kansas City on the Great Plains

For a child who grew up on a farm on the Great Plains in the 1950s, Kansas City in Missouri was a place full of not only original sin, corruption, evil, and poverty, but also elegance, refinement, worldly wisdom, and sophistication. In his imagination, the neon lights of the city flickered with a decadent and warm halo, they were dangerous and enchanting. It was a city with Elvis Presley style that provided all sensory freedom and spiritual experimentation. Although it had a bit of the flavor of rural life compared with New York or Chicago, it actually made him feel close to a child from the Great Plains. His first adventure in the city might be the visit to an auto show with his father. Every man from the Great Plains in the 1950s dreamed of possessing a new car of his own to replace his old Dodge covered with dust, which seemed that every former cowboy hunted for a good horse and released excess hormones while galloping. In 1951, Ford Motor Company built the Ford Kansas City Assembly Line in Claycomo, a suburb of Kansas City. Thus shining Ford Fairlanes went off the assembly line in the city one after another. The cars produced on the Great Plains seemed a bit unwieldy, but they were sturdy and durable. However, when a modern hall was filled with them, and the attractive figure of an auto salon girl dressed in thin clothes was reflected on their surface, all of these could make the rural child confused and excited. He, together with his father and friends, went to the church every Sunday.

S. Hou, *Wallless Cities*,
https://doi.org/10.1007/978-981-97-7827-0_6

Under the cover of darkness, they would quietly sneak into the lobby of the Muhlbach Hotel. What attracted them was not to visit the ancient landmark or to pay tribute to the president of every term who stayed there, but to the authentic Bourbon Whiskey in the lobby bar. The lights from the nightclub not far away illuminated the entire night sky of the city, and the sound of the saxophone sobbed in the distance. For the child from the countryside of West Kansas, what a day it was! When he left by train in the early morning, behind him was everything that was familiar to me, a simple church and an honest farmer. Suddenly, not just one, but innumerable New Worlds of different kinds unfolded before his eyes, each brought him a new spiritual liberation.

For many years to come, the experience of the day will resonate in his mind. He will connect those fragments that have already become blurry in his mind, so as to make the auto showroom exceptionally bright, the model girl extraordinarily graceful, the whiskey in the bar particularly strong, the lighting in the nightclub all the more ambiguous, and the sound of jazz especially sobbing. But what truly lingers in the mind of that ten-year-old child may still be the food there. Every thick, well-textured steak that totters under the bright red sauce, such decadent food can stimulate every taste bud on the tongue; it is simply the paradise of the original sinner. It is a city renowned for its steak and Kansas City BBQ. Some of these rotisseries are run by blacks, such as the famous Arthur Bryant, while others are owned by Italian immigrants and gastronomes. They brought their secret recipe from their hometown, but once they arrived in Kansas City, the hometown of steak, the taste of KC entered their barbecue sauce and blended strange spices with the unique flavor of tomatoes, thus giving birth to the best barbecues in America. Even to the present day, the city still supplies the most juicy and flavorful "KC strip steak" from just a few hundred miles away from it although stockyards were officially closed in 1991. A housewife who had experienced the difficult times of the Great Depression and sandstorms on the Great Plains would never have placed such luxurious food on the dining table. Although the 1950s were already the time when historian David M. Potter of Stanford University wrote *People of Plenty: Economic Abundance and the American Character*, an elite and professor of private university could not clearly imagine the frugality and simplicity of ordinary farm families on the Great Plains.[1] The "other half," who the middle class in New York, San Francisco, and even Kansas City could not understand, not only lived in the slums of New York during the Progressive Era, but also on the prosperous Great Plains in the 1950s.[2]

A large number of Irish immigrants arrived here earlier than blacks and Italians, and they were also the earliest large-scale immigrants. In addition, Eastern Europeans, especially Polacks, arrived here earlier than blacks and Italians as well and along came not only more Catholic churches but also exotic food, which amazed a young man born into a simple Scottish immigrant family. At that time, a large number of Latino immigrants flooded into Kansas City, and they opened a large number of Mexican restaurants along the railway to entertain dark-skinned, dark-eyed railway workers who had crossed the southern border to get here. How can a child who had oatmeal for breakfast and meat loaves for lunch and supper every day imagine that there was such food in the world, corn cake wrapped in red spicy beef, mixed with some rice, added with soft beans, and coated with butter jam. The Country Club Plaza was another scene. It, as the first suburban shopping center in the world, provided a large parking lot since it was built. It achieved great success after opening in 1922 and followed by later urban developers. Such success benefited from J C. Nichols's design, whose architectural style was imbued with Spanish style. That was the place the wealthy upper middle class could afford to spend, but the boy could at least catch a glimpse of the French restaurants, Italian restaurants, British and Irish bars, and even Chinese restaurants with red upholstered chairs and shiny goblets as they passed by, all of which had

[1] David Potter, *People of Plenty: Economic Abundance and the American Character*. Chicago: University of Chicago Press, 1954. In the book *Shrinking the Earth: The Rise & Decline of American Abundance*, Worster had a wonderful discussion about the book. See Worster, *Shrinking the Earth*, p. 206.

[2] Jacob, Riis, *How the Other Half Lives: Studies among the Tenements of New York*. New York: Charles Scribner's Sons, 1890. The documentary album recorded the existing situation of the slums in New York in the 1880s, which initiated the "Muckraking" Movement in the American press circles during the Progressive Era and a series of urban reforms aimed at improving the slums. Compared with the wide sympathy for the slums in the inner city, mainstream public opinion in the United States paid little attention to the "redneck" on the Great Plains. In their description, these were just a group of ignorant and conservative evangelical fundamentalists. However, it is impossible to understand the huge divergences in politics and society in the United States in 2020 if we don't try to understand what these people thought and we only think about the disruptions in the city.

never even heard of in the Western Great Plains. There were drinks on the menu of each restaurant, which children growing up in West Kansas couldn't believe. Although the nationwide Prohibition ended 20 years ago, Kansas to the west of the state border remains the "driest" and devoutest state of Prohibition. No one drank alcohol during meals, and few people knew the taste of red wine, cocktails, vodka, or gin. In the city, they appeared so freely, generously, and openly on people's dining tables. German immigrants' contributions couldn't go unnoticed as they added draft beer and grilled sausages to the exotic food map.

There has been a perception for a long time among the intellectual elites in the United States that Americans, especially Westerners, have a distinct anti-urban tendency. However, such assertions neglect the fundamental fact that Americans constantly build new cities and endlessly recreated them. Just like Kansas City, the natives of the city love it although non-native population often snaps theirs fingers at it. In almost a century since its establishment, it always hovered on the edge of anarchy. Most of municipal and state governments were weak and powerless, and few communities had cohesion. Of course, the strong local aristocracy didn't insist on urban order and planning. Waves of new immigrants came here from all directions. Those from the countryside were often poor, had low levels of education and knew nothing about the various norms of urban life. However, the new life that unfolded before their eyes to some extent generated a strange excitement in everyone's mind, an original motive power for adventure, as felt by that child from the Great Plains. Standing at the confluence of the Missouri River and the Kansas River, they looked westward into its eternity, which might be the reason why the city first appeared on the skyline of the Great Plains.'

2 "Westport"[3]

A traveler started from Boston, crossed New England and New York, then entered Pennsylvania, and continued going westward, passing through Ohio, Indiana, Illinois, and Missouri, and finally reached Kansas City, the

[3] Some contents in the chapter were published in the Collected Papers of History Studies, No. 2, 2021, and greatly adjusted and expanded when they were included in the book. Hou Shen: *The Entangled History between Nature and Cities: The Evolution of the City of the Grass Sea—Kansas City*, Collected Papers of History Studies, No. 2, 2021.

core of the Midwest in the United States, after he traveled 2200 km in total. If his destination was San Francisco on the Pacific coast, he was still required to continue going westward and marching nearly 3000 km so as to breathe the salty sea air when he arrived in Kansas City. No matter what mode of transportation he adopted, on a horse, in a caravan, or by train, plane, or car, the second journey meant he should cross Nebraska, Wyoming, Utah, Nevada, and even California. He would cross the middle dimensions of the western United States, through boundless prairies, high mountain ranges, dry valleys, and deserts interspersed with shrubs. If he were a modern traveler, he would also cross innumerable towns, fields, and pastures of all sizes.[4]

Boston and San Francisco are both port cities open to the world, with the former facing the Atlantic and the latter embracing the Pacific. However, whether departing from the East Coast or traveling from the West Coast, Kansas City is the big city in the United States which is farthest away from the ocean. However, it is also a port, a port of rivers, and its starting point is called "Westport," which clearly defines the self-orientation of the city. Nevertheless, what the port yearns for was not the vast ocean or distant exotic civilizations, but the interior of continental North America, and the fertile soil, forests, prairies, minerals, and animals on it. It would become a city quite different from Boston and San Francisco, and a city that would provide unexhaustible food for the booming country. Thanks to its geographical location, the boosters of the city repeatedly used phrases such as "the heart of America" and "the capital of the west" to call the attention of the east and west coasts, and to remind them that there was a city on the mainland that boomed with the lifeblood of the country.[5]

[4] Before the opening of the Transcontinental Railway in 1869, about half of the people would choose to take a ship from a port on the east coast, head southwards to the Caribbean Sea and arrive in Panama, then walk through the Isthmus of Panama, take another ship along the Pacific coast, and head northwards to San Francisco to avoid the sufferings of crossing the continent.

[5] The Kansas City metropolitan area belonging to both Missouri and Kansas, is bordered by the Kansas River, to the east of it is Kansas City, Missouri, and to the west of it is Kansas City, Kansas. Kansas City, Missouri, developed early and had a large population. It is the center of the city, while The Kansas City, Kansas, was built relatively late and mainly a working-class residential area in the early years, but later developed into a prosperous middle-class community. When it comes to Kansas City in the chapter, most of the time it refers to the metropolitan area, which will be noted in special circumstances, and when it comes to Kansas without the word "city," it refers to the State of Kansas.

In 1950, 100 years after the official establishment of Kansas City, the Kansas City Star, one of the largest newspapers in the city, published a special issue titled *The City of Today Rises from the River Town of Yesterday*. The cover of the special issue consists of two colored pictures that are nested. In the smaller picture below stands a pioneer holding a rifle, and beside him is an aborigine wearing feathers and a caravan drawn by two oxen. In the picture, only his back is firm and straight, in contrast to him, the aborigine beside him is so short-statured and humble. He looked at the Missouri River flowing into the distance, with a small gabbart nearby and a steamboat emitting clouds of smoke heading westwards to the open, free, and undecided land. The riverbank is covered with luxuriant, fragrant grass, and seven or eight two- to three-storied small buildings are arranged in a relatively flat area, with a trace of human habitation in sight. Behind the buildings is a hill with a green scene. The smaller picture is embedded within the larger one that occupies the entire page. In the big picture still stands the back of a man, still with a river running towards the distance, the long grass swaying on the embankment, but a bridge spanning the river from north to south. To the north of the broad river, there is a huge, towering city with gleaming towers. The hill that used to be filled with walnut and oak trees has disappeared, but replaced by high-rise buildings. A modern American city can only be proudly called a metropolis when these buildings outline the city's skyline. The man isn't set off by the aborigine on his side. He is confident and calm, with his hands in the pockets of his suit pants. He looks no longer at the wild and hopeful west, but at the city where the deep chasm has been turned into a thoroughfare by the Hannibal Bridge. The steamboats on the river have disappeared, and the caravans have long been a thing of the past. The former river port, along with the construction of the Hannibal Bridge, has become the center of a dense railway network in the central United States. Moreover, a plane flying towards the city in the sky in the picture marks the arrival of a new era of transportation and seems to declare the end of the era of a river port and cow town.[6]

As the Great Acceleration approached in 1950, the 100-year-old Kansas City was brimming with ambition and life. The natives of the city yearn to tell the history of their city, just like all those who fervently praise their hometowns. They vividly recall the sufferings and glory of

[6] "The City of Today Rises from the River Town of Yesterday,"Centennial, *The Kansas City Star*, June 4, 1950: p. I.

the past, and how they endured great hardships in pioneer work, kept on fighting in spite of all setbacks and created today's prosperity with the intelligence and wisdom of ordinary people. In their narration, a story of victory gradually took shape, with a progressive tone, a fast pace, a democratic temperament, and a free faith. It will be a typical Western story and also a distinctive American story. In the writings and mouths of those who have a deep sense of identification with a city, an individual's urban biography is similar to an individual's biography, while what people are willing to record and spread is the story of successful persons, so are historians, and even more so are ordinary people. Indeed, it is a story about success to a large extent. A small fur trading town that used to have only a few dozen people grew into a metropolis stretching across two states and having a population of nearly 3 million. Over the past century and a half, it provided not only opportunities, hope, and wealth for countless people who had come here one after another but also food, the most indispensable energy source, for America's growth.

However, the story of victory is not always the entirety of history, and the end of travail does not inevitably lead to the final glory. Replacement of victory or progress is not simply decline, but more likely, what we see will be a complicated non-linear story. Many people are already telling this new story. They see how some people pay the price that they should not afford for the success of others, and tell the injustice, setbacks, and failures suffered by the previous individuals and groups. The efforts of these historians have made the history of cities a mosaic diagram, which shows different landscapes when seen from a distance and at hand. It has a complete structure, rich details, and bright color.[7] However, just as

[7] The relevant works concerning Kansas City are concerned about the conditions of blacks, cattle fences as well as workers in meat processing workshops. Works concerning the racial history of Kansas City include Sherry Lamb Schirmer, *A City Divided: The Racial Landscape of Kansas City, 1900–1960*. Columbia: University of Missouri Press, 2002; Kevin Fox Gotham, Race, Real Estate, and Uneven Development: The Kansas City Experience, 1900–2010. Albany: Sunny Press, 2014; G. S. Griffin, *Racism in Kansas City:A Short History*. Traverse City, MI: Chandler Lake Books, an import of Mission Point Press, 2015. There are many articles on the situation of stockyards and black workers in meat processing workshops, the representative ones include Darryn Snell, "Meat Packaging, Race, and Unification in Kansas City, 1880–1904," in *Global Humanization: Studies in the Manufacture of Labour*, ed. by Michael Neary (London: Mansell Publishing, 1999), 127–63; John Herron, "Making Meat: Race, Labor, and the Kansas City Stockyards," in *Wide-Open Town: Kansas City in the Pendergast Era*, eds. by Diane Mutti Burke, Jason Roe, and John Herron, Lawrence: University of Kansas Press, 2018.

history is not just about the success stories of a certain person or some group of people, it is also not just about the past of "personal" success or failure. The cost of success is often borne by non-human organisms and the environment that they rely on for survival, and their changes in turn can affect the way human beings, which they rely on for survival, and their creations, such as the survival mode of cities, historical trajectories and power relationships.

The growth process of Kansas City is largely a replica of Chicago on the Great Plains. It is 800 kilometers away from Chicago and, like its big brother Chicago, tells its story by returning to the Great West, tall grass and shortgrass prairie, corn and wheat farms, as well as stockyards and meat processing workshops. The story to be told in this chapter is about a city located 800 kilometers away from Chicago, which almost shares a common hinterland and some similar experiences. But in the story of Kansas City, what we will see is not a grand metropolis that used capital as a powerful tool to crush and conquer the ecology of its hinterland, or rather more than just such a story; In the past ecological experiences of Kansas City, the power of nature didn't eliminate, and its changes always influenced the changes in urban power and economic form.

Kansas City, was born and grew in the central part of the vast continent, is located at the confluence of the Kansas River and the Missouri River. The latter with the nickname of "the Big Muddy" is the longest river in North America. It originates from the snowy summit of the Rocky Mountains, and winds southeast with sand and mud carried before drains

Open Town: Kansas City in the Pendergast Era. also discusses issues related to other marginal population, such as women and Hispanic immigrants. It is worth noting that in 1939, the Urban League of Kansas City (ULKC) issued a report on black workers, which provided important records of the working environment at that time. The Urban League of Kansas City, as a branch of the National Urban League, was established in 1919 to promote the development of civil rights, education, survival and economic conditions for African Americans. Urban League of Kansas City, *The Negro Worker of Kansas City: A Study of Trade Union and Organized Labor Relations.* Kansas City: Urban League of Kansas City, 1939. In addition, for research on gender history in Kansas City, see Amber R. Clifford Napoleone, *Consulting Kansas City Jazz: Gender, Performance,and the History of a Scene.* Lincoln: University of Nebraska Press, 2018. Amahia Mallea's book *A River in the City of Fountains: An Environmental History of Kansas City and the Missouri River* is an environmental history work about Kansas City, it mainly focuses on urban water pollution and its governance. However, in the book, Amahia specifically discusses the issue of environmental justice. Amahia Mallea, *A River in the City of Fountains: An Environmental History of Kansas City and the Missouri River.* Lawrence: University of Kansas Press, 2018.

found three deer swimming in the river, which were hunted by the accompanying indigenous hunters. When another group of hunters returned, they brought eight more deer and two wolves, one dead and one alive. On the first day after they arrived, Clark saw a large group of now extinct parakeets. He also recorded the situation of the "bellicose" Kansa, who were hunting bison on the Great Plains at that time, and the members of the expedition also saw bison while hunting deer. Either side of the Missouri River and the Kansas River was lined with mountains and cliffs, but the diary didn't describe the vegetation on them. Clark only pointed out that it was a good place to build forts and unload goods.[10]

There were no clues in the Journals of the Lewis and Clark Expedition that could inspire people's imagination about the future of the land. After three days of rest, they traveled up against the current and continued their exploration of the Missouri River basin. During the long journey from the origin of the Missouri River to the Mississippi River, they saw countless tributaries, big and small, converge into the rushing torrents. However, what was the special significance of the Kaw Point? The expedition did not inquire or answer the question. In 1804, the young United States just kicked off the Western stage, and no one among them knew what potential the flat land unfolding their eyes contained and what possibilities it would bring to the country. For President Thomas Jefferson who completed Louisiana Purchase and his contemporaries, and even for the next two generations, regardless of North or South, their world is a humid, verdant, and forested eastern North America. However, the undecided areas in the west of the Mississippi River, especially those in the west of 98° W, were desolate, arid, and treeless but unsuitable for farming. A ghost of "the Great American Desert" began to emerge above the imagination of the public of the United States and solidified into a basic understanding of the region. The transformation from the "Great American Desert" to the "American Garden" required decades of exploration and understanding. Therefore, what followed the Louisiana Purchase was

[10] *Journal of Lewis and Clark Expedition*, June 26–28, 1804 https://lewisandclarkjournals.unl.edu/item/lc.jrn 1804-06-28, which was logged in on August 16, 2020. Kaw Point has now become Kaw Point Park, located on the Lewis and Clark National Trail. The military fort proposed by Clark was not put into practice here, but Fort Osage was built about 50 kilometers away from the place in 1808 to control the local indigenous tribes. However, in the 1820s, the fort gradually became abandoned with the retreat of the Osage.

not the agricultural empire that President Thomas Jefferson yearned for, but a huge commercial wave.[11]

Kansas City is the first to be affected by the wave of collision between capital and nature. It is here that the Missouri River no longer flows southeast, but takes a big turn and runs eastward and joins the Mississippi River. The turn means that over the next half century, men and women who aspired to travel southwest would find it the most convenient place to abandon ships and change their mode of transportation. Initially, it meant that it was a natural depot for people who headed to Santa Fe, New Mexico. Those Eastern adventurers who hoped to trade with aborigines and Mexicans took a boat to Kaw Point, where they purchased caravans and livestock, crossed the Great Plains, then headed southwest, and finally arrived at Santa Fe. The famous Santa Fe Trail was formed with the blessing of caravans and rifles. Soon, a new vision emerged, with the Oregon Trail leading to the Pacific becoming a path for more people to travel westward, but they shared a common starting point, Kansas City and Independence City adjacent to Kansas City. Especially since the start of the gold rush in 1849, more caravans carrying people's yearning for the future, wealth, and freedom set out from here to the west so as to achieve the "Manifest Destiny" of the country and individuals. The wave brought waves of immigrants towards their dreamlands, until the Transcontinental Railway was completed in 1869, and the procession of caravans slowly disappeared from the Great Plains. Before this, at least 400,000 people traveled westward through the Oregon Trail, they performed countless tragicomedies that were told, praised, and analyzed by people of all kinds. And what was the ecological role played by Kansas City, a small town with a population of only 500 in 1855?

It is not just the best geographical transfer site, but equally important is its grassland ecology. Whether these immigrants chose Santa Fe or Oregon Trail to explore the wilderness in the west, their caravans would need to use thousands of cows, horses, and mules, which in turn required food. The prairies surrounding Kansas City were their natural pasture for reorganizing, resting, and wintering. James R. Shortridge, the most important biographer of Kansas City, believed that these were one

[11] The construction and deconstruction of the myth of the "American Desert" has always been an important topic in Western history, but so far, the most exciting discussion is still Webb's classic work: *The Great Plains*. Walter Prescott Webb, *The Great Plains*. Boston: Ginn, 1931.

of the three most important geographic features that made Kansas City a metropolis, and also the one that was neglected easiest.[12] However, what Shotridge and other historians or geographers who studied Kansas City also neglected was that the prairies of Kansas City belonged to a vast plain that stretches westward, and the vast "sea of grass" that stretches south to Texas and west to Colorado, including Kansas, Nebraska, Oklahoma, and Wyoming, was the most important ecological feature of the city.

3 Looking Out to the Hopeful West

Half a century after the Lewis and Clark expedition, the image of the "Great American Desert" began to fade away. Although it was still regarded by the people at that time as a wilderness with bison and aborigines, it had the potential to become a "garden." What needed to be done now was how to tap into its potential and make the "desert blossom like a rose." In 1854, the Kansas-Nebraska Act was signed with a clear political purpose that the Missouri Compromise was put to an end and newly added Western states were turned into slave states, ultimately leading to a series of sanguinary conflicts between supporters of slave states and of free states on the Kansas-Missouri border.[13] However, regardless of the political stance and moral demands of the opposing sides in "Bloody Kansas,"

[12] James R. Shortridge, *Kansas City and How It Grew, 1822–2011*. Lawrence: University of Kansas Press, 2012: pp. 6–7. Shotridge believed that the other two features were the geographical location that Kansas City is located at the confluence of two rivers and the geologic features of its limestone formation.

[13] In 1820, there was a debate in the United States over whether to introduce slaves to new states and the Missouri Compromise was passed at last, which stipulated that slavery was not allowed to enter north of the 36° 30′ parallel, except for Missouri. But after the end of the Mexican–American War in 1848, California applied to join the Union in 1849, which sparked a new round of debate between free states and slave states. In the end, in 1850, the Congress passed five relative acts, collectively known as the Compromise of 1850. These acts allowed adding California to the Union as a free state, but the United States began implementing stricter Fugitive Slave Act. The Compromise of 1850 did not appease the dissatisfaction of the Southern states, which hoped to continue expanding slavery to the west, and they, in 1854, impelled the Congress to pass the Kansas-Nebraska Act, which stipulated that 36° 30′ parallel no longer applied, and newly added States could decide voluntarily to become slave states or free states. The Act sparked anger among pro- and anti-slavery advocates of Northern free states, and multiple armed conflicts erupted at the border between Kansas Territory and Missouri, which is known as "Bloody Kansas". Eric Foner, *The Fiery Trial: Abraham Lincoln and American*, translated by Yu Liuzhen, Beijing: The Commercial Press, 2017.

they shared a common belief that the future President Lincoln expressed in his Peoria Speech made in 1854. While refuting the logical fallacies and moral absurdity of the Kansas-Nebraska Act, he said, "The whole nation is interested that the best use shall be made of these territories. We want them for the homes of free white people..."[14] In fact, they also became the home of some free black people after the end of the American Civil War. As for how to make full use of them and make them become the home of free white people, they were the questions answered by President Lincoln in the Homestead Act of 1862.[15]

In Kansas City, commerce was still the basic means of creating wealth, but at that time, what commerce relied on were no longer furry creatures bred by nature. In the greed of human nature, the support of capital, the lure of money, the stimulation of whiskey, and the consummate hunting and fishing skills of aborigines, those animals favored by hunters for their beautiful and warm fur, for instance, beavers, were hunted to the death. The ambitious and determined country needs a more solid material foundation, and its business couldn't continue to rely on the products that nature made carelessly and inefficiently. Human hands shall intervene in nature so that what nature will produce and in what manner, speed and cycle the products are made shall be decided by human beings. It is the basic logic of agriculture and animal husbandry. In a traditional agrarian society, farmers, after paying taxes, provided their products to meet the needs of themselves, their families, and their communities. However, the agriculture and animal husbandry that was about to emerge in the Great Plains would be the world's first agriculture and husbandry which focused on commerce and market from the beginning. The capital chain would be embedded between the labor of farmers and the will of the country as well as human demand and land supply. It would change the agricultural ecology and urban landscape of the Great Plains, where Kansas City is just located.

[14] As cited in Eric Foner: *The Fiery Trial: Abraham Lincoln and American*, p. 81.

[15] During the stalemate of the Civil War in 1862, the Homestead Act, which played a decisive role in the fate of the American West and the lives of thousands of men and women, was enacted. Afterwards, with the signing of a series of related acts, nearly 1/10 of the territory of the United States was eventually granted free of charge to nearly 1.6 million homestead farmers, thus ensuring their pursuit of freedom and wealth. A large amount of land was gifted to railroad companies to encourage private enterprises to join the country's camp and work together to achieve the so-called "Manifest Destiny" of the State, a continental empire that comprehensively developed nature.

Compared with other American cities, Kansas City stood at the forefront of the Westward Expansion of the United States and was the city which was most clearly defined by the transformation of the ecological and economic roles of the Great Plains. The formation and development of Chicago and St. Louis, especially during their takeoff phase, was more closely related to capital and markets in the Eastern United States and even foreign countries. The construction of the Erie Canal and the dense railway network that emerged first in the Eastern United States closely connected Chicago with New York, which laid the foundation for Chicago becoming the largest city in the Central United States. As Cronon said, Chicago, a global metropolis, had to rely on the exploitation of the natural wealth of the vast Western United States at last. However, the development of the Great Plains emerged after the formation of Chicago, which was already a growing city with clear and easily accessible prospects at that time. In the competition for the largest city in the Midwest, St. Louis was Chicago's strongest rival. It is located at the confluence of the Mississippi and Missouri rivers, so in the age when transportation was still based on waterways, no inland city could match it in terms of geographical advantages. Anyone with common sense of life would decide to settle here even though they didn't have much wisdom or foresight, so did the Cahokia residents of that day, and Europeans who came here later were no exception. When the fur trade began to shift towards the interior of North America, Saint Louis immediately became the center of inland trade and the stronghold of the fur dynasty established by the Chouteau family in France. Before the development of the Great Plains, Kansas City was established by marginal men of the Chouteau family through their commercial network. However, as people began to eagerly explore the new wealth of the Great Plains, St. Louis did not have such an intricate relationship in the east as the one with New York that was established by Chicago through the construction of the Erie Canal and was too far from the Great Plains in the west. The emergence of railways made its advantage in waterways diminish. Philadelphia, its alliance in the east, fell into decay with each passing day. Thus St. Louis was no longer unable to compete with Chicago.[16]

Kansas City occasionally wanted to compete with Chicago, but most of the time, it willingly followed Chicago, learned from and imitated

[16] William Cronon: "Gateway Rivalry: Chicago and St. Louis", Section IV, Chapter VI of *Nature's Metropolis: Chicago and the Great West*, pp. 415–436.

Chicago with it as a model, and enjoyed the protection and support of the latter. In addition, it often envied and occasionally hated Chicago. Kansas City was different from Chicago whose rapid rise first relied on the market of the Eastern United States, the development of agriculture in the Western United States was the key to the transformation of Kansas City from a trading hub, a minion of the fur trading dynasty of St. Louis, to the largest city in Missouri (including its parts in Kansas). In the 1850s, boosters of Westward Expansion were well aware of that, and the *History of Kansas and Nebraska* published in 1857 voiced support for the possibility of the New Westward Expansion. The book had a popular lengthy subtitle at the time: *Describing Soil, Climate, Rivers, Prairies, Mounds, Forests, Minerals, Roads, Cities, Villages, Inhabitants, and Other Subjects Relating to that Region*, with a Correct Map, and the title page was the publicity calling for people to immigrate to Kansas City. In the book, the author Walter Sloan painstakingly deconstructs the image of the "Great Desert" in Kansas and the Great Plains from all angles such as local soil, rivers, climate, and the richness of flora and fauna to attract pioneers from the East. He wrote, "The climate in Kansas is similar to Missouri, Kentucky and Virginia, but slightly drier. The soil here can produce perfectly all grains, forage grass, vegetables and all the fruits that originally came from the central states. ... there is an abundance of excellent wood along the banks of the river for more than 100 miles from the west of the Kansas River estuary. ... Many of the oaks are five or six feet in diameter, and the cotton-wood is often even larger than that." In a chapter dedicated to Kansas City in the book, the author wrote: "This ambitious little city, now striving to her utmost powers for the whole of the emigrant, Santa Fe, and California trade..."[17] The author of the book foresaw the bright agro-pastoral prospects of Kansas, Nebraska and the Greater Plains to the west, but obviously, he still did not have a clear understanding of the ambitions of the small city with an overwhelming desire.

Robert T. Van Horn, one of the early leaders of the city, looked forward to a far brighter future. Van Horn arrived in Kansas City from Ohio in 1855. He was right in his early thirties, and took a heroic posture, while Kansas City just recovered from the cholera epidemic of 1850 and

[17] Walter B. Sloan, *History of Kansas and Nebraska: Describing Soil, Climate, Rivers, Prairies, Mounds, Forests, Minerals, Roads, Cities, Villages, Inhabitants, and Other Subjects Relating to that Region, with a Correct Map*. Galesburg, ILL: Boishel, Kuhn, & Co., 1857: pp. 10, 12–3, 17.

was back in the saddle. At that time, its population was only 478 people, but boosters of Westward Expansion had endorsed the small town as a starting point for exploration in various ways. Encouraged by the New England Emigrant Aid Society, painter F. Buckeridge created a panoramic view of Kansas City, which is believed by later generations to be a true portrayal of the city at that time, and also be the scene that Van Horn saw after arriving there.[18] Half of the picture is occupied by the broad Missouri River, and the most conspicuous thing on the river is naturally the steamboat, the main means of transportation on the Missouri River and a symbol of prosperity and progress. But even people at that time were well aware that they also brought plagues and death. The more developed the commerce and the denser the population, the faster epidemics spread. The case was seen especially in river port cities, and Kansas City was no exception.[19] Van Horn himself temporarily helped his relation by marriage with the operation of steamboats, so like most Eastern immigrants at the time who hoped to enter the West through Oregon, Santa Fe, or California Trails, he took a steamboat and headed westward along the Missouri River to Kansas City. Unlike most people, he wanted to run his own business here from the beginning, so he successfully bought a newspaper firm that was just established and published the first newspaper in the town: Kansas City Enterprise.[20] Looking south from the steamboat, one could see Kansas City. Most of the buildings

[18] The nested picture on the cover page of centennial review of the Kansas City Star 1950 was also designed on the basis of the picture and published in the second edition.

[19] Charles P. Deatherage, *Early History of Greater Kansas City, Missouri and Kansas, the Prophetic City at the Mouth of the Kaw*, Vol. 1. Kansas City. Interstate Pub Co., 1927, pp. 371–5. Deatherage originally intended to write three volumes of the book, but only the first volume was published at last, and the other two volumes still remained manuscripts.

[20] When the newspaper started its publication, it was named *Kansas City Enterprise*. In 1857, after Van Horn took control of the newspaper, it was renamed *The Western Journal of Commerce*. In 1858, it became *The Kansas City Daily Western Journal of Commerce*. It was later renamed several times, and in 1938, it was renamed *The Kansas City Journal* again. After the establishment of *Kansas City Star* in 1881, these two newspapers competed for a long time, and *The Kansas City Journal* ultimately lost and ceased its publication in 1942. The fact that the newspaper continuously changed its name implies a shift in the positioning of Kansas City in a sense, namely, it turned to a comprehensive large city from a commercial town entirely dominated by trade at first. During the Civil War, due to the Republican background of Van Horn, the newspaper firmly supported the Union. Van Horn himself was elected as the mayor of Kansas three times in the 1860s, which had a significant impact on the early development of the city.

along the river bank were warehouses, which indicated the port position of the city. Behind the harbor stretches the cliff several hundred feet high, there was a dense forest on it, but by that time several residences could already be seen.[21] In the coming decades, as Kansas City continued to expand its hinterland westward, it would also cut down its forests and level most of the cliffs to build a city. Charles Deatherage, the author of *Early History of Greater Kansas City, Missouri and Kansas, the Prophetic City at the Mouth of the Kaw*, was himself a local timber dealer. Therefore, his works extensively describe the forest conditions here.

Deatherage might have been born in 1852 as a native of the city. Since founding his own timber firm in 1878, he was a beneficiary of the city's economic development, a developer of resources, and a steadfast booster of Kansas City.[22] He witnessed the rise of the city, so when he retired in 1919 and wrote the history of the city, he recorded it as a witness rather than a researcher. It was a continuation of the tradition that the British gentry wrote local history, in other words, they retired and began writing history after achieving success in their careers and living the life of Riley. Such a situation was quite common in the New England region, and it was more or less inherited in the Midwest region. These kind of works largely played the role of local chronicles.

In *Early History of Greater Kansas City, Missouri and Kansas, the Prophetic City at the Mouth of the Kaw*, Deatherage cited several pages of an address titled "The Spirit of the Times" made by Van Horn at a Christmas dinner in 1857, in response to the toast: "Railroads and the Press: Twin Brothers in American Progress and Development." Van Horn made his address passionately in exquisite wording. The message he conveyed was a voice which the people at that time were familiar with: the railway would sound the horn of progress and implement democratic systems. "Railways involve a philosophy in the progress of the world ..." he told his audience that in ancient countries that heavily relied on human and animal labor for transportation, "courts built cities, and the decrees of despotic rulers obliged whole empires to pay tribute to their licentious capitals. ... Here, the people establish their own commercial

[21] For the description of the picture and the situation of Kansas City in 1855, see *Early History of Greater Kansas City, Missouri and Kansas*. Deatherage, *Early History of Greater Kansas City, Missouri and Kansas*, pp. 384–5.

[22] The character information comes from the introduction of the collection of Charles P. Weather Book Manuscripts at the Kansas City Public Library.

capitals, ... New York, Cincinnati, Chicago, St Louis and New Orleans are the metropolitan cities of our people." God fixed the centers and marts of commerce by the ranges of mountains and course of rivers. "it is by studying these tracings of the Almighty's finger that the pioneer of trade and the herald of civilization has selected the site of these gigantic cities of the Republic, and which has fixed upon the rock-bound bay of the Missouri and Kansas as the last great seat of wealth, trade, and population in the westward march of commerce toward the mountain basins of the Mississippi and Pacific," that was Kansas City. He spoke forcefully, "Since the days of Columbus, commerce and enterprise has been seeking the west. West, west, has been the watchword over the Atlantic, up the Potomac, across the Alleghanies, down the Ohio, over the Mississippi, up the Missouri. It is found at last. Kansas City stands on the extreme point of western navigation—it is the West of commerce: beyond us the west must come to us overland* I say again—the west at last is found." After the speech, the audience broke into enthusiastic and prolonged applause throughout the room.[23]

In fact, similar words and phrases almost repeatedly appear among city boosters in the cities west of the Mississippi River, such as St. Louis, Chicago, Omaha, and Denver. But Van Horn did not just harp on the same old tunes, on the contrary, he keenly saw the unique location of Kansas City at a time when the railway era was approaching—it was located in the geographic center of the United States. Based on that, he told people that when railways were open up to Kansas City from all directions, Kansas City would be "a mine of wealth unsurpassed by any city in the world, bringing within seventy hours of each other the cotton, sugar and stock of Texas, the robes and furs of the plains and mountains, the manufactures of the East, and the lumber and copper of the Mississippi and Lake Superior." Finally, he listed the miracles created by Chicago and confidently declared, "Let the world know of as it did of Chicago, that here is the commercial center, fixed by the laws of nature herself, and the capital of the world will stretch out its iron arms for our commerce... Let us work westward—that is the word [westward] for Kansas City...." Once again, he received enduring cheers from people.[24] On Christmas

[23] Deatherage, *Early History of Greater Kansas City, Missouri and Kansas*, p. 573.

[24] Deatherage, *Early History of Greater Kansas City, Missouri and Kansas*, p. 574.

Eve in 1857, the business elites in the city gathered together to celebrate. Although the rift between free states and slave states was deepening, Kansas City was deeply trapped in the gap and had felt the shadow of civil war through sanguinary conflicts. However, they still saw a rich and fertile land full of hope when they looked westward. They firmly believed that God's plan and the laws of nature would make a city rise, which was their "Manifest Destiny."

4 When Capital in the East Encounters Cattle in the West

So what could people see when looking westwards from Kansas City? Whether on that day or now, what people see is a vast plain, which is so flat that someone later compared it and found that the terrain of the Great Plain is even flatter than the surface of corn muffins under a microscope. There were no lofty mountains, peculiar canyons, lush forests, or running waterfalls, but it had a vast open sky. At the moment of sunset, the sky was filled with strange hues from red to violet in the spectrum of light. On a thunderstorm night in the summer, narrow silver lightning flashed into the dark wilderness from the dark blue sky and the light on the horizon shone with soul-stirring beauty. Today, standing at the confluence of the Kansas and Missouri Rivers, you can see a crisscrossing railway network and Interstate 70 that runs through continental North America, and on which 16 Wheelers speed in an endless stream. You can also see Johnson County, a streetcar suburb that formed after Kansas City spread to Kansas. The County with flat lawns and single-family mansions has long been listed as the wealthiest county in the United States. Going further west, you can see large, neatly-rowed corn fields as well. Under the scorching sun, maize leaves are slightly yellowed and withered, and small planes are hovering above them to spray pesticides. In the famous film *North by Northwest* directed by Alfred Hitchcock, Cary Grant narrowly escaped machine-gun fire in such a landscape. Further west, there are more irrigated farms and pastures, which have the same green color in the breeze. The sturdy and similarly-sized herds of cattle are almost at rest, where crops and cattle create a unified, repetitive, and monotonous industrial landscape. Of course, you can also see more suburban areas of various sizes, with wide and narrow highways at all levels. It is the "Manifest Destiny" that today's American Midwest, as well as the elites in Kansas

City and innumerable Easterners who advocated Westward Expansion, hoped to achieve on the eve of the Civil War.

However, such a western region only appeared in the vague imagination of a few political and commercial elites in the 1850s, and even they had no way of predicting what would happen there. When Van Horn and his companions looked westward, the residential area of the upper middle class, which was known today as Johnson County, remained the reservation of Shawnee who retreated from the Mississippi River basin, but soon they would also permanently left there and headed towards the more desolate Oklahoma. No railway line arrived here, let alone Interstate 70 that ran through the East and West, but the Santa Fe and Oregon Trails were still busy at the time, which lit up the shining business opportunities in the eyes of traders in Kansas City. Most importantly, they saw such vast and fertile prairies on the Great Plains. Indeed, aborigines of various ethnic groups meandered and lived there for thousands of years, but they had never become the keystone species in the ecosystem of the Great Plains. Here or there, they burned forests and prairies, cultivated small plots of corn fields, and pitched several tepees, but their labor and livelihood did not change the natural rhythm and the composition of the diversity of species here.[25] The seemingly monotonous prairies of the Great Plains were home to hundreds of different grass species, such as buffalo grass, sleepy grass, rough bent grass, blue stem, purple three-awn, downy wild rye, and plain muhly, which vary in height, shape, and color. As Worster, the historian of the Great Plains, once said, "Complexity, adaptability and beauty are the ways in which all the organisms in nature meet the challenges on the Great Plains."[26] Although fur traders had founded profitable companies in St. Louis and Kansas City and fur hunters shuttled through the grass leaves on the Great Plains, Millions of bison still lived on the Great Plains in the 1850s. While their companions with lush fur were killed by rifles and traps, these huge animals were still chewing on the tender and juicy bluestem grass. Their rough fur couldn't

[25] Whether the indigenous tribes on the Great Plains are ecological is a pseudo question, it isn't discussed here. What the book emphasizes is that they were only one of the participants in the local ecosystem before they were in contact with each other, rather than the keystone species. As for whether there is the "Pleistocene extinction" as geologist Paul Martin said, no final conclusion has yet been reached on this matter in the academic circles. For the discussion on "ecological Indian," see Shepard Krech III, T*he Ecological Indian: Myth and History*.

[26] Donald Worster: *Dust Bowl: The Southern Plains in the 1930s*, pp. 85–88.

match the delicate skin of noblewomen and men in cities, but only the belts of the machines in the factories in the East needed their thick and rough leather which used to serve as teepees to shelter aborigines lived on the Great Plains from wind and rain.

Of course, these overlords of the Great Plains would eventually disappear, which was not because the profit of their fur and flesh was meager, but because they had to make way for their distant relatives from distant continents. Although the Great Plains was vast, it could not accommodate the coexistence of bison and domestic cattle. How can bison or American buffalo be as pleasing to the ear as cattle? The latter sounded so similar to capital, and it also proved that it would not disappoint the holders of capital. Moreover, the place would become not only a vast pasture, but also a bread basket for the world. Ploughing up the land of the Great Plains and weeding out the native grass seeds would make it become a home for wheat and corn. As the boosters said, Kansas City, located in such a position, was "destined" to become an agricultural capital. In the introduction of the book *The Leading Industries of Kansas City: A Review of the Manufacturing, Mercantile and General Business Interests of the "Gate City of the West" To Which Is Added a Historical Sketch of Its Rise and Progress* published in 1882, the author enthusiastically repeated the praises that can be seen everywhere in the city's publications in the nineteenth century: "At the entrance of a glorious territory, there is a vast territory, with not only incomparable fertility and productivity but also superexcellent and beneficial climate, as well as close to the sum of the territorial area of several countries in the Old World, in Kansas City, the centropolis of our great republic. It is situated on continuous hills and overlooks the magnificent area, inviting the best nations and people in the world. It is almost located in the geographic center of the United States. The center distance obtained by the precise calculations from any direction is only a few miles. Moreover, an undeniable fact is that it happens to be located in the agricultural belt of the continent. Its neighbors have the best livestock territories in the world. In addition, there are famous lead, zinc, and coal mines in Missouri and Kansas, as well as lead, coal, iron, silver, and gold mines in New Mexico bounded by Colorado."[27] In the annual report compiled by the Board of Trade

[27] *The Leading Industries of Kansas City: A Review of the Manufacturing, Mercantile and General Business Interests of the "Gate City of the West." To Which Is Added a Historical Sketch of Its Rise and Progress*, Kansas City: Reed & Co., 1882: p. 3.

of Kansas City from 1878, the same information appeared in the annual report in relatively simple language: "In the past year [1878], the weather was good for the crops all the year round, and farmers had a bumper harvests. Their poultry and cattle were increasing, the area of fertile land we cultivated was also expanding. Immigrants, which were estimated to have hundreds of thousands of strong and energetic hardworking people, were seeking opportunities on the fertile prairies to add their wealth, strength and efforts to the productivity of the existing community. This year, we also fully shared the dividends of the construction of American railway.... Although our market prices were brought down, goods were flowing freely and money was quite adequate. In the city [Kansas City], there is also a satisfactory situation, because the relationship between Kansas City and the New West is so close that they acted in cooperation with each other. Whatever affects one side, the other will quickly be affected in the same way."[28]

With the vast hinterland, the boosters of Kansas City had the backbone to lobby the tycoons who controlled capital in the East and attract their wealth. Chicago had already provided a template for how to develop, so Kansas City was very clear about the significance of railroads for modern cities. The best position of its waterway was still its advantage to brag about, but its practical value had greatly decreased in the railway era, especially under the condition that the Kansas River was not a suitable river for steamship navigation, so Kansas City's advantages in waterway were limited. If you go up the Missouri River, like Lewis and Clark in the past, the Great Central Plain rich in milk and wort has little to do with the city. Therefore, only the railway was the foundation for the rise of any small town in the Midwest. Van Horn's vision in 1855 was interrupted by the long sanguinary conflict within Kansas City and the Civil War which subsequently broke out nationwide. But as soon as the civil war ended, they began to go around advocating the incomparable geographical location of Kansas City and the firm and resolute wisdom and spirit of cooperation of commercial elites in the city. After all the rivals located in Kansas and Missouri were defeated, Kansas City joined hands with the capital of New England to build the Hannibal Bridge, the first permanent bridge on the Missouri River. It was expected and designed to stand

[28] William H. Miller, *Second Annual Report of the Trade and Commerce*, Completed for the Board of Trade, Kansas City KC, MO.: Peter H. Tiernan, Printer and Binder, 1879: p. 9.

forever at least at that time. However, a tornado destroyed its middle section in 1886 (so it was replaced by the Hannibal Bridge which was built 60 meters upstream later), which officially launched the process of Kansas City becoming the center of the railway network on the Great Plains in the future.[29]

The construction of railways directly triggered a similar process as Chicago. The city would become a hub for the transfer, processing, and distribution of livestock and food. Unlike Chicago, wood was of little importance in the city's commerce. Although there were still many forests along the Missouri River basin and even in the west along the Kansas River when Kansas City was first founded, the essence of the Great Plains was prairie, so tall trees were not a dominant species here. However, for the city, the huge profits from livestock and food were enough to make its business elites complacent. Thus, the driving force of economic development came from capital once again. In 1869, the same year as the completion of the Hannibal Bridge, Charles Francis Adams Jr. began to ponder in his mansion in Boston about where his money to be invested should be invested, Kansas City was selected at last. In 1871, he built the stockyard of the city, which was then only 13 acres and located in the West Bottoms of the Kansas River, west of the state border. From then on, not only the political, economic, and social landscape of the city changed fundamentally, but the ecology of the new western region closely related to it also changed.

Few people would associate Adams with those smelly, furry, gentle, and bulky guys. His great-grandfather was John Adams, the Founding Father, the second President of the United States, and one of the authors of the Declaration of Independence. His grandfather was John Quincy Adams, the sixth president and a political, economic, and cultural elite, who came from a family of Boston Brahmins. Although the United States does not have a nobility, there is no doubt that the Adams family exists just like a nobility. Shortly after Charles graduated from Harvard University, the American Civil War broke out. After the end of the War, he hesitated for a short time, but soon became clear about what he should do, and what was "the most developing force and largest field of the

[29] For the construction of the Kansas City Railway and its relationship with the development of the city, see Charles N. *Glaab, Kansas City and the Railroads: Community Policy in the Growth of a Regional Metropolis*. Lawrence, KS: University Press of Kansas, 1993.

day," whose answer was Railroads, and he determined to attach himself to it. He not only served as a director of several railroad companies, but also wrote numerous articles related to railroads and canals as a historian. He saw the infinite business opportunities in the West and knew that the "most developing force" in his eyes would be ample scope for giving full play to his abilities in the West, but it did not hinder his disdain for the cultural and economic status in the West. In 1860, on his first journey to the west of Niagara, he complained about the backward transportation, the crude facilities and the rough service. Apart from the sight of torches illuminating the night along the upper reaches of the Mississippi River, which made strong impression on him, he wrote, "Kansas was the most interesting of the regions we visited."[30] But he didn't explicitly state what made him interesting. It is hard to imagine that he would be tempted by the setaria viridis still swaying in the wind at that time, but perhaps the breathtaking sunset he saw on the Great Plains, or the golden sunflowers blooming in the vast green made him fall into a short trance. Adams was not without the ability to appreciate the beauty of nature. On the contrary, he has a deep affection for the mountains, rocks, and trees around Boston, and played a crucial role in the establishment of the Boston Metropolitan Park System. He once served as the chairman of the Massachusetts Parks Commission.[31] In this autobiography written 55 years later, he did not mention the landscape of the West. However, when speaking of Kansas, he placed in the same category the sanguinary conflict between slave states and free states that Kansas experienced after becoming a quasi-state and the recent drought here, writing: "Kansas, the always either bleeding or starving. ...in 1860 but sparsely settled, and the unfortunate inhabitants were suffering under a prolonged and very destructive drought. There seemed no end to their misfortunes."[32] Perhaps the passage was just a copy of the journal he wrote then, but when he reread it in 1915 and looked back at the land and people that had been changed by the stockyard he and his capital had built, what did he really feel?

[30] Charles F. Adams, *An Autobiography, 1835–1915*. Boston & New York: Houghton Mifflin Company, 1916, p. 63.

[31] Namely, the Boston Metropolitan Park System discussed in Chapter III of the book.

[32] Charles F. Adams, *An Autobiography*, p. 63.

Adams did not conceal his complacency in his achievements in Kansas City. He wrote,

One great business success I did achieve; and it is the only one on which I can fairly plume myself. Going into it at its inception, in 1869, I have for over forty years been at the head of the Kansas City Stock Yards Company, directing its policy and development. When I became President, it was a concern of 100,000 capital, earning, perhaps, 20,000 a year gross. From this I stage by stage have built it up, always its President, until to-day [1915] it is capitalized at above ten millions, and earns annually over 1,200,000 gross. The second largest institution of the kind in the world in all these years it has missed but a single quarterly dividend; and that owing to a cataclysm—the Kaw Valley flood of 1903, which in three days swept away at least 600,000 of the values of enterprises of which I was the originator, and in which I was personally most largely interested. Financially, it was for me as a brick falling on the head as I walked along a familiar street. The loss was never recouped. Managed in a broad, liberal spirit, the Kansas City yards have been a great public benefit as well as a considerable commercial success. ......[33]

Like many entrepreneurs like him, especially those in Boston, Adams sincerely believed that the company he invested in was a great good business, which created such large profits for this ambitious city and provided jobs and bread, possibly even meat loaves, for thousands of poor immigrants who came to the city to seek means of livelihood. He never recorded whether he had ever been to the damp depression, the stockyard built beside the rumbling railroad, and every small cowshed was jammed with cows, pigs, and sheep which waited to be slaughtered. In the era when animal power was not completely replaced by motor vehicles, there were still a large number of horses and mules in the area. Of course, he didn't visit the places where the workers who benefited from his charity worked day and night. The air around the stockyard always emitted the unique odor of feces, sweat, and cattle. Compared with his confidence in the public welfare of the company, what made him feel blinder was his repeated emphasis on the decisive role of his own efforts in the construction of the second largest stockyard in the United States. Adams was more willing to be identified as a historian and leader in public affairs. Obviously, he had the same mindset on how to treat the rise of a business

[33] Charles F. Adams, An Autobiography, pp. 187–8.

chaired by him as all entrepreneurs: the power of the mind of a great individual which guided the intelligent flow of capital was the key to success. However, the soil, water, grass, cattle, and thousands of immigrant labor in the West played at least equally important roles in the management of stockyards.

Anyway, as Adams said, it was a great success. In 1891, after the 20th anniversary of the establishment of the Stock Yards Company, the annual report on commerce in Kansas City reviewed the development of the city's animal husbandry. Due to the fact that it was the first report written by the Commercial Club of Kansas City, which took over the Board of Trade, after the annual report was discontinued for three years, the content of the report was particularly detailed. The report states: "The animal husbandry in Kansas City began 23 years ago. In 1868, it received a total of 13,000 pigs and 4200 cows.[34] ... Compared with any other industry, the animal husbandry has made the greatest contribution to the incomparable development of Kansas City. The business is not only in sync with other industries, but always in a leading position. Now, Kansas City ranks second in the country in the animal husbandry, second only to Chicago, which holds a leading position in the global livestock market." A vast and mighty body of cattle continually arrived in Kansas City by train, it was so large that the report continued: "If all the livestock brought into Kansas City were loaded into a continuous train, the collection would extend from Louisville, Kentucky to San Francisco, and then back to Louisville. If the train ran at a speed of 50 miles per hour, it would take 14 days to pass through the same location. If all these livestock were placed in one row, these living line would be 39,376 miles long, or could make one circle a half around the Earth. If the row were made to move forward day and night at a uniform speed of four miles per hour, it would take one year and 45 days to completely pass through the same location." At that time, the stockyard here had already covered an area of 100 acres.[35] In the 1920s, the livestock received by

[34] The passage of livestock through Kansas City was a turning point before the establishment of the Stock Yards Company, but the number of livestock received was very low from 1868 to 1871.

[35] *Twelfth Annual Report of the Trade and Commerce of Kansas City*, being the first under the auspices of The Commercial Club of Kansas City, for the commercial year ending June 30, 1891. Kansas City: Press of Hudson-Kimberly Publishing Co, 1892, pp. 47–8.

stockyards reached a new height. In a good year like 1923, 2,631,808 cows, 2,736,174 pigs, 377,038 calves, 1,165,606 sheep, 42,987 horses, and mules arrived in the city.[36] By 1945, the "livestock inn" covered an area of 238 acres, with a total of 4200 cattle houses (stockyard), 700 pigsties and 450 sheepcotes.[37]

Not all livestock were passersby of the inn, full up, they would board the next train and head to markets in other cities. In fact, for most of them, the stockyards in Kansas City were not a "hotel," but a janitor of "slaughterhouse." Along with the rise of stockyards in Kansas City was the meat packing industry here. Taking 1923 as an example, 1,194,527 cows, accounting for 45% of the total, were left in the local area and directly entered the disassembly. 1,857,143 pigs and about 48% of calves were left behind, while the proportion of sheep was even larger, nearly 62%. The vast majority of horses and mules were kept in Kansas City, their fate was not to be slaughtered, but to be sold to farmers in the Great Plains to continue their fate of being ordered about and dominated.[38] Next to these meat packing plants and stockyards was a nine-storied livestock trading hall, where people trade livestock that were about to leave Kansas City. The highest bidder will return with fruitful results. These animals, which had been destined since birth, had a good time chewing on juicy and delicious grass in the vast western sky, but in the eyes of buyers, sellers, and brokers, they were just moving goods, not the life that could grow, breathe, eat, shit, and piss from the moment they were driven into the train.

The sincere cooperation between meat packing plants, stockyards, and livestock trading hall built the prestigious reputation of "Cowtown" for Kansas City. Of course, Kansas City was not just a cowtown, it was also a "barn." Starting from the Union Elevator, huge barns defined the scenery along the railroad in the West Bottoms in a more monotonous and dull way than stockyards but in a more orderly and highly mechanized way. Wheat, corn, oats, barley, hay, and alfalfa flowed endlessly into Kansas City like livestock. In the report on the post-war development of Kansas City made by the City Plan Commission in 1945, they wrote, "Kansas

[36] *Forty Fourth Annual Report of the Trade and Commerce of Kansas City*, 1924.

[37] "Your Kansas City at War: The Story of the Stockyard," March 11, 1945. Archives of the Kansas City Public Library, scattered files with the source not indicated.

[38] Ditto.

City ranked first in normal years. In the production of mill feeds, and for a number of years has been consistently second in the U. S. in the production of wheat flour. This is related to the fact that Kansas City has for some years ranked first in the United States, If not in the world, as a winter wheat market." Like the relationship established between stockyards and meat packing plants, barns, and mills, as well as various other food factories such as bread, pasta, and candy, appeared side by side. Of course, the presence of a grain trading hall was indispensable here. From the moment these grains were dumped into the barn, their connection with their place of origin was completely severed. People could no longer distinguish whether it came from a wild turkey farm in Kansas or a Johnson Family Farm where they used to grow in Nebraska. They were simply graded to wait for the next buyer to arrive.[39]

In the logic of capital operation, the Great Plains became a hinterland under the control of the commodity economy, while Kansas City became a true "gateway city." Like Chicago, it, facing the west, gathered the abundant output there, and leading to the east, produced food for the markets there. As the City Plan Commission identified the orientation for it, "Kansas City performs several distinct and valuable services in the national economic pattern. Into her rail and truck terminals pour livestock and grains, milk and dairy products, poultry, eggs, vegetables and fruit from a rich farming area. There they are moved to packing houses, storage plants, wholesale markets, processing plants, retail outlets, and exchanges. Some are consumed in their raw state; others are first processed and then used in Kansas City. Some move on through the city without even leaving the cars or trucks and are used in distant centers."[40] Everything seemed beautiful. One side was responsible for production, the other side buying and selling, and another side consuming. The addition of the buyers and sellers hid the connection between consumers and producers as well as the connection between the city and its hinterland. Consumers only cared about whether the food on the shelves in the grocery store or supermarket was sufficient and whether there was a price increase. As for the situation

[39] City Plan Commission, *Agriculture, Greater Kansas City*, 1945: p. 14.

[40] City Plan Commission, *Agriculture, Greater Kansas City*, 1945: p. 3.

of the place of origin of these foods and the status of producers, they neither cared nor had channels to know the information.[41]

However, the concealment of connections did not mean that it was cut off. The operators of the urban agricultural economy, who are the buyers and sellers, were actually aware of their connection with their hinterland from the beginning, and the connection was constantly emphasized in economic planning. In the same report, the Commission once again clarified that the relationship between the city and its hinterland was inseparable, and called on people to recognize that "Greater Kansas City must be considered as an economic unit" and "the region's dependence on its agricultural hinterland." "The metropolitan area of Kansas City exists because of a very productive agricultural hinterland which is tied to it by a spider-web of rail and truck lines."[42]

In 1870, Ukrainian immigrants brought strong and drought-enduring Russian winter wheat. It seemed that people suddenly could see the dawn at the end of "misfortune" of the residents of Kansas. Although the people who opened up virgin soil here, in reality, had bumper harvest years and famine years. However, the continuous influx of new immigrants, constant reclamation of new land, ceaseless use of new machines, and uninterrupted introduction of new systems encouraged people to continue their efforts to convert prairies into capital. At the end of World War I, agriculture in the Great Plains also attained its own summit when the demand for wheat in the world reached a maximum.[43] In response, Kansas City also strengthened its belief in its own growth during its ups

[41] In the United States, Europe and metropolises in China, the situation has changed. With the boycott of food safety and fast-food culture, more and more urban consumers appealed for food traceability. In California, New York and other places, "farm to fork" movement, which aimed at rebuilding the organic connection between city and its surrounding hinterland, emerged with the expansion of organic agriculture. However, the change in urban culture was still the demand of the niche and upper-middle class on the whole, and most people still bought food from Wal-Mart or RT-Mart that they didn't know where it came from.

[42] City Plan Commission, *Agriculture, Greater Kansas City*, 1945: pp. 1 & 3.

[43] The study of agricultural expansion in the Great Plains is a major research focus of Western American history. The first work that truly reflects on the relationship among capitalism, agricultural expansion and dust storms was *Dust Bowl: The Southern Plains in the 1930s* by Donald Worster. Since its publication, it has revolutionarily challenged the people's existing impressions of agriculture on the Great Plains and become a classic work in the study of environmental history and Western history. A discussion on agricultural development of the Great Plains can be found in Part II of the book.

and downs. In the report of 1891 there is a thought-provoking paragraph, "over the past year, the internal development and trade of Kansas City were greatly affected by the poor harvest in 1890, which greatly intensified the regional and overall depression of western cities. All industries [animal husbandry and agricultural trade] mentioned earlier, as well as trade, bank and various types of industry, were deeply restricted and depressed by that reason. The harsh conditions of the winter of 1889 caused damage to winter wheat and the impact of the severe drought in 1890 on corn made average yields reduce to the minimum in the history of the region, which is the lame excuse for the widespread depression in the agricultural and commercial circles." However, these brave and optimistic people had no reason to be unable to recover due to it. In the following pages, they wrote: "In the years to come, there will still be years of poor harvests, and an overall financial and commercial depression, Kansas City cannot stay outside of it. Other cities in the position similar to us have shown us that there will always be spasmodic attacks when developing in certain special directions, a period of adverse reactions will follow a significant progress until the situation gradually become stable. However, such a situation will soon give way to a steady and upward movement, which moves forward with an indefatigable and increasing force."[44]

5 Whiskey, Jazz, Drought, and Floods

Interestingly, in the report 1945 of the City Plan Commission, nature, a term that frequently appeared in the nineteenth century, never appeared. Both the earliest boosters of Kansas City and the early writers of reports of the Board of Trade repeatedly used the term to remind people that the fertile land and products of the Great Plains were a generous gift of nature, Kansas City, located at the confluence of two rivers, was an excellent location selected by nature. However, in 1945, the elites of the Commission saw the hinterland and wisely reminded the city's planners that Kansas City could not exist independently from its hinterland, but did not see the nature that still existed behind the hinterland and did not follow the logic of capital. Cronon did not either. Although he described the establishment of a "nature's metropolis," in the end, nature

[44] *Twelfth Annual Report of the Trade and Commerce of Kansas City*, pp. 4–5 & 13.

also became the hinterland, the "Second Nature," an artificial landscape influenced by capital. The hinterland, a product of human planning, was gentle, orderly, and predictable. Although occasional "spasms" occur, it always returned to the fixed planning path and moved forward. But nature is an untamed, spontaneous, and independent strength, and when human beings step into it with a posture that they strive to adapt to it, there may be relatively stable cooperation between them, while they are in a state of confrontation only for the purpose of pure conquest and exploitation. Sometimes, technology and capital will prevail, making the "desert blossom like a rose." But when such exploitation reaches a certain limit, the antagonistic force of nature will burst out. Even if the force does not always bring earthshaking changes, it will still destroy existing plans, bring huge disasters for human beings, and force the capital, technology, and urban power that conquered it to adjust. Kansas City, Chicago and thousands of other cities are no exception.

In the 1930s, Kansas City, along with other cities mired in the depths of the Great Depression of the capitalist economy, was in difficult times. It has a special label, the Pendergast Age, in the political and social history of the city. Like many American cities, Kansas City was also controlled by the politics of big city bosses. James Francis Pendergast (Big Jim) and Thomas Joseph Pendergast (also known as Boss Tom), the two brothers, were in charge of Kansas City for nearly 40 years, from government elections to gang fights, from the police department to small bars, from the mayor of the city to Truman, who would be serving as President Roosevelt's deputy in Washington D.C. Their likes and dislikes deeply infiltrated into the spinal cord of the city's culture. However, from another perspective, the culture of the city shaped their likes and dislikes. Tom is a godfather-like gang boss who has a clear-cut view on kindness and hatred, is cruel and bloodthirsty but pitiful for the poor and weak and generous in aiding needy people. He is a nostalgic setting sun at the gradual conclusion of the past events of the Kansas City frontier.[45]

Today, it is hard to imagine that the city, which follows the rules and regulations and half of which is located on the religious belt, was a notorious "wide-open town" during the Prohibition period, just like its elder brother Chicago and the distant yet inseparable New York. Since the 1920s, the United States has implemented Prohibition nationwide,

[45] Jason Roe, "Thomas Joseph Pendergast," Biography from *The Pendergast Years: Kansas City in the Jazz Age & Great Depression*. Kansas City Public Library, 2018.

which was strictly enforced in Kansas, and alcoholic drinks were still rare there until the 1950s. However, in downtown of Kansas City, whether black settlements on both sides of the Paseo Corridor, the classic parkway designed by George Kessler during the City Beautiful Movement,[46] or the Quality Hill built by the city's white elites on white limestone cliffs, were lined with bars of all sizes, expensive or cheap, as well as gambling-houses and brothels. In its early years, the former was the center of Kansas City, with neat streets and alleys as well as exquisite houses on both sides. At that time, it became the main settlement for blacks who later moved to Kansas City. In 1900, the black population here was only 1331 persons, accounting for 29% of the community's population. By 1920, whites had moved out, and the black population had reached 75%.[47] The latter was an upscale community built on a 200-foot-high cliff, which overlooked the confluence of the Kansas and Missouri rivers running eastward. Tom built his mansion here and took charge of the operation of the city. Surrounding the city, there were numerous moonshine workshops, and even in times of economic depression, the city still had enough corn for them to brew Bourbon Whiskey. Its high-quality barley and German immigrant tradition also ensured the brewing of barrels of Kansas-style beer with a rich taste. It was probably an exceptional advantage of the Capital of Grain. After Prohibition was issued, countless cities could only operate underground bars and try their best to obtain channels to purchase moonshine, thus Kansas City became a producer of moonshine.

The interesting thing about a "wide-open town" was that when it could tolerate things that were excluded and despised by a normal society such as gangs, gambling, brothels, and bibulosity, it often had a stronger tolerance for multiculture. The urban role attracted all manner of people

[46] In order to break free from the label of "Cowtown", Kansas City began to re-plan itself in the early 1890s, pioneering the "City Beautiful Movement." Its chief designer is George Kessler, a German-American designer. He was deeply influenced by Olmsted, so compared with the key personnel of the later City Beautiful Movement, he paid more attention to the preservation of nature in the city. His design was a complete park system that includes urban parks, squares and avenues, which laid the foundation for the basic pattern of the landscapes of Kansas City. The later Country Club Plaza continued the spirit of the City Beautiful Movement. Relevant research can be found in the works of William Wilson, the founder of City Beautiful Movement history. William Wilson, *The City Beautiful Movement in Kansas City*. Columbia: University of Missouri Press, 1964.

[47] James R. Shortridge, *Kansas City and How It Grew*, p. 87.

to come here again. There was no doubt that Kansas City had established a hierarchical society determined by assets and education levels since its establishment. With its growth and the continuous influx of new immigrants, the hierarchy became increasingly strict. However, in the 1920s and 1930s, the sluggish economy and bleak sky made everyone feel a sense of impending doom, the boundaries between evil and justice, poverty and wealth, and even blacks and whites became less distinct for a time. Next to the bootlegger, there lived business managers swinging their sticks on the golf course to discuss business. The black community rapidly expanded, with underground bars and jazz ballrooms mixed among numerous small churches and slums. At that time, the city already had a large number of Catholic immigrants from Ireland, southern Europe, and the southern border, and in the early twentieth century, a large number of Polacks had also moved in and become the main labor force for stockyards and meat packing plants in the West Bottoms. A community of working people devoutly going to Mass every week formed near the gambling houses and bars where people lived a gay life every day. Not far away from it, the tycoons who earned a lot from livestock and food built their garden mansions defended by cast-iron fences.

On a certain level, Kansas City is not unfamiliar with this kind of openness. When it was just "Westport," it was a completely open town on the restless frontier of America. Fur traders who spoke French, Spanish and English rushed from all over North America, followed closely by farmers with the Bible on the left hand and a hoe on the right, who searched for land to feed their descendants. Soldiers shuttled through the town and spent a lot of time in bars. A small building that was on the verge of collapse might burst into flames, and it was probably a masterpiece of some drunkard. The declining indigenous tribes were deprived of their livelihoods, wandered the streets, sometimes begged for half a glass of whiskey or a cigarette. The priests of the church tirelessly resisted these evils, strove to make their devout believers turn a blind eye to sin and a deaf ear to filthy speech. In such a city, municipal order was not easily established. There were no city walls or sacred land here, and the streets were constantly renamed, and always under construction and demolition.

Compared with the rural communities that were continuously built westward on the Great Plains, the population composition of Kansas City was greatly different as Kansas City was home to a large number of blacks. Some of them had already arrived here before the American Civil War, and some were free blacks and some were slaves. Throughout the 1850s, the city vacillated to the left and to the right in the confrontation between free states and slave states. Some arrived in succession after the American Civil War and began a new free life in a non-plantation environment. Some of them moved westward and became homestead farmers; they hoped that the former free states could accept them with a more inclusive attitude. The persons who stayed in Kansas City gradually formed Blacktown. Although they endured segregation and discrimination, they also built their own communities, churches and commercial districts, including bars and nightclubs clustered on 18th adjacent to "the Paseo Corridor" and the Vine Street. It was here that they added a new title to the business cards of Kansas City's "Cowtown," "Barn," and "Wide-Open Town"—"Cradle of Jazz" (of course, just one of the cradles, other cradles include New Orleans—the birthplace of jazz, Chicago, New York, Philadelphia, Pittsburgh and St. Louis). The "Wide-Open Town" connived in the willful performance of jazz in a completely open style. In the smoky bar on 18th, accompanied by the sound of dices and weights, a new jazz style, known as "bebop," flew out of the saxophone. Bebop, a meaningless sound word, was just like the jazz style named after it. It had no theme, meaning, fixed melody or distinct emotions. It was an impromptu product, and almost jumped out of the performer's control and had its own rhythm in the interpretation of the performers at their own will. Charlie Parker, a native of Kansas City, was the genius founder of Bebop.[48]

And what is the relationship among all of them, the land and rivers? The "Wide-Open Town" was built on the foundation of multiracial immigrants, with the most unique group being blacks. From 1900 to the late 1920s, the black population here increased by 119%, up to 38,574 persons in total. Like many immigrants of other races who arrived at the same time or before them, they had unique social, political, and economic reasons behind them, but had the same ecological reasons—ambient pressure. In order to meet the needs of the rapid industrialization and urbanization of northern cities after the American Civil War, southern plantations made unprecedented efforts to invest in cotton production. The former slaves became tenant farmers, but they were still penniless.

[48] Ted Gioia gave a wonderful description of the transformation of jazz style in Kansas City and the main characters. Ted Gioia, *The History of Jazz*, translated by Li Jianmin, and reviewed by Li Wan, Beijing: The Commercial Press, 2020, Section III "Kansas City Jazz" of Chapter V "The Swing Era".

The fertility of the original land gradually decreased day after day without rotational tillage and care. Visitors who had never appeared in the plantation flew over the southern border and entered Texas to begin their "Reverse Sherman's March to the Sea." Some people referred to boll weevil (namely, cotton boll weevil) as the most expensive insect in American history, an incubus for southern cotton plantation owners, and a plague of the cotton economy that the people in the southern region relied on for survival. According to statistics, by the end of the twentieth century, the cotton boll weevil had destroyed tens of billions of pounds of cotton, which caused a loss of up to US $ 1 trillion. They were not the last straw that broke the camel's back, but a series of powerful bombs that ravaged the already fragile ecosystem of plantations wherever they went. When they were combined with other driving forces, nearly 2 million blacks had to give away the land they had lived on for centuries and become a home interwoven with blood and tears.[49] Some of them came to Kansas City, a city that was extremely strange to them. There, they did not find an urban paradise, but at least found a city that could accommodate them and immerse them in their own "blues" world most of the time.

At a time when many black immigrants arrived, the whole world was just in the Great Depression of 1930s, and the tone of the city was as gloomy as the sobs of jazz, and its rhythm was as disorderly as the bebop's. Compared with cities in the East and West that were both in the Great Depression, Kansas City suffered a double whammy. The hinterland, which they were proud of, and had driven the growth of the city over the past 80 years and provided it with "stronger development" during economic depression and speculative failures, finally ceased to produce in the way expected by urban capital, but swept away all hopes with a black storm that covered everything. Drought, extreme

[49] For the works discussing great immigrants from the perspective of environmental history, see Brian McCammack, *Landscapes of Hope: Nature and the Great Migration in Chicago*. For the entanglement among cotton boll weevil, southern plantation culture and ecology, see James C. Giesen, *Ball Weevil Blues: Cotton, Myth, and Power in the American South*. Chicago: University of Chicago Press, 2011. For a comprehensive examination of the environmental history of Southern United States, see Paul Sutter & Christopher Manganiello eds., *Environmental History and the American South: A Reader*. Athens, GA: University of Georgia Press, 2009. If you are interested in the environmental history of Southern United States, you can follow the book series of "Environmental History and the Southern United States" published by the Publishing House.

drought, nagged most parts of the country throughout the 1930s. For the Great Plains, drought was the norm. Looking back at the annual business reports of Kansas City from 1877 to 1939, it can be seen that droughts continuously occurred, thus leading to crop failures and a decline in profits on a portion of Kansas City's agricultural industry that year. However, people seemed to be always optimistic and looked forward to the "coming year." The problem was that drought did not necessarily lead to sandstorms, which are another type of disaster caused by the fact that people uprooted perennial weeds but planted annual crops to replace them. These grasses produced food for human beings and were cultivated by farmers year after year. The part on the ground appeared to be as strong and lush as native weeds while the part buried underground were quite different. Their roots are too short. When drought began and the fierce wind howled, the soil, which didn't fix on the perennially entangled root one or two meters under the ground and which nurtured innumerable lives and were the food source of human beings, turned into dust in the next moment, thus dust storm began.

Donald Worster, in his classic work *Dust Bowl: The Southern Plains in the 1930s*, told the story of how dust storms formed. The work, published in 1979, constitutes the classic narrative of the study of environmental history: capitalism is not just a mode of production and economic relations, but also a political and social culture that views nature as capital at the ecological level. Capitalism extremely exploited nature to achieve maximum output, thus dust storms were the great calamity created by the culture. The grass, water, soil, and the people who created and suffered a lot in the dust storms that Worst cared about in the Great Plains, but he did not attempt to connect the Great Plains with the cities it serves and to reflect on the interaction between nature and cities. Cronon found another direction, namely, starting from the city, the opposite direction of the Great Plains. He also gave an account of the capitalist culture of plunder of nature, but in the end, what such culture achieved was Chicago, a nature's metropolis. However, despite his solicitude for the wilderness and disgust at the city, Cronon still did not consider the rise of Chicago as a complete ecological disaster. Moreover, what he cared about was only the process of how nature is commercialized, and neither droughts, locust plague and floods, nor subsequent dust storms appeared in his story of urban construction. It is now time to combine the issues that are pondered in these two classic works and rethink the meaning of nature to cities, rather than just the hinterland to cities.

The dust storms of the 1930s destroyed the agriculture and animal husbandry on the upgrade in the southern Great Plains, caused soil erosion, death of crops due to drought, and death of livestock as a result of thirst and hunger. Worster said, "The story of the Southern Great Plains in the 1930s was fundamentally the one of dust storms. At that time, even the soil went crazy. During most of the ten years, it occurred not once or twice, but again and again: such situation repeated day after day and year after year that the sand made windows creak and dust stuck to people's lips. Beautiful youth turned into the despair of young people, poverty devoured people's confidence."[50] The area engulfed by dust was the most important hinterland of Kansas City. On the annual report of the city, we can clearly see the changes in the income from food. In 1900, Kansas City received a total of 46,638,250 bushels of grain, including 34,775,450 bushels of wheat and 8,334,250 bushels of corn. In the following twenty years, although the quantity of different grains changed, it was on the rise as a whole. In 1921, the bumper harvest of wheat after World War I also brought new prosperity to Kansas City. That year, the total income was 139,629,550 bushels of grain, including 110,204,550 of wheat and 15,495,000 of corn. Although it occasionally fluctuated in the following decade, it remained relatively stable. By 1931, the cities in the East were still in the midst of an economic crisis, but the food industry in Kansas City remained prosperous. That year, it received a total of 152,949,350 bushels of grain, including bushels of 125,032,000 wheat, and bushels of 22,044,000 corn. In 1932, drought began to affect the Great Plains, and the total amount of grain decreased to 91,699,559 bushels, 80,294,400 bushels of wheat and 7,630,500 bushels of corn. In 1933, a dust storm broke out, the total amount of grain decreased to 72,751,650 bushels, including 49,115,200 bushels of wheat and 18,831,000 bushels of corn. In 1934, the total amount of grain increased to 73,136,000, including 44,057,000 bushels of wheat and 25,236,000 bushels of corn. In 1935, the total amount of grain increased to 73,351,400 bushels, including 49,700,800 bushels of wheat and 18,054,000 bushels of corn. In 1936, the total amount of grain slightly rebounded, up to 87,105,400 bushels in total, including 63,744,000 bushels of wheat and 17,481,000 bushels of corn.[51] Such

[50] Donald Worster: *Dust Bowl: The Southern Plains in the 1930s*, p. 7.

[51] Board of Trade, Kansas City, MO, *Annual Report*, 1937: p. 9.

situation continued until 1939 when it gradually returned to normal. In the report of the City Plan Commission, the author stated that "the territory immediately surrounding Kansas City, (Kansas, Missouri, Nebraska, Iowa, Arkansas and Oklahoma), furnished almost two-thirds of the agricultural raw materials for the Kansas City market in 1939. ... In Kansas City, essentially dependent as it is on the surrounding agricultural area, the 'reason-for-being' of all occupational groups is directly or indirectly related to agriculture." The City Plan Commission did not fail to see the impact of dust storms on the economy of Kansas City, it pointed out that in 1939, Kansas City ranked in the upper seven of this country's each livestock market, while going back to 1932 and 1933, "before the extremely hot, dry years, the city's (Kansas City's) rank is even higher among major livestock markets than that in 1939."[52]

Under the dual pressure of the Great Depression and dust storms, Kansas City, like his fellows on the Great Plains, turned to Washington for help. When the federal government entered, especially during World War II, a large amount of military industry, including military aircraft manufacturing, was moved inland, and Kansas City was once again redeemed. This time, it did not rely on its dusty hinterland, but on Franklin Roosevelt, the charming angel of Washington, and Harry S. Truman, the steady and tenacious native son of Kansas City, beside him.

In 1950, Kansas City had a history of 100 years. Although in the eyes of a metropolis like New York on the East Coast, it was still a rural area located in a flyover region of the Midwest, boosters of the city had the confidence enough to laugh at such ignorance. They established the Kansas City Centennial Association to review the glorious history of the city and look forward to its bright future. They wrote, "Progress in commerce and industry, as well as growth in civic and cultural stature, were to place Kansas City in the cavalcade of progressive cities that combine to make up a great nation. And Kansas City, with one hundred years of growth from a pioneer, seedling community to a sturdy city flexing its commercial, industrial, civic and cultural muscles, now strides confidently forward, toward industrial power ... industrial productivity and industrial leadership." The author of the article hoped that the city would become a city with more industries, but it was undeniable that agriculture was the foundation of everything, and agriculture

[52] City Plan Commission, *Agriculture, Greater Kansas City*, pp. 4, 14, & 19.

itself directly promoted various industries. After dust storms and World War II, the Great Plains regained its former prosperity, and Kansas City once again became the center of livestock and food industries "the first ranking winter wheat market; 2nd greatest grain center…"[53] In terms of resilience, nature and cities shared a certain commonality. The business elites of Kansas City were eager to make a contribution to a new round of growth in 1950.

However, a new accident occurred. In 1951, after many days of rainstorms, the Kansas River and Missouri River were in spate, and the Kansas River could not flow into the Missouri River. The flood broke out, broke through the bank from the West Bottoms, overflowed its banks, and caused an explosion, thus destroying the stockyard that had long been the symbol of the city's identity. Afterwards, it continuously declined and eventually disappeared from the city's territory in 1991. The former trading hall became a market for antiques and handicrafts, where strange little things of all kinds were sold. Tourists who come to Kansas City, in hopes of visiting the historical relics, go there and buy a hamburger made of Kansas flour with the famous Kansas beef patty in the bread, and a glass of local iced beer, then fancy the unique smell and prosperity of the place.

In fact, floods are to Kansas City what droughts are to the Great Plains, which shall be originally the norm of the city's prudential response to them. In 1845, when the city did not exist, a flood destroyed the commercial settlement that just emerged. By 1903, flood broke out again in the Kansas and Missouri river basins. Compared with the flood in 1845, people had a much clearer memory of the flood in 1903. Most young witnesses still lived in places that had been swept through by the flood. While the turbid waves were still crashing against the limestone cliffs on both sides, someone decided to document the flood and wrote a book. In the first sentence of the preface to the book, Author W.R. Hill wrote, "Believing that the great Hood of 1903 deserves a place in the history of this country, the writer herein attempts to give graphic details of the catastrophe." He encouraged people to "fancy two mad rivers rushing down upon a populous city of 250,000 souls, including suburbs. Fancy water twelve feet deep racing like an express train through the Union depot and

[53] "Midwestern Titans: Kansas City Industries Work for the World," *The Kansas City Centennial Association*, The Kansas City Centennial. 1950, p. 72.

out over the surrounding territory...." Obviously, people at the time had experienced it, feared it, mourned it, but forgot it at last.[54]

Not only were the flood forgotten, but to a large extent, people turned a blind eye to these two rivers that carried Kansas City in the era of railroads, automobiles, and the coming aviation. The steamboats that once ran on the Missouri River in an endless stream had long disappeared, and the role of these two rivers in the city seemed to have disappeared, at least in the cognition of ordinary people. For urban planners, they were still flowing water works which provided drinking water, and like most cities, urban wastewater and garbage were directly discharged into rivers for a long time.[55] They were another kind of hinterland of the city. After the Railroad Era, only in the moment of recalling the past and comparing it with the present can their existence be noticed by people. However, the occurrence of the flood of 1951 once again reminded people that floods were nature's rivers. No matter how many dams and embankments were built, their ancient wild power would be awakened again when enough rainwater flooded into their channels; floods would break through all the shackles of human beings and declared their existence.

6 Power of Nature

These American heartlands were far more unpredictable than their creators had expected. They had different types of soil and vegetation, wildlife species, evolutionary histories, fluvial morphology, hydrological environment, and energy resources (such as oil, natural gas, and coal), needless to say, different climate types and nutrient cycles. All these differences were never set in stone. Much of the time, their changes stemmed from variations that occurred over a long evolutionary history of their own. However, after the beginning of agricultural and pastoral civilization, many changes were the result of the interaction between human beings and the environment where they lived. If what the developers of the West, the boosters of Kansas City, and thousands of people who came to the city in search of a comfortable and stable life had hoped

[54] W. R. Hill, *The Great Flood of 1903: Being a Graphic Story of How Two Mad Rivers, the Missouri and the Kaw, Deluged Kansas City and Its Suburb*. Kansas City: Enterprise Publishing Company, 1903.

[55] Amahia Mallea, *A River in the City of Fountains: An Environmental History of Kansas City and the Missouri River*.

for was the eternal land of Canaan, obviously, they were destined to be disappointed time and time again. Every bubble was uncertain and can resist city builders or grand or subtle designs. People should not expect that once the hinterland was built, it would always be in a stable, docile, and sustainable state, and it might be disturbed or scattered. Sometimes, bubbles could break down. For Kansas City, the dust storms of the 1930s were the fragmentation of a huge bubble, but after all, it was just the most extreme catastrophic event. Sometimes, bubbles disappeared. When the original resources (such as oil fields, coal mines, gold mines, and forests) of a place were completely consumed, their former hinterland functions no longer existed. More often than not, their changes were gradual and subtle. They constantly reminded their creators and successors to adjust their plans. Sometimes, such case even meant complete change in the original intention of the plan or abandonment of the entire plan. Throughout our history, much of the time, history, especially environmental history, is a story of failure, a story about the disillusionment with a certain group of people, a certain city, or even the great plan of economic development of a country in the face of being misunderstood and underestimated even in unstable natural environment.[56] Fortunately, history is also telling about the adaptation of bubbles to their ecological reality.

No one can deny that cities are ever-changing, but the change of cities does not only mean growth or decline. For most modern American cities, their changes are more about readjusting their relationship with their

[56] Such examples can be found everywhere in history, from the relatively age-old examples listed by Jared Diamond in *Collapse: How Societies Choose to Fail or Succeed* to the failure of highly modern totalitarian state control over nature explored by James. C. Scott in *Seeing Like a State: How Certain Schemes to Improve the Human Condition Have Failed*. The relationship between the agricultural era and the environment is not as harmonious and beautiful as that people fancied. Whether John Mark Dutton Elvin's reflection on the reasons why China fell into the "High-Level Equilibrium Trap" in *The Retreat of the Elephants: An Environmental History of China* or Richard Hoffmann's exploration of the degradation of aquatic ecosystems in medieval England forces historians to reflect on the "failures" in history. Jared Diamond, *Collapse: How Societies Choose to Fail or Succeed*, translated by Jiang Ying et al., Shanghai: Shanghai Translation Publishing House, 2008; James. C. Scott, *Seeing Like a State: How Certain Schemes to Improve the Human Condition Have Failed*, Beijing: Social Sciences Academic Press, 2019; John Mark Dutton Elvin, *The Retreat of the Elephants: An Environmental History of China*, translated by Mei Xueqin et al. Nanjing: Jiangsu People's Publishing House, 2014; Richard Hoffmann, *An Environmental History of Medical Europe*. New York: Cambridge University Press, 2014.

hinterland and learning to adapt to new environments, rather than simply plundering, learning to restrain themselves instead of endlessly expanding. It means a cultural change of a city, a reflection by its residents and policymakers on what is truly important to the city, its residents, other species and the ecosystems they rely on. It also means that they realize the limit which nature allows their bubbles to expand, and once such limit is exceeded, the city will be faced with new failures. Perhaps it will still recover, just like Kansas City's rebirth from dust storms and floods, but it will inevitably require a high cost paid by one or even several generations, who are often ordinary people living in the inner city of Kansas and on the farms on the Great Plains. It is equally inescapable that the ecosystem, which was once lively, rich, and diverse, is the most innocent victim of disasters. If American cities have always relied on disasters or exhaustion to make necessary changes in the past two hundred years, it is now time to restrain their ambitions, confront their own fragility directly, and learn to coexist with nature which is equally protean and fragile before a new disaster arrives.

More than sixty years later, when the child of the Great Plains wandered for many years in alien lands, traveled around the world, witnessed countless cities, hinterlands and wilderness, and returned to the city that once made him so excited and confused, could he still recognize the mystery and degeneration that mixed with the stinking smell of stockyards in the air? It may be difficult. What he will see is a livable city which is tidy, clean, elegant, and secure. It remains an international city, even more so than ever before. It has more food and restaurants from different countries and regions, and its culture has become more diverse and inclusive. It is no longer the last frontier town, the Cowtown a hundred years ago, or even the city that was still somewhat chaotic but ambitious in declaration of the spirit of Kansas City to the United States more than sixty years ago. It has gradually evolved into a larger and more complex economy. However, its basic hinterland is still the Great Plains, and agriculture still occupies an important place in its economy.

After the dust storms in the 1930s, the Great Plains revitalized once again and continuously exported food and beef to the world. Its agriculture has become more mechanized, modernized, and commercialized. The discovery of the Ogallala aquifer has temporarily made it have no worry about water shortage. But what it will be afterwards? The aquifer does not guarantee that the water source is perpetual. It will disappear in the coming decades, up to a hundred years, due to increasing water intake. The increase in water intake largely stems from the desire of more people for juicy steaks and the desire of ranchers for lucrative profits from raising cattle. As Worster reviewed the changes in the Great Plains after the 25th anniversary of the publication of *Dust Bowl: The Southern Plains in the 1930s*, he wrote, "Like the legendary copper and gold mines in the Western American, the livestock industry that dominates the Great Plains today is also a mining economy. It extracts meat rather than minerals. Like all mining economies, its outcome is inevitably a ghost town, abandoned home, uncultivated farms and collapsed businesses."[57] What will Kansas City do when dust storm comes again? Or people may say that it now has a large-scale manufacturing, high-tech industry, and health services, let alone the fact that it can turn to the federal government for greater help. The latter helped it out of its past difficulties. In the present day, the federal government is so powerful that it will offer even more generous support to Kansas City. However, Kansas City still needs the energy to support the operation of its machines, the food and water to feed its people even if it is as powerful as the country. Even if Wall Street business elites and Silicon Valley tech geeks have minds that are a hundred times more savvy and powerful than the one that they possess now, they still need the products from land, water, and fresh air to breathe freely without purification, and the wild world preserved on the edge of civilization. In that world, non-human life, including coyotes on prairies, lizards in deserts, and the birds, insects, and plants that have evolved on their own courses over billions of years in the Amazon Rainforest, everything has their own right to survival.

As for Kansas City, some things have been lost, lost in the midst of change, and cannot be restored. What has not been lost is the hot and humid days and nights in summer, which is a climate type shared by the Mississippi River basin and its tributaries. The boy once came to the city for the first time in a sultry season, and when he went back, the sultry weather remained the same. He might feel a bit distracted and nostalgic, but he knew that nothing was eternal and unchanging. Does the repeated revival of the city tell a new story of victory? Has it learned how to get along with its limits and its ecological reality? It is a part of the Great Plains, as Worster said: "The fact that the United States rules over the Great Plains for only one or two centuries cannot serve as a basis for

[57] Donald Worster: *Dust Bowl: The Southern Plains in the 1930s*, p. 330.

predicting the probability that any society or mechanism can exist for a long time. On the basis of such a short stage, no historian, environmental historian or anyone else can choose a future winner."[58]

[58] Donald Worster: Dust Bowl: The Southern Plains in the 1930s, p. 330.

PART IV

Ideological Landscape of Cities

CHAPTER 7

City, Wilderness and Intermediate Landscape

1 Prologue: The Birth of Urban Thought

In the book *The Brown Decades: A Study of the Arts in America, 1865–1895* (1931), which established him as a master of urban research in the twentieth century, Lewis Mumford referred to Frederick Law Olmsted as the "first urbanist" in America.[1] Before Olmsted, Americans had thought about cities in a limited and vague manner. When they left the British Isles and crossed the Atlantic to the continental North America, they brought not only wheat, plantain herb, livestocks, and bacteria, but also the anti-urban mentality formed in British society from the early modern era.[2] They were concerned about the country, democracy, freedom, and law, but rarely discuss the city as an independent unit, what it should look like, what kind of relationship it should have with the people living in it, how it should coexist with nature and rural areas, and how they could make urban life better. Even if they occasionally discussed the issues like those, they often criticized those as a source of pollution, plague, and moral decay. The person who completely changed the ideological point of view was Olmsted, who clearly declared that the urban age had come, and every person and every place couldn't escape.

[1] Lewis Mumford, *The Brown Decades: A Study of the Arts in America, 1865–1895*. New York: Harcourt, Brace and company, 1931.

[2] Morton and Lucia White, The Intellectual versus the City: From Thomas Jefferson to Frank Lloyd Wright. Cambridge: Harvard University Press, 1962.

S. Hou, *Wallless Cities*,
https://doi.org/10.1007/978-981-97-7827-0_7

In terms of landscape renovation in the United States, perhaps no one else's influence could surpass that of Olmsted. His works could be found across 25 states across the continental North America and several provinces in Canada, with innumerable city parks, national parks, schools of all sizes, and private houses. More importantly, his design philosophy dominated the landscape design in the entire second half of the nineteenth century and early twentieth century, which is a time when urbanization was flourishing in the United States. Until the present day, although Olmsted's authority is constantly facing various challenges, his ideas have infiltrated every corner of the United States, and no challenger can withdraw them from it. In any city, town, state park, or national park in the United States, people may encounter Olmsted, either the landscape built by the master in person, or the legacy of his ideas. Olmsted became a part of the American landscape.

Olmsted grew up in the soil of New England in the nineteenth century, which was immersed in transcendental ideas. Olmsted's aesthetic taste was purely romantic, and the ideals he pursued in his design expressed his faith and pursuit of democracy. In his view, it is the new cities born in the nature of the New World that could become places that not only realize their social ideals but also convey their aesthetic appeals. He firmly believed that in the New World, there is nothing that is too beautiful for ordinary people to possess: comfort, health, education, art, especially the beauty of nature. As discussed in the previous chapters of the book, the idea was truly implemented when he went back to New York in the east from the wilderness in the west and fully embarked on his career as a landscape designer. While most of his contemporaries were still immersed in the old dreams of pastoralism, feeling sorrow for its disillusionment, Olmsted had already begun actively exploring and reconstructing the relationship between city and civilization—defining the role of nature in the urban age and the position of cities in the vast ecosystem of nature.

As early as 1870, Olmsted read an article titled *Public Parks and the Enlargement of Towns* at a conference of the American Social Science Association, which was a proclamation announcing the arrival of the urban age. In the article, Olmsted keenly perceives that urbanization, or as he puts it, "a strong drift townward," has become an irresistible trend in the development process of civilization. He rejected the resentment of seeing cities as a "moral epidemic" or the sentimentality of a backset toward "rural simplicity." On the contrary, in his view, the process of urbanization is full of "elements of human progress," and the growth of cities is closely related to "the dying out of slavery, feudal customs, of priestcraft and government by divine right." He praised "the multiplication of books, newspapers, schools, and other means of popular education and the adoption of improved methods of communication, transportation, and of various labor-saving inventions." He asserted, "no nation has yet begun to give up schools or newspapers, railroads or telegraphs, to restore feudal rights or advance rates of postage. King-craft and priestcraft are nowhere gaining any solid ground." He confidently prophesied, "It would seem then more rational to prepare for a continued rising of the townward flood than to count upon its subsidence." It is worth noting that although Olmsted firmly believes that urbanization represents a trend of progress, he is not blind to the "diseases and hardships," "corruption and sin" of urban life. He further wrote, "If modern science didn't explore the causes of the unique evil things that torment human beings in many cities and handed over our methods of resisting them, the prospects of civilization would be bleak."[3]

Among the evil things, as Olmsted stated, was undoubtedly the pandemic plague that still threatened the lives of urban dwellers in the nineteenth century. By the time he read the article, the germ theory had already begun to develop, new water-works technologies were becoming increasingly mature, urban public health construction on the other side of the ocean had stood out conspicuously, and the outline of Central Park in New York had gradually become clear. Most importantly, with the end of the American Civil War and the numerous obstacles to the reconstruction of the Southern United States, the entire United States, from the East Coast to the Great Plains, and even to the Pacific coast where Olmsted left five years ago, was revitalized with a new atmosphere and vitality. Just like Whitman, who never stopped singing for the New World, wrote the preface to *As a Strong Bird on Pinions Free*, his poetry collection, in 1872: "The mighty present age! To absorb and express in poetry, anything of it—of its world—America—cities and States—the years, the events of our nineteenth century—the rapidity of movement—the violent contrasts, fluctuations of light and shade, of hope and fear—the entire revolution made by science in the poetic method—these great new underlying facts and new ideas rushing and spreading everywhere;—truly a mighty

[3] Frederick Law Olmsted, "Public Parks and the Enlargement of Towns," *Civilizing American Cities: Writing on City Landscape*, ed. by S. B. Sutton. Cambridge, Mass.: MIT Press, 1971: pp. 56–7, 64–5.

age!"[4] In Olmsted's view, it was indeed an extraordinary era, where all dangerous, dirty, and terrifying beings in nature were expelled and destroyed in the unceasing update of modern science, and all decadent, corrupt, and unjust things in human society were also suppressed and eliminated in the constant improvement of modern democracy. There was no doubt that the reason why the optimistic and confident ethos of the Victorian era permeated the bone marrow of the generation of Olmsted and Whitman was that they cherished the belief, in near veneration, in the progress brought about by the magnificent science and technology in the nineteenth century. Just like Whitman's unconstrained poem: "The sum, concluding all we know of old or modern learning, intuitions deep/of all Geologies—Histories—of all Astronomy—of Evolution, Metaphysics all/ is, that we all are onward, onward, speeding slowly, surely bettering."[5]

Of course, Olmsted is not Whitman. Whitman longed to completely shake off all the old yokes and disciplines imposed on himself, and to stand naked in the center of the New World like a neonate. For Olmsted, the traditional moral fiber in New England always constrained his thoughts and actions. The knowledge and rationality necessary for him as a professional landscape architect also made his observation and reflection on society and the city not only rich in the romantic imagination of artists, but also with the sober analysis of sociologists. As Whitman joyfully ran on the continent of North America, every imaginary and real leaf in his eyes, including men and women, factories, cities, sky, oceans, and the great plains, were freely unfolding and blooming in a New World without hindrance. In Olmsted's design, he deeply perceived the threat that industry and cities pose to nature on the path of immoderate pursuit of benefits.

However, how similar are the two in step with the spirit of their time and beyond the limitations of their time. They all put great stock in the power of science, firmly attained the ideals of democracy, abhorred slavery, and called for a just and healthy society. However, unlike most people with the same faith in their time, they clearly felt (for Whitman) and realized (for Olmsted) the importance of nature to the civilization being built in the New World. In Whitman's poetry, the city has never

[4] Walt Whitman, Preface to *As a Strong Bird on Pinions Free*, *Leaves of Grass*, translated by Zou Zhongzhi: Shanghai: Shanghai Translation Publishing House, 2019, p. 1.

[5] Whitman: Going Somewhere, from Leaves of Grass, p. 575.

been separated from nature. It is from the wildness and richness of the latter, and from the resilience and vitality of leaves that the civilized city, the cause and the people he loves, each of them draws vibrant, original, and enduring strength. In Olmsted's view, the beauty of nature is not only his concern, but also the inspiration for art and the necessity of urban civilization. His grand urban blueprint came from his understanding of the evolution of the human ecosystem: cities will inevitably become a more civilized stage in the system and a residence for civilized people, and science will make the new human residence more livable. But in his optimistic vision of urbanization, he always maintained a cautious vigilance. He worried that it was possible to separate nature from cities after these dual concepts interwove, and that the alien threat might arise in the environment created by technology where humans are fully immersed. In his view, whether it was Jefferson's ideal agrarian society centered around yeomen or the plantation economy in the Southern United States was out-of-date, urbanization was destined to the development direction of civilization. However, the development of civilization does not mean the separation between nature and cities, let alone the death of nature. In the urban age, human needs for nature are not merely limited to physiological needs for the sources of clean water, clean air and safe food, and the understanding of nature should not be limited to knowledge. In fact, as civilization shows human beings the way forward, the aesthetic awareness of and demand for nature become stronger. As scientific understanding of nature is deeper and more comprehensive, civilization shall test its blind confidence and revere the rules and rhythms of nature. When Olmsted designed parks for American cities, his thinking was not limited to the boundaries of the parks, but focused on the plan for the entire city and the civilization it embodies. During the period of rapid rise of American civilization, he hoped to naturalize the city by integrating natural gardens into the city and preserving wilderness outside the city, thus making the city more civilized.

Up to the present day when cities become increasingly huge and crowded, pollution turns more severe, climate gets warm every day, biodiversity steadily disappears, and the scope, speed, and frequency of pandemics gradually increase, the confidence in progress in the era of Olmsted and Whitman seemed somewhat ridiculous, even incredible. However, they worked hard for the progress they aspired for, while we did not.[6] They directly responded to the impact of nature on their thoughts, and wrote their thoughts not only on paper, but also built them into the landscapes that innumerable people would experience. They lived in an era where the discussion of nature, art, and beauty was despised. That era has not only its own innocence, prejudice, and blindness, but also the ability of constant introspection. It is precisely from Olmsted that generations of urbanists examined the reflection on nature and civilization in their ideological origins, integrated the favor of nature and the rejection of cities in American thought into the reflection and transformation of urban civilization, and promoted the evolution of cities in the ecological paradox of cities.

2 Climbing Romantic Mountains

In 1836, Ralph Waldo Emerson anonymously published a small book titled *Nature*. At the beginning of the article, Emerson called out, "Why should not we also enjoy an original relation to the universe? Why should not we have a poetry and philosophy of insight and not of tradition, and a religion by revelation to us, and not the history of theirs [foregoing generations]? Embosomed for a season in nature, whose floods of life stream around and through us, and invite us by the powers they supply, to action proportioned to nature, why should we grope among the dry bones of the past, or put the living generation into masquerade out of its faded wardrobe? The sun shines today also. There is more wool and flax in the fields. There are new lands, new men, new thoughts. Let us demand our own works and laws and worship." It was a new Declaration of Independence, 60 years after the publication of the *Declaration of Independence*, it heralded the arrival of an era in which the United States sought independence of thought. Young Emerson keenly noticed that new ideas were closely connected to new land, and that a new culture would burst out in the United States when the new ideas collided with the power of nature in the New World.

Compared with the European continent where Romanticism emerged, the relationship between the natural environment of the New World more agrees with the aesthetic imagination. Although the natural landscape of

[6] When E P. Thompson said he hoped to condescend to rescue those untimely marginalized groups from the contempt of their descendants, today's historians may be more willing to patronize these former intellectual elites.

Conrad where Emerson lived had already greatly changed by that time, there were still large mountains and valleys around the small town which were in their wild state. Even though they had been damaged by agricultural, industrial, and commercial impacts, they still maintained a certain degree of biodiversity and lush natural scenery until the mid-nineteenth century, thereby stimulating civilization's imagination of nature. Similarly, it was impossible to sever the relationship between the Puritan tradition in the New England and the subsequent cultural development. Although Puritanism, by its nature, often treated the relationship between man and nature from a utilitarian perspective and do its utmost to extort goods from them, when the seeds of romanticism were sown on New England soil, it couldn't avoid being irrigated by the water of Puritanism. As one of the earliest and most active cities in the United States, Boston in the mid-nineteenth century foreshown the advent of a new urban age. Emerson and his successors needed to seek new position and moral judgment for nature and their relationship with nature. It is precisely in such a specific natural, ideological, and contemporary environment that transcendentalism, a romantic American expression, sprouted and grew in Concord, a small town several miles away from Boston, and became an important basis for a large portion of people to interpret the relationship between the natural world and human society in the second half of the nineteenth century.

From the remarks of Confucius that "the virtuous love mountains and the wise love waters" to "the clear river is the same as silk," "Carefree I see the southern hill," the ones put by the people of the Wei and Jin dynasties, the Chinese people's enjoyment of mountains and rivers is almost natural and logical. However, for modern Europeans, the course of aesthetic development from "mountain gloom" to "mountain glory" was as steep and rugged a road as crossing the Alps.[7] From ancient Greece and Rome to the Renaissance, mountains were viewed as a stain on the works of God and a cancer on the landscape of the earth. The towering and obscure peaks are hiding-places for evil and heretical forces, a blasphemy of Christian faith, and a place where civilization and reason flinch but reluctantly struggle to conquer. It is often associated with desolation and barbarism, as well as with well-known and unknown wildlife,

[7] Marjorie Hope Nicolson, Mountain Gloom and Mountain Glory: The Development of the Aesthetics of the Infinite. Ithaca: Cornell University Press, 1959.

including wild man. For them, mountains mean wilderness, a derogatory term that existed before modern times, whether from a utilitarian, religious, or aesthetic perspective.

Therefore, when whites who inherited Western traditions came to the continental North America, although the abundant resources and vast land made them feel that they were God's favor and a reappearance of the lost Garden of Eden. But they also felt that the desolate and lonely wilderness was a God's test for them. Its existence itself was odious, and their mission was to conquer and turn the wilderness into farmland, orchards, and cities. From the sixteenth century to the early nineteenth century, the manifestation of American colonizers on the wilderness accurately conveyed the mainstream attitude toward nature in Europe for thousands of years. In the book *Wilderness and the American Mind*, historian Roderick Nash provides an extremely accurate summary of it. He believed that two components figured in the American pioneers were hostile to wilderness. On the one hand, from a practical perspective, wilderness constituted a formidable threat to his very survival. In order to obtain food, shelter, and necessary comfort and safety, pioneers felt an urgent need to fight against the wilderness. They also had to face the danger of succumbing to the wildness of his surroundings and reverting to savage himself. Therefore, their attitude toward the wilderness was completely utilitarian, and full of confrontation, conquest, and being conquered. On the other hand, the attitude of pioneers toward the wilderness of North America reflected their European traditions. They "imagined wild country as a moral vacuum, a cursed and chaotic wasteland. As a consequence, frontiers men acutely sensed that they battled wild country not only for personal survival but in the name of nation, race, and God. Civilizing the New World meant enlightening darkness, ordering chaos, and changing evil into good." Therefore, the conquer of wilderness will not only guarantee their survival, but also make them moral and religious heroes.[8]

But when did the Western world sweep away the shadow of the "mountain gloom" and let the brilliant rays of the "mountain glory" shine into the aesthetic mind of Europeans? It should be mentioned that the Romanticism that emerged in the late eighteenth century opened up the new relationship between man and nature. In the present day

[8] Roderick Nash: Wilderness and the American Mind, pp. 24–25.

of postmodern era, the term "romanticism" has lost its previous charm, and to a certain degree, it seems to have become the synonym of flamboyant and exaggerated sentimentalism, or have been blind to or even hypocritically concealed the so-called "real world." A romanticist may be seen as somewhat foolish, irrational, feeble, and anti-scientific. However, the understanding originates from ignorance of the history and connotations of Romanticism, as well as a misunderstanding of our reality. Before we truly criticize it, we shall first answer the question of what exactly romanticism is.

To sum up, Romanticism is a thought and culture movement that emerged in Western Europe from the late eighteenth century to the mid-nineteenth century. Some historians believe that the movement has swum with the tide and continued to the present day. The Romantic movement, by its nature, firmly believed that individual emotions—subjective feelings and emotions were reasonable and important, and people should trust it and allow it to run its own course, rather than suppress or question it by dull rational logic—scientific or economic logic. However, more importantly, romanticists extended their emotional objects to non-human beings so as to express their love and respect for nature and all living beings in nature. It regarded nature as a vibrant whole, and tried to appreciate it through emotional experience so as to seek inspiration for art, literature, and thought from nature. It was under the influence of Romanticism that Europe rediscovered the beauty of mountains and rivers, as well as the need of human mind and life for nature.

Scholars have different opinions on the origin of the rise of the movement. In the mid-twentieth century, some scholars believed that the rise of Romanticism was closely related to the introduction of Orientalism, especially Chinese landscape design.[9] Some scholars also deemed that although Romanticism sharply criticized the rational logic of science, the scientific interpretation of nature, especially the revelation of the operation of nature's economic system, played a decisive role in the rise of Romanticism. It was science that broke the fixed mindset of theology, allowed people to see the intimate connection, and even the familial relationship, between man and nature, and that broke the myth that all things

[9] For instance, the paper "The Chinese Origin of A Romanticism" published by British scholar Arthur Lovejoy in 1933, and the book *Tides in English Taste* by American scholar B. Sprague Allen published in 1937 link the rise of romanticism with the introduction of art and ideas from China and India.

in nature were centered around humanity and began to examine the relationship between man and nature with a more equal perspective.[10] However, more scholars held that the rise of Romanticism in Western Europe in the late eighteenth century was closely related to the economic and cultural development of Western European at that time. It was the result of various social, economic, and cultural changes. Romanticism largely stemmed from criticism of the material culture of capitalists and rebellion against the shackles of religion and social customs. Environmental historian Donald Worster referred to romanticists as "the first great subverters of modern times," which aimed to overturn at least "established concepts formed by science; the constantly expanding values and structures of capitalism; and the traditional prejudices of Western religions against nature."[11] Therefore, romanticism is by no means a shallow sentimentalism or simple hedonism. What it does is a profound critique and reflection on human nature, civilization, and nature.

Rousseau tends to be regarded as the pioneer of the movement to liberate emotions. He attempted to tell his contemporaries who were going through capitalist transformation how the civilizations they flaunted and were keen on were decadent and how to suppress and distort people's emotions. Compared with the well-organized Palace of Versailles, Rousseau yearned for the towering and steep peaks of the Swiss Alps, where others could truly feel freedom, nature, the sudden burst of inspiration, and even moral cleansing. Since Rousseau, romanticism has largely been a pursuit of a fresh aesthetic concept, that is, a redefinition of what is the beauty of literature, art, music, and scenery. They often rejected traditional aesthetic tastes—straight and rigid lines, highly artificial landscapes, precisely measured spaces, limited visions, and landscapes that boast of wealth and privilege. Contrary to the old concept, what romanticists admired and pursued was a more natural form—organic flowing forms, asymmetrical and less decorative landscapes, infinite space, wilderness, and piedmont, as well as the dwellings of "ordinary beings" and even "savages." The flat and regular landscape became stiff and mediocre. What people's aesthetic emotions require was the contents which went beyond simple "goodliness" or "beauty," in other words, a stronger and even

[10] See Marjorie H. Nicholson, Mountain Gloom and Mountain Glory: The Genealogy of an Idea.

[11] Donald Worster: Nature's Economy: A History of Ecological Ideas, p. 81.

daunting impulse. Just because of this, high mountains were transformed into marvels of creation from blisters of herpes on smooth surfaces, and became a representative of "sublime." And "sublime" was the highest realm of beauty. In the face of sublime landscapes, people would feel small and fragile, and awake to quite closeness to death and tremble with fear. It is at that moment that romanticists realized the existence of God.

Therefore, Romanticism is largely a religious movement, or rather a spiritual movement. In some cases, romanticism attempted to awaken traditional religions, for example, Catholicism, and to inject stronger emotional experiences into them. In other cases, and even more generally, it attempted to create a new religion, a more direct religion based on the worship of nature. Many former devout men and women were no longer able to continue their faith in traditional Christian doctrine. The endless debates over theology, norms, and dogmas made them feel tired, and true piety had been lost in the midst of expanding material civilization. Faced with the invasion of Western civilization by rough and indifferent materialism, the resistance of traditional religions appeared particularly weak and feeble.

It was in such a case that many romanticists turned to nature to seek faith conversion and moral Sukhavati. They climbed high mountains, strolled along the seaside, wandered among the dense forests, where they found the great beauty of nature. It was precisely the natural beauty that nourished and enriched their spiritual world, which the Bible or church couldn't provide. There was no doubt that Romanticism subverted the disregard for, and even hostility toward, nature, from the Western religious tradition, by viewing nature as a masterpiece of God and using nature rather than the Bible as a model to recognize the power and revelation of God. However, the God here has also become an uncertain concept. Sometimes, he refers to God in traditional Christianity, but in other cases, God becomes a tremendous force that exists in the universe but free of a name, so it is called God or nature for the time being.

In fact, the new religion that Romanticism aims to establish is largely one of the oldest religions, or pantheism, a religion worshiping nature or wilderness. From the reconstruction of "natural piety" as Wordsworth mentioned to Byron's passionate and unrestrained lines, "I love not Man the less, but Nature more," the natural religion of Romanticism echoed the ancient voice of nature worship. However, from the perspective of the development of Western civilization, it is a new religion indeed. Although romanticists draw a lot of inspiration from primitive religions, such inspiration is not a simple repetition of pantheistic worship in the stage of savagery. Its foundation is based on the reflection and questioning of Western religious traditions and modern civilization, as well as the nostalgia for and return to nature in isolation from man in the era of industrialization and urbanization.

It should be noted that what Romanticism aims to subvert is not science, but to view reason as the only way to understand nature, and science as the only authoritative concept for explaining truth. One important theme of Romanticism is its "fascination with the study of biology and the organic world. Romantics found that the world of science was a way to understand the intuitive ideas of past heresy. The intuitive ideas held that the entire nature was alive and jumped with the strength of energy and spirit. Moreover, it was precisely at the core of the Romantic view of nature that future generations would propose an ecological perspective, that is, an exploration of the concept of wholeness or interrelationships, an emphasis on interdependence and correlation in nature, and a strong desire to restore humanity to a position closely related to the vast organisms that make up the Earth."[12] All of these have elevated the natural religion of Romanticism to a new level, the one that is far more complex and profound as well as strongly critical and forward-looking.

In addition, what Romanticism aimed to subvert also includes the construction of materialistic value which constantly expanded in the capitalist system, the calm weigh of gains and losses, and the endless pursuit of wealth and money. Their critique of capitalist values extends to the criticism of industrial civilization, the product of capitalism. The farms, cities, and factories that are built by industrial civilization and capital accumulation make them feel bored. The sense of boredom comes not only from their contempt for the ideas and power behind these artificial products, but also directly from the exhaustion of sensory organs, resistance to and hatred for rigid lines, cramped spaces, noisy sounds as well as deep and dead environments. On the contrary, nature grows with vigor.

Romanticism is still a movement to get away with norms, traditions, customs, and conventions. After it was spread to the United States, the subversion was particularly intense. In the new land, romanticism encouraged people to put society and dogma behind them, but engage in candid self-expression, thorough-going individualism, and self-pursuit. In

[12] Donald Worster: Nature's Economy: A History of Ecological Ideas, p. 109.

the midst of such rebellion, romanticists passionately sang the praises of youth. They questioned why in a young country, young people were bound by outdated customs and rules? Why couldn't they conduct new explorations and experiments, try new ideas, create new art, and write new chapters? It was under such questioning that American romanticism emerged. It was not a replica of the thought of Old World, but a new life that grown in the cultural and natural soil of the New World. It would change the attitude of Americans toward nature and civilization which inherited from their European ancestors, and also reshape the natural landscape of North America in practice, the new life is known as Transcendentalism.

3 "I Wish to Know an Entire Heaven and an Entire Earth"

After Henry David Thoreau lived in a wooden house by Walden Pond for two years and two months, he went back to his parents' home in Concord. On the surface, he seemed to have estranged himself from the nature he once lived in, but in fact, in the last decade of his short life, the relationship between Thoreau and nature became unprecedentedly close and profound while Thoreau wandered dozens of miles around Concord. Thoreau, who built his own wooden house on Walden Pond between 1845 and 1847, was the recluse who people are well acquainted with. He resisted civilization but sang about nature with exquisite diction, radical actions, and sharp thoughts. But it was in the last decade that he was neglected by later generations that Thoreau became a true thinker, who was contemplating the evolutional process of nature. At that time, what he saw in nature was no longer the model that could be depicted in a harmonious and stable form by the Walden pond, but a system similar to a civilization not far away that was experiencing turbulence and change. The transformation of the system largely originated from external forces, which we call "civilization." Thoreau was born too late, and he had no way to witness the process of the change. Instead, he could only search for its brilliance of the past from the incomplete nature. Therefore, in his pursuit of natural history, Thoreau felt an unprecedented sense of loss and pain. He felt lamentable and wrote, "I take infinite pains to know all the phenomena of the spring, for instance, thinking that I have here the entire poem, and then, to my chagrin, I hear that it is but an imperfect copy that I possess and have read, that my ancestors have torn out many of the first leaves and grandest passages, and mutilated it in many places. I should not like to think that some demigod had come before me and picked out some of the best of the stars. I wish to know an entire heaven and an entire earth."[13]

The place that made Thoreau heartache and headache was his hometown—Concord, an agricultural town located only 20 miles northwest of Boston. Like most small towns around Boston, it was a product of Puritan tradition and Boston culture, and also served as an interior for Boston's development. In the 1850s, the small town still seemed like a rural scenery, but the tranquility of Sleepy Hollow was shattered by the sharp whistle of trains, and machines suddenly entered the garden. For Thoreau, even before the train passed through Concord, the nature here had become fragmented under the agricultural cultivation of the Puritans, but the tentacles of industrialization made the remnant nature disappear more quickly. Equally important was that the small town, like other New England towns, was constrained by rigid dogma, stiff customs, and immutable lifestyles, thus causing freedom and ego to be lost in the midst of daily norms for petty profit. But it was precisely on such a typical land of New England that transcendentalism, a heresy which rebelled against its Puritan traditions and customs, emerged. Its founder was Ralph Waldo Emerson, around whom a transcendental cultural circle formed, including Margaret Fuller, Amos Bronson Alcott, and Thoreau.

In the following years, and even throughout the nineteenth century, transcendentalism became the mainstream ideology in America, and even to a large extent, a representative of American culture. However, it cannot abandon its European origin. In fact, it was a component of European and American romanticism, and had all the romantic characteristics as mentioned previously. The word was first used by Immanuel Kant. In Kant's view, the highest truth, that is, the truth known to humanity about God and the natural world, does not come from physiological senses

[13] Henry David Thoreau, *Journal*, in *The Works of Henry Thoreau*. Vol. 8. Concord: 1906: pp. 220–222. The most thorough and comprehensive study of Thoreau's ecological thought can be found in Part II of Nature's Economy: A History of Ecological Ideas by Worster. There are numerous studies on Thoreau's view of nature, and classic studies on the relationship between Thoreau and Romanticism include *Thoreau in the Context of International Romanticism* by Perry Miller. Perry Miller, "Thoreau in the Context of International Romanticism," *The New England Quarterly*. Vol. 34 (June, 1961): pp. 589–617.

such as vision and smell, or purely empirical knowledge, but from intuition. Human beings are born with the ability to perceive these truths. No knowledge accumulation or scientific speculation can create such truths or confirm them. The only way to know them is through intuition, therefore, human beings should believe in and rely on it. For the Transcendentalist in New England, the way to touch his own intuition is to step into nature to have an aimless and unplanned wander for pure experience of the beauty around them, thus awaken their intuition. You, anyone, and I will discover hidden truths and virtues in the existence of nature.

However, what cannot be ignored is the distinct American characteristics of transcendentalism, even the Boston characteristics. Although these transcendentalists always complained about and criticized the small town they live in, they did not deny the charm of Concord and the enlightenment of the environment there on their thoughts. If Thoreau lived in New York, or even in Boston, he would have never been able to write the *Walden* or thousands of pages of *Journal* we see today. He, like Emerson, was deeply attracted by the natural beauty of the small town. Although the beauty was already shattered, it still retained many wild and natural qualities. Meanwhile, the geographical proximity between Concord and Boston also had a great impact on their thoughts. Compared with most cities in the United States, Boston was a city that venerated philosophy, poetry, and spiritual pursuits. They needed a city that could provide them with a market to listen to their thoughts, a vibrant cultural atmosphere, audiences that could accept theological liberalism or even natural religion, a new religion, and an open world for thinkers.

Although the religious tradition with Puritan origin in Boston was the main object that Transcendentalists criticized, they couldn't deny the influence of the tradition on their thoughts. When Puritans moved away from their hometowns to the continent that they were unfamiliar with, they were rebels of English orthodox culture. The thoughts of Winthrop et al. were filled with the rebellious spirit of their time, which became a unique cultural factor in Boston, and grew into different forms of rebellion against the new orthodoxy. These rebellions not only included the Quakers' rebellion against Puritan dogma, but also the American Revolution's rebellion against the crowns of England, as well as the rebellions of Emerson, Thoreau, and others against religion itself, social customs, and institutional concepts.

More importantly, compared with European romanticists, American transcendentalists represented by Emerson and Thoreau had a much stronger sense of moral self-discipline and the democratic consciousness of egalitarianism, and the inheritance relationship between these two points and Puritan traditions was evident. However, unlike Puritanism's attempt to cleanse society with Christian dogma and emphasize the idea of man as the wisest of all creatures, transcendentalism sought a moral purpose by getting close to nature. The equal democratic consciousness developed, in Thoreau's view, into the one that all living beings should be treated with an equal perspective, and democracy was expanded to nature, which is beyond the realm of human society.[14]

Although both belong to transcendentalism, there is a great difference between Emerson and Thoreau. Throughout his life, Thoreau never diminished his reverence for Emerson, but he always deviated from social norm from his thoughts to his life practice. He hated all injustices and inequality of human society and the natural world. He always satirized and took a pessimistic view of the development of civilization itself and ceaselessly argued with American society. From his ecological perspective, Thoreau went further than any of his contemporaries did. He observed the ever-changing nature with a new science from a holistic, interconnected, and organic perspective although the science, at that time, had not yet been given its future name—ecology. The science was fundamentally different from laboratory work, as it studies the relationships between living individuals in the organic world and their environment. It requires its researchers to step into nature and be exposed to every living being there.

Today's environmentalists see Thoreau as a prophet, but at least in the nineteenth century, Emerson's influence on American culture, including the relationship between Americans and nature, was more profound than

[14] The relationship between Puritan tradition (more broadly, Protestant tradition) and transcendentalism can be found in Perry Miller's collection of papers: *Errand into the Wilderness*. Cambridge: Belknap Press of Harvard University Press, 1956), Especially the article "From Edwards to Emerson"; Mark Stoll, *Protestantism, Capitalism, and Nature in America*. Albuquerque: University of New Mexico Press, 1997; Donald Worster, "John Muir and the Roots of American Environmentalism," Worster, *The Wealth of Nature: Environmental History and the Ecological Imagination*. New York: Oxford University Press, 1993.

that of others, including Thoreau. Emerson was born of unitarian clergyman family in Boston in 1803.[15] But his father passed away when he was 8 years old, so Emerson was brought up by his mother and other female relatives in his family. His formal school education was almost spotless. Boston Latin School and Harvard University where Emerson studied are, both then and today, the best school in the United States. There is no doubt that Emerson is a beneficiary of the wealth and refinement of Boston, a rising commercial city where education has always been respected. After graduation, he followed his father's footsteps and became a unitarian clergyman. However, when his first wife passed away at the age of twenty, Emerson began to question his faith. Living in the city that provided him with education and even a foundation for his thoughts, he felt constant pressure to follow social customs and the expectations of others. As a symbol of rebellion, he moved out of the city where he had lived for thirty years and came to Concord, a small town in the west of and twenty miles from Boston, where Emerson would embark on a journey of independent thought. He yearned for freedom.

To a large extent, Emerson indeed broke free from the limitations of New England society and religion, won the freedom he had expected. However, interestingly, when he was alive, he became a philosopher of New England, even an "American philosopher." Even though living in Concord, Emerson was always a Bostonian, yet he also shook off the constraints of the city on his thoughts and found his own voice. More importantly, the new generation of young people in the United States listened to and needed his voice. He became a leader of ideas rather than a follower of anyone.

"Self Reliance" is Emerson's most famous article even though later critics often did not consider it his best work. In the article widely read by everyone in the United States, Emerson said that everyone must adhere to their independent spirit, thoughts, and speech, and never follow others blindly and echo the views of others. He believes that the United States is a country constantly striving for independence. During the revolution, it gained independence from Britain and won political independence. Now, the United States should throw off outdated European traditions and fight for ideological and cultural independence. The time in which Emerson lived just required such a voice, and Americans made him one of

[15] Unitarianism is a relatively free and open branch of Puritanism.

the most influential thinkers in nineteenth-century America. Even to some extent, Emerson's ideas still lead American culture until the present day even though his works may not be read as extensively as before. However, sometimes it is difficult to determine to what extent a person's influence extends when his voice spreads too widely.

But what exactly did Boston give Emerson? An ideological basis, a necessary training for careful and meticulous thinking, and a writing and speaking ability to express his thoughts clearly and passionately; an age-old tradition of challenging authority that was passed down from Puritans; an ethical concept that emphasizes moral enthusiasm and social responsibility. Meanwhile, he also provided security and affluence. But in the city that rose through commerce, there was always a counterbalance of materialism, and the idea of disdaining the single-mind pursuit of money, which believes that individual's noblest pursuit is not commercial profit but philosophy, religion, beauty, and spirit. The architecture in Boston expressed the traditional definition of beauty in the city, which is often simple, serious, chillax, and free of magnificence and embellishment. It is precisely Emerson that experienced the hustle and bustle of the world, yet still stepped into nature to seek beauty because beauty there was a blend of elegance, simplicity, and practicality. Just like simple white wooden churches and spires, they were a model of artificial beauty in his eyes even though they fiercely challenged the religions in those churches, the reason was that they appeared so harmonious and beautiful after they integrated into the forests and lakes of New England.

However, in any case, Emerson was a rebel in Boston, not a convert. However, his rebellious voice became an authority newly established in Boston, New England, and even the United States, which would help reshape the culture and land of the city and country. Unlike the authority of Puritanism, Emerson emphasized a theme all his life: believing in himself. He told his followers to believe in their deepest instincts and their own minds, but not to blindly believe that they also had the ability to think critically; to search for individual independence and freedom, test every established norm and tradition, and make their own judgments on whether to accept or not. The best way to achieve the theme was to step into nature, where the spirit and practice of "self-reliance" were everywhere, and where individuals could stay away from the various constraints of the city and society. In nature, a person could find the freedom to think and find himself. In nature, a person could find the courage to follow his own moral compass.

4 The Virtue of Restraint

By the last 20 years of the nineteenth century, Emerson's voice had spread throughout the intellectual space of America. Today's readers are amazed by their moral preaching and questioning when reading American works of that period. However, it was a collective characteristic of the American intellectual elites deeply influenced by transcendentalism at that time, and an important aspect that set them apart from their European Romantic pioneers. No matter how seriously and intensely they criticized theology, they were still largely descendants of Protestants, and emphasized the importance of moral code in maintaining the development of civilization. However, the moral teachers they found were not the *Bible* or priests, but rather nature. Their nature was different from that in the works of British Romantic poets such as Byron or in the eyes of Whitman, author of *Leaves of Grass*. There, people could indulge all their desires and instincts, and reject social, religious, and traditional constraints and taboos. In the minds of urban elites in the United States at that time, the exchange of nature and human spirit would elevate people's moral realm and maintain social harmony.

The combination of nature and morality was not discovered in the late nineteenth century, nor was it even the creation of Emerson, their revered philosopher, and the transcendentalist master, but rather the revival of pastoralism in the New World of America. For whites who just arrived in America, the vast wilderness presented to them often aroused their complex thoughts and feelings, that is, fear, loss, strong desire to conquer, and a sense of wonder to see the Garden of Eden again. They generally believed that the pastoralism lost on the European continent would bloom and bear fruit in the process of transforming every acre of wilderness into farmland. What they practiced on the vast forests and plains of America was not only the methods and habits of farming operation that originated from Europe, but also the integration and reconstruction of pastoral ideal. In the pastoral ideal, nature was harmonious and orderly. It was a moral model higher than human civilization that divine providence showed to the world. People were required to understand and learn in the simple life of farming so as to cleanse and reflect on their own morality. The ideal of pastoralism was thoroughly elucidated and even practiced by several founding fathers in the late eighteenth century, especially Jefferson et al. Before the 1860s, it occupied the mainstream position of the imagination of nature and morality in American society.[16]

Even in the late nineteenth century, when American society had gradually entered the urban age, the interpretation of nature and morality from pastoralism still had a lasting impact. In the view of historian James Macher, the core of urban reform in the late nineteenth century was to mix cities and countries with the ideas of pastoralism. Although these ideas were criticized time and again in the twentieth century, they still resonated.[17] However, it is worth noting that among the reformers of urban naturalization, both nature and morality, as well as the connection between nature and morality, diverged from traditional concepts. Urbanization and industrialization overturned the social foundation on which pastoralism relied for existence, but the most intense theoretical impact on it came from the spread and acceptance of Darwinian evolution in the New World on the other side of the ocean. The sacred natural system believed in pastoralism completely collapsed in the merciless revelation of evolutionism, and harmony and order were replaced by endless struggles and chaos for survival. What was exposed to people was a world full of sharp teeth, blood, and slaughter, which ran its own course. As Darwin said in his scientific notes, "It's hard to believe that a terrible yet quiet war of organic beings is taking place in the tranquil forests and smiling fields."[18] The kindhearted natural world designed by the almighty and merciful Creator collapsed, and people couldn't continue to seek the moral care of nature for civilization in the murky and wild jungle. They could only fall back on humanity itself, where they had discovered science, technology and the new order established on them.

Consequently, a large number of specialized talents who received strict scientific training stepped into the mountains and forests which struggled to survive between city and country. They began to build roads, lay rails,

[16] The most famous work among them is *Letters from an American Farmer* by Hector St. John de Crevecoeur, a French immigrant. The most classic work on the tradition of pastoralism in the United States is *The Machine in the Garden: Technology and the Pastoral Ideal in America. New York by Leo Marx,* Oxford University Press, 1964.

[17] James H. Machor, *Pastoral Cities: Urban Ideals and the Symbolic Landscape of America*. Madison: University of Wisconsin Press, 1987.

[18] *Darwin's Notes on the Evolution of Species*, Part 4, p. 113. As cited in Donald Worster: *Nature's Economy: A History of Ecological*, p. 162. The impact of Darwinian evolution on the ideal of pastoralism is brilliantly discussed in "Part III: The Dismal Science: Darwinian Ecology" of *Nature's Economy: A History of Ecological.*

dig canals, repair embankments, as well as protect forests, plant trees, regulate rivers, distribute water sources, and conserve soil. Their pursuit was a dual rationalization, namely, not only to eliminate various disorderly and terrible factors of nature itself, but also to correct the traditional abuse and waste of natural resources in the United States. The notion of inexhaustible natural resources in the old dreams of pastoralism gradually faded away, while the endless treasure of technology, the new creed of the industrial age, was being established. For Americans living in the Victorian era where science was looked up to as the standard, the degraded nature would be redeemed in the dawn of science and successfully incorporated into the tide of civilization at last.

However, how will morality survive? As mentioned earlier, progressive reformers felt an unprecedented moral panic in the curtain call of pastoralism and the advent of the urban age. Since the nature's moral structure collapsed, the morality of the urban age seemed inevitably to be built on the consciousness and confidence of human civilization. After the impact of Romanticism on knowledge and reason, the great development of scientific theory, the decline of theological dogmas, the rapid growth of technological strength, the significant improvement of medical conditions, and the general rise in living standards in the Victorian era made the authority of science re-established and stronger than ever. At that time, society was not without different ideas. In fact, among the intellectual elites living in the era of industrial cities on both sides of the Atlantic, there were continuous doubts and questions about the direction of modern civilization and its cornerstone, namely, science. However, such ideas were often full of contradictions and could not resist the arrival of a far more powerful scientific age.[19] People generally believed that science would raise the moral standards of society by improving educational standards and living environments, ensuring the health of citizens, and various other means while the people's mind was enlightened.

[19] In the late nineteenth century, a wave of reflection on the direction of civilization progress emerged on both sides of the Atlantic. Some intellectual elites believed that Western society had become overcivilized, so they called for a return in various different meanings. For the issue, see T J. Jackson Leers's classic work, *No Place of Grace: Antimodernism and the Transformation of American Culture, 1880–1920*. T. J. Jackson Lears, *No Place of Grace: Antimodernism and the Transformation of American Culture, 1880–1920*. New York: Pantheon Books, 1981.

If the government and citizens actively participated in the transformation, science would help establish rational individuals and society, thereby maintaining moral balance.

The active urban naturalization reformers in the late nineteenth century were also firm believers in science and guardians of civilization. In their calls and actions to protect the wilderness and build urban parks, there was not only a passionate emotion but also a rational spirit and practical considerations. From the text they left behind, we can easily find that they know well the evolution. Not only did the profession of landscape designers like Olmsted need cooperation with nature but also these designers are required to eliminate uncertainties in nature with artistic perspectives and scientific methods so that their profession more agreed with the existence of civilization. Even in the national parks they strive to preserve, they inevitably bring machines into the wilderness and use planning to restrict nature in order to meet the leisure and entertainment needs of most people, which is a typical feature of the treatment of nature in the urban age.

However, there were subtle differences between them and other progressive reformers in the way they treat nature, or even further, the relationship between wilderness and morality. Olmsted as well as his followers and supporters still discovered a mysterious connection between nature and morality. In the report prepared by Olmsted for the establishment of Yosemite State Park, he pointed out that "it is unquestionably true that excessive and persistent devotion to sordid interests cramp and distort the power of appreciating natural beauty and destroy the love of it which the Almighty has implanted in every human being, and which is so intimately and mysteriously associated with the moral perceptions and intuitions, but it is not true that exemption from toil, much leisure, much study, much wealth, is necessary to the exercise of the aesthetic and contemplative faculties."[20]

The source of the mysterious connection, in the eyes of those urban green reformers at that time, first came from human demand for natural beauty. Undoubtedly, the appreciation of natural beauty had a mystical connection between human body and mind and nature, which is the remaining influence of Romantic thought in the late nineteenth century. In the meanwhile, it couldn't be denied that they still retained a strong

[20] Olmsted, "Typed Transcription of Draft of Preliminary Report upon the Yosemite and Big Tree Grove.".

demand for pastoralism in their consciousness, and had a persistent pursuit of the serious morality brought about by simple life although they lived in cities and believed that cities were a more noble and beautiful stage of civilization. However, for these urban reformers, the connection between natural beauty and morality laid more emphasis on the thirst as a result of the great pressure to the physical and mental health of urban population brought by overcrowding urban environment and the intense modern life. It regarded the demand for natural beauty as a fundamental physiological requirement that human beings were born with, a fundamental condition for the existence of goodness and a fundamental link between the urban age far away from the land and nature.

Furthermore, the connection between nature and morality also originated from the pursuit and retention of a free, aggressive, and persistent frontier spirit. The advent of the urban age marked the disappearance of the free land of the vast continental North America. In 1890, the US government formally declared the border closure, thus officially ending American Frontier era which lasted for more than two centuries. At the World's Columbian Exposition of 1893, Turner published the famous paper *The Significance of the Frontier in American History*, which directly reflected the American understanding of and confidence in local culture and morality at that time, irrespective of the enduring debate it sparked in the historian circle. Turner wrote at the end of the article: "and yet, in spite of environment, and in spite of custom, each frontier did indeed furnish a new field of opportunity, a gate of escape from the bondage of the past; and freshness, and confidence, and scorn of older society, impatience of its restraints and its ideas, and indifference to its lessons, have accompanied the frontier."[21] Although the frontier spirit was still reverberated with the long-standing inquiry of the relationship between nature and freedom shared by various cultures, it was no longer a continuation of European pastoralism on the New World, while cultural and moral orientation with distinct American characteristics nurtured in the American wilderness.

Therefore, in the urban age in which American frontier faded out, the preservation of wildness held an equally important position as natural beauty. Once wildness was driven away by man-made buildings such

[21] Turner, "The Significance of the Frontier in American History.".

as railways, highways, markets, and residences, places like the Adirondacks would become a combination of mountains, forests, and lakes. The common state of busy life would be in sight, and they would still be attractive, but no longer a wilderness. The call for wildness that modern population makes in the wilderness means their rediscovery of the freedom lost in industry, urban norms, and hierarchy. When wildness disappears, the fundamental significance of urban population entering these mountains and wilderness will no longer exist. As Thoreau shouted when the urban age faintly appeared on the shores of Walden pond, the reformers living in the late nineteenth-century urban age also sought to tell people: "In Wildness is the preservation of the World." An instruction that people can listen to their inner moral compass, stay away from authority and constraints, and break free from the unchanging norms and petty profit of daily life.

It was in the call and action to actively preserve the wilderness that urban reformers in the late nineteenth century began to interpret the "virtual of constraint," the third meaning between nature and morality The morality of the urban age began to reflect on the excessive individualism in American culture, and attempted to suppress it with the freedom that was referred to as collective by Charles William Eliot. It was the spirit of the entire progressive era, and the spirit that people had to follow when urban public spaces were constructed. Regardless of whether a person was willing or not, constraints would inevitably become a necessity in human attitudes toward nature. In the new urban age, Americans warned that what needed to be preserved was the simple, robust, and equal qualities of the frontier spirit, rather than its recklessness toward nature and hatred toward the wilderness. People searched for the wilderness, stepped into the wilderness, and protected it with civilization because they needed freedom. However, such freedom was not behavioral indulgence but spiritual liberation. In order to achieve it, a city or even a country had to recognize the boundaries of civilization's expansion into the wilderness, and people had also to learn the "virtue of restraint." It emphasized the responsibility of each individual toward others, the group and descendants, and ultimately emphasized the respect and obligation of human toward nature.

5 Reflection on Megacities

In 1895, Olmsted officially retired, and Mumford was born. He grew up in New York City, where was full of Olmsted's works everywhere. In the year of his birth, urbanization and industrialization that Olmsted was afraid of had become prominent, and the natural landscape was shattered by the invasion of cities. A nature preservation movement launched from the middle class also shone brightly on the political and social stage of the United States as a strong modern demand, and has not come to an end. Olmsted was undoubtedly a pioneer of the movement, but Mumford did not stand at the stage of the movement. More importantly, he internalized it as a profound reflection on the relationship among civilization, cities, technology, and nature.

Although Mumford, who grew up in the twentieth century, still had confidence in the future of civilization, he was more anxious about and fearful of it. In the 1930s, when his ideas were mature and efficient in writing, the United States had fully entered an urbanized society. The hustle and bustle of the 1920s had fallen silent under the heavy blow of the global economic crisis, but the megacities, which Mumford was anxious about, spread like mollusks, and the social " mechanical system" dominated by machine alienated human nature and threatened the environment in which organisms lived. In his view, the finance, politics, and architecture of megacities all threatened the individuality that each city should possess and that was built together by unique community cultures, as well as the organic connection between city and nature. If cities continued to follow the existing path, the big cities of the New World would suffer the fate of the collapse of ancient Roman cities.

Many of Mumford's contemporaries had similar fears with him. In the 1930s, pastoralism revived again, it responded to the ubiquitous monopolistic economy in cities and its global collapse at that time. Compared with the "back to nature" movement in the late nineteenth century, the new generation of pastoralists had a clearer anti-urban intention. They refused to seek a middle road between prosperous urban economy and natural beauty, and insisted that their rejection of modern capitalist society should start with "back to the land" for labor and livelihood.[22] Their denial of cities was not moral concerns at all, or the ways to seek methods for reconciling emotional breakdown. Although both haunted in their minds, the denial came from the disillusionment with the capitalist economy and the accompanying institutional confidence. Similarly, the new pastoralism was not a simple reversion to the ancient ideals of Jefferson's yeoman although Jefferson's influence always existed like a shadow. These advocates of "back to the land" were also different from Thoreau, Muir, or subsequent Ed Abbey, who firmly believed in the power of wilderness and sought the redemption of civilization in the wilderness. Although most of them were pursuers of Thoreau's "simple life" and faithful observants of Emerson's "self-reliance" philosophy. Their ideological origins were mixed with various ideas from the America and new ideas from across the Atlantic during entire nineteenth century. They laid different emphases for different individuals. However, their common reaction is to escape from the shadow of capitalist economy in cities.

One of their representatives is Ralph Borsodi, a Manhattanite. He began to reflect on the issues of American society, and even the civilization of the entire industrialized city, in the 1920s which was prosperous and tumultuous. In the 1920s, he, together with his wife and two sons, lived in a rented apartment in Manhattan, New York, and maintained a modern nine-to-five life as a low-level white-collar worker. The city they lived in had the most exquisite apartments, the brightest buildings, the most delicious restaurants, and the most active cultural life. However, when a family still struggled for a living, how did they enjoy all the urban charm that people called? Moreover, he deeply felt the lack of "real beauty—the beauty which comes only from contact with nature and from the growth of the soil, from flowers and fruits, from gardens and trees, from birds and animals?"[23] Finally, when their landlord sold the apartment they rented but didn't inform them of it, they faced the common problems of thousands of urban tenants and began to search for a place to stay again in the

[22] In the movement of anti-mainstream culture in the 1960s and 1970s, "back to the land" became an influential social movement. A large number of young people declared that they broke away from mainstream culture by leaving industrialized cities, returning to the land, establishing communes and practicing self-sufficient agricultural lifestyles.

[23] Ralph Borsodi, *Flight from the City*. New York: Harper & Brothers, 1933: p. 2. In the Chapter I of *Flight from the City*, they described their own experience in vivid detail.

vast and crowded city. At that moment, he and his wife made an important decision—to escape from the city. In their original imagination, they still hoped to find a place not far from the city where they could enjoy the comfort brought by modern electricity and water-works, as well as the joy of direct exposure to the land. However, in the end, their budget only allowed them to purchase a seven-acre farm located more than an hour's drive from the city, with a house without any modern facilities on it. From then on, they began to cultivate in person on the land, a completely self-sufficient life that fundamentally rejected capitalism's control over human life through the market.

Ralph Borsodi published his book *This Ugly Civilization* in 1929, and *Flight from the City* in 1933.[24] In these two books, he repeatedly emphasized that the current problem was not a problem of a city or a country, nor is it a temporary spasm of capitalist economy, but rather a chronic illness of industrial civilization that "is either on the verge of collapse or of rebirth on a new social basis." "Men and women who desire to escape from dependence upon the present industrial system and who have no desire to substitute for its dependence upon a state controlled system, are beginning to experiment with a way of living which is neither city life nor farm life, but which is an effort to combine the advantages and to escape the disadvantages of both."[25] How to combine them? and how to escape? Borsodi suggested escaping from cities and markets by taking possession of land, and escaping the inconveniences of rural areas by using modern machines to adding modern comfort to a self-sufficient life.

In Borsodi's time, his voice was not lonely. The advent of the Great Depression made it difficult for many people who lived in cities to continue making a living and returned to the country which they had fled from. At that time, there was no longer the one in western America in the time of the Homestead Act, and there were no longer large areas of unowned land for people to freely choose and develop. Moreover, the land there was completely involved in market production, and in the process of industrial development, it soon fell into a huge ecological and economic crisis—a dust storm. However, there were many abandoned small farms near the big cities in the Eastern America. If people chose to

[24] Ralph Borsodi, *This Ugly Civilization*. New York: Simon and Schuster, 1929. In addition, Borsodi also published *The Distribution Age* in 1927.

[25] Borsodi, *Flight from the City*, p. xiii.

survive on the land rather than made profit from it, those places could still provide livelihoods, safety, self-reliance, and "real beauty" that Borsodi believed. Later than the publication of *Flight from the City*, Maurice Grenville Kains directly, in 1935, provided a survival manual for such a life: *Five Acres and Independence*.[26] For Kains, ordinary people living in cities, especially tenants, were always in an unsafe state of livelihood. However, more importantly, they were unable to be truly productive in their urban life. While rural areas provided them a sense of security, they could also make people feel the joy and satisfaction of production, and receive the liberal education for physical and psychological health. However, compared with Borsodi, Kains' strategy of "back to land" was closer to Jefferson's Agrarian Vision of the United States since its founding. It lacked deeper reflection on both capitalist economy and its culture, but was only a expedient measure for seeking livelihood during the Great Depression.

On the back-to-the-land road, Helen and Scott Nearing went far than anyone else. They also fled from New York during the Great Depression and went to a farm in Vermont. Unlike Borsodi, Kains, or millions of people who left cities to make a living in the countryside during this period, Helen and Scott Nearing returned to the land not forced by livelihood but for a self-selection life, which they called "simple living."[27] Helen and Scott both came from wealthy families on the East Coast so they weren't worried about their livelihood. After obtaining a PhD in Economics, Scott Nearing taught at the Wharton School of the University

[26] Maurice Grenville Kains, *Five Acres and Independence*. New York: Greenberg, 1935.

[27] According to the US Population Census statistics cited in Borsodi's book *Flight from the City*, Americans always migrated between the city and the country. However, from 1910 to 1929, the number of people who migrated to the city remained higher than those who migrated from the city to the country, the case lasted until the economic crisis broke out in 1929. In 1930, for the first time, the number of persons leaving farms for cities exceeded the number of persons arriving at farms from cities. In that year, 1.74 million persons arrived at farms from cities and 1.723 million persons left farms for cities. The former was 17,000 persons more than the latter. By 1931, the gap between farm population and city population had significantly increased. 1.683 million persons arrived at farms from cities and 1.469 million persons left farms for cities, and the difference was up to 214,000 persons. In 1932, it reached a new peak, with 1.544 million persons arriving at farms from cities, 1.011 million persons leaving farms for cities, with extra 533,000 persons arriving at farms from cities. In Bosodi's view, such large-scale migration indicated that persons were not satisfied with both urban and rural life. Borsodi, *Flight from the City*, p. xii.

of Pennsylvania around 1910. However, his radical criticism of capitalism annoyed the Board of Trustees which was composed of big capitalists, and he was notified in 1915 not to renew his appointment contract after it expired. The event sparked a debate on academic freedom throughout the entire University of Pennsylvania and even the academic circles on the East Coast, and Nearing built his reputation as a radical public intellectual. His own political tendency also kept turning left and became a believer in communism. In 1956, he and Helen came to China again and the former Soviet Union. After returning from China, they both borrowed the title of the famous dystopian novel *Brave New World* and added the article "the" in 1958 to write their book *The Brave New World* for recording their travels to the Soviet Union and China. In the stalemate age of the Cold War, it was believed in the book that the two communist countries mentioned above were "peaceful socialist giants."[28]

Their political conception directly affected their living practice. In 1932, Helen and Scott purchased an approximately 750 acre forest and a small-sized farm in Winhall, Vermont at the price of $2.5 per acre, and moved their family there from New York. It undoubtedly marked their disillusionment with capitalist culture and practical implementation of their belief in "simple living." In their most famous work *Living the Good Life: How to Live Sanely and Simply in a Trouble World*, which was published in 1954, they explain clearly the reasons for leaving New York and heading to Vermont farm. The book had a huge impact on the movement of anti-mainstream culture and the "Back-to-the-Land" movement that emerged in the 1960s and 1970s. They wrote:

We left the city with three objectives in mind. The first was economic. We sought to make a depression-free living, as independent as possible of the commodity and labor markets, which could not be interfered with by employers, whether businessmen, politicians, or educational administrators. Our second aim was hygienic. We wanted to maintain and improve our health. We knew that the pressures of city life were exacting, and we sought a simple basis of well-being where contact with the earth, and home-grown organic food, would play a large part. Our third objective was social and ethical. We decided to liberate and disassociate, as much as possible, from the cruder forms of exploitation, the plunder of the planet;

[28] Regarding his personal activities and political beliefs, see his autobiography *The Making of a Radical: A Political Autobiography*. New York: Harper and Row, 1972.

the slavery of man and beast; the slaughter of men in war, and of animals for food.[29]

It is precisely on the third objective that Helen and Scott Nearing, surpassing the "Back-to-the-Land" movement represented by Borsodi, Kains, and others, began to reflect on the ethical relationship between human beings and the Earth they live on as well as other living beings on it. At the moment they turned around and left the city, they bid farewell not only to an economic system or political structure, but also to the human-centered morals which had a firm foundation.

In 1952, Helen and Scott Nearing left their forest farm in Vermont because the area nearby their farm would be developed into a ski resort. They moved to the remote and desolate town of Brooksville in Maine to continue practicing their simple life. At that time, the little town on the coast of Maine became the place where many East Coast elites who were tired of the big city and yearned for a more honest, simple, and free intellectual life. As mentioned in Chapter IV, the mount desert island in Maine was favored by many people of the upper class in New York and Boston as early as the late nineteenth century, and villas were built there, including Charles Eliot's father, and Charles William Eliot, the President of Harvard University. The behavior also became the reason why Eliot deliberately protected the coast of Maine and ultimately made it the first national park on the East Coast. After the outbreak of the Great Depression in the 1930s, more intellectual elites from New York and Boston began to seek new ways of life in Maine which was sparsely populated. However, at that time, many of them no longer saw it as just ornaments for their luxurious lives, but as the foundation of their lives. For example, the famous writer E B. White and his wife, Katharine Sergeat, one of the founders of *The New Yorker*, also left New York in 1937 and moved to Brooklin, another small town which was only 12 miles away from the town where Helen and Scott Nearing moved later. Four years ago, they bought a farm and a house built in the late eighteenth century, and finally decided to move their family. There, White wrote the book *One Man's Meat* (As Sweet As the Shepherd's Purse) to record his life and work on

[29] Helen and Scott Nearing, Living the Good Life: How to Live Sanely and Simply in a Trouble World. New York: Schocken Books, 1954, p. ix.

the farm. Of course, he also wrote the popular *Charlotte's Web* and many other works on that farm.[30]

"As Sweet As the Shepherd's Purse" may accurately reflect the attitudes of Scott Nearing, White and others, self-selection in a complete sense. The selection was not unique to the urban elites of the 1930s in the United States. The time and space of various civilizations such as China, Britain, and France reverberated with "Ah, homeward bound I go! Why not go home, seeing that my field and garden with weeds are overgrown?" From Tao Yuanming (a famous Chinese poet from the six dynasties period) to White, the nature's mystery they felt in nature, the spiritual freedom they pursued by returning to the rural areas, are not completely unrelated as a result of the remote spatio-temporal distance. However, in the 1930s when the tentacles of modern urban civilization were ubiquitous, when the Great Depression ended unprecedented material prosperity, and when these intellectual elites on the East Coast of the United States were disappointed with the democracy they believed in and pursued throughout their lives in the urban power with strict hierarchy, their selection remained strong in their pursuit of social justice and equality, in addition to the pursuit of personal freedom. They hoped to tell Americans that if they were willing to live a simpler life in the country, it was possible that they still found a better, healthier, more independent, and equal life.

Researchers have pointed out that Helen and Scott Nearing had never been completely "independent" as they themselves claim because both had inherited a large amount of legacy. However, their way of survival didn't deviate from their political conception, life principles, and life philosophy. They truly worked on the land but didn't rely on the services of child servants or tenant farmers. They always led a vegetarian life because they didn't believe in the need to slaughter other lives to satisfy their own appetite. They never amassed wealth. When they were preparing to leave Vermont, the price of land there skyrocketed. The price of the woodland which once sold as the price of US $2.5 per acre was US

[30] E. B. White, *One Man's Meat*. New York: Harper & Brothers, 1942. The book has a Chinese translation, and the translator translated the title as *Everyone is Different*, which is an accurate translation in Chinese. However, White's book title is derived from the English proverb "One man's meat is another man's poison," which reminds me of the Chinese saying "Who says that the sowthistle is bitter? It is as sweet as the shepherd's purse?" Therefore, it is translated accordingly.

$8000 per acre in 1952. They could earn US $6 million from their forest farm due to the rising price of the land. However, both of them chose to present it as a municipal forest to Stratton. Compared with White, their understanding of freedom and equality did even better. White always cared about persons and human society, while Helen and Scott Nearing extended their moral reflections on slavery, plunder, and slaughter to other living beings and the Earth. If their practice of "back to the land" and simple living was just a reconstruction of Utopia in the twentieth century, at least their astonishing longevity proved the rationality of their pursuit of health. Scott, who maintained strong creativity throughout his life, was born in 1883 and passed away in 1983. 12 years later, Helen died at the age of 91.

Compared with these anti-urbanists, Mumford still stood on the side of the city. Like Olmsted, he did not want to wallow in any form of old dreams of Utopian, pastoralism, medieval cities, or back to the land of his contemporaries. Although nostalgia was constantly expressed in his works, it never drowned out his rational thinking. He was well aware that modern cities had gone too far, and the complex problems and ecological dilemmas of cities couldn't be solved only by individual emotions, local knowledge, or personal practice. Compared to the cities in the age in which Olmsted lived, and even the cities in the 1930s that hesitated about which way to go under the heavy impact of the Great Depression, the technological level of cities in the 1960s exceeded the imagination of any science fiction writer in the nineteenth century. Cities became the great achievements of human culture, but cities and the driving force behind their technology and human migration still stemmed from the natural world that they attempt to escape from. Nature was not an isolated space far from human society, but a force that not only was changed by human technology but also constantly shaped the thought, society, and space of human beings. Mumford, deeply influenced by ecology, incorporated cities and the technologies that support their existence into the historical course of human society to analyze their mutual influence with natural history. He saw in the historical evolvement, how cities sprouted within an ecosystem, grown under its nourishment, recklessly plundered the nature, the mother they relied on for existence, when they grew strong enough, and how cities severed the psychological bond between them and the system through technological self-deception and deception. Mumford did not probe further into how nature set limits on the growth of cities, but the long history he witnessed made him clearly point out that the end-result of urban development was still the ecosystem which the city depend on. Regional planning of the city, in combination with new biotechnology, would make the life of the city and nature extend together.

From Olmsted to Mumford, cities were expanding and growing step by step, and the knowledge that shaped cities constantly produced fission. However, they always manifested the pragmatic wisdom of the American intellectual circles, which was full of idealism. The pragmatic wisdom made them face the complexity of the world and be ready to perceive changes at all times, adjust strategies, and make corrections, rather than restore and process them in a simple way. They all had faith in modern science, but it was a trust in Arcadian ecology, an organic, holistic, and historical ecology. For Mumford, the new pragmatic wisdom came from such ecology. In his view, modern science had developed into two paths: science of mechanism and science of organism. The former was accompanied by technology from the industrial age, while the latter corresponded to technology from the zoic age. Mumford did not intend to completely negate the former. In fact, while criticizing technological dominance, he had the due respect for technology. However, "This reawakening of the vital and the organic in every department undermines the authority of the purely mechanical".[31] The new machine system would adjust its operation on the basis of the purpose of life, not only human life, but also the entire natural life, so as to achieve a dynamic balance between the city and nature, and make the transformation from a hygienic city to an ecological city.

The period from the late 1860s when Olmsted began designing American cities as a landscape architect to the 1960s when Mumford ended writing about American cities as an architectural critic was a century for American cities to achieve modernization. However, the practice of modernization in the United States has never been a forced transformation of rigid concepts, nor is it a linear process of a guiding ideology going forward, but rather a matter of innumerable twists and turns, a complexity with its own characteristics in every period. Modernization implemented in American cities not only means the development of natural resources with modern technology, changes in the material landscape of cities, efficient operation and growth of cities, and the construction of the overall

[31] Lewis Mumford, *Technics and Civilization*, translated by Chen Yunming et al., Beijing: China Architecture Industry Press, 2006, p. 327.

order and discipline of cities, but also a profound reflection on the development direction of cities, and even the civilization age represented by cities, with modern science as its knowledge structure and the modern spirit of challenging authority as its ideological background. It is the latter that serves as the brake to prevent the high-speed train from derailing in modern cities.

In the dual sense of humanism and ecologism, Olmsted and Mumford were the two most important urbanists in the nineteenth and twentieth centuries of the United States, respectively. For their time, they both possessed prophetic qualities that transcend their time, but the knowledge structure of their time still laid the foundation for their reflection on urban civilization. If there was a consistent attitude toward the challenge of authority in the American experience, modern scientific knowledge was by no means rigid and unchanging, it was always in a state of self-challenge and change. The transformation of the ideal of urbanism from Olmsted to Mumford conveyed information about changes in American environmental knowledge.

In Mumford's view, the change manifested what he called urbanism. It was first and foremost a humanistic spirit; on the one hand, it was the open and inclusive temperament that Olmsted saw in modern cities, which broke the democratic concept of the original rural social hierarchy. On the one hand, it was the principles of human nature that should be followed in city planning, as Mumford expected. Subsequently, urbanism also expressed an ecological ideology. When Olmsted was active in American urban landscapes, "ecology" as an independent discipline had just emerged on the domain of human knowledge. However, behind it lay a long tradition of natural history, which views nature as an organic whole. Darwin's *On the Origin of Species*, which was published in 1859, brought an idea storm to the European and American intellectual circles. By the early twentieth century, the explanation of evolution had already entered a deeper level of knowledge structure in Europe and America. From then on, people could not imagine nature as a perfect and predetermined God's design any longer, but saw its long evolutionary process and the competition and turbulence within it. Olmsted, who was active in American cities in the period from the 1860s to the 1890s, was an early practitioner who incorporated organic, holistic ways of thinking about nature into city planning to stress the indispensable role of nature in urban civilization. On the basis, Mumford, who wrote in American cities in the period from 1930 to 1960s, incorporated the historicity of nature into his reflection on the development direction of urban civilization. In *The Myth of the Machine: The Pentagon of Power*, his final important work on cities, he claimed, "Today, all valuable ideas should be ecological."

But what was the significance of those anti-urbanists? Any society requires some form of Utopia and idealism, at least they provide imagination for the possibility of another life. From Thoreau to the Helen and Scott Nearing, their practical significance for simple living lay first and foremost in that. However, perhaps more importantly, the selection of these persons who escaped from the city and confronted it was also a reflection on the relationship among city, countryside, and nature. They attempted to explore another path of civilization evolution, and hoped to question the fate of modern civilization growing in the crowded and poisoned cities dominated by capital and technology through their own reflection and practice, and force those who still chose to dwell and live in cities to fundamentally reflect on the path of urban civilization, and also force urban reformers to naturally and unquestionably make progress and growth a prerequisite for a healthy, beautiful and good life. Furthermore, their thinking transcended human self-care, constrained the recklessness of human behaviors, endowed freedom and ethics with new meanings, and made people realize that other living beings which live together with humanity on the planet also had the right and freedom to survive.

CHAPTER 8

City by the Bay

1 Prologue: "I Left My Heart in San Francisco"

Urban landscapes and natural landscapes share a common charm, namely, they often can arouse the inspiration of musicians. Many American cities, big and small, trembled with the sensitive mind of artists at a moment. At that moment, the musician might compose song a new song to either passionately chant nostalgia for and attachment to the city, or bestow praise on and take pride in it. For an urban environmental historian, these songs were far more than just their enchanting melodies, but provided special clues to convey certain cultural and environmental implications. For example, the song "New York, New York" in the musical film of the same name directed by Martin Scosese in 1977 became popular for a time and almost became a New York endorsement song when Frank Sinatra sang it with his magical voice two years later. It made innumerable homeless persons stop their wandering footsteps, yearn for a new beginning, and hope to blend their own figure into the bustling crowd of New York. However, the song belonging to New York didn't have the water flow of the Hudson River, the birdsong of Central Park, the surging waves at the feet of the Statue of Liberty in the distance, and the swamps that disappeared in the urban sprawl. It even freed of physical form of the city's architecture. Giant skyscrapers stretched out on both sides of narrow streets, one after another, they resembled deep corridors, thus making Manhattan one of the most densely populated places on Earth's surface. The New York described in the song was the place where innumerable

S. Hou, *Wallless Cities*,
https://doi.org/10.1007/978-981-97-7827-0_8

people hoped to achieve great success and look down on creation. "I want to wake up in a city that doesn't sleeps," the lyrics say, "and find I'm king of the hill." If an ambitious American/immigrant "make it there," then he "make it anywhere." The city, the city of the American dream, was defined by self-reliant individualism which was appealed for during two hundred years, the city that fulfilled everyone's great aspirations, and the city was ultimately chosen by thousands of tough-minded and enterprising individuals who yearned to escape the drowsy towns, rural life and outmoded conventions of the place. "These little town blues are melting away/I'll make a brand new start of it/in old New York."[1] Like many providers of the American dream, the song promises those who come here: in the city, as long as they have clear goals and work hard, they will ultimately reap their rewards due to their talents.

Of course, the promises of the songs, like the speeches of politicians, are often only their appealing voices. Innumerable people left their hometowns, lived in New York and surged with human flow through the dark corridors of the city that never sleep. However, when they woke up in the morning, they found themselves still empty-handed and penniless, even abandoned and forgotten in the huge city, and found themselves still keeping their hearts elsewhere. Among all the urban songs originating from New York, the most famous one is not "New York, New York," but "I Left My Heart in San Francisco," the song about a distant West Coast city. Two San Franciscans, who had already passed their thirty years of age and resided in New York, left their hometowns during World War II. After the war, they, like thousands of drifters, arrived in New York where they hoped to realize their dreams. Eight years later, George Cory (composer) and Douglas Cross (lyricist) co-wrote hundreds of songs, but none of them brought them success, fame, or money. Or on a noisy but lonely winter night, or on a brilliant yet desolate summer day, recollections of San Francisco overwhelmed their lives in Brooklyn apartments, thus the city song of San Francisco was born.

Of course, these two "New Yorkers" in dire straits would have to wait another nine years. The song was in favor after Tony Bennett, another Italian American singer, sang it at Fairmont Hotel, a landmark building in San Francisco. It would take another 22 years until 1984, when the song,

[1] John Kander & Fred Ebb, "New York, New York," 1977. In the film, the song was first sung by Liza Minnelli. Frank Sinatra deeply loved the song and first recorded it in 1979, and recorded it again at the age of 78 in 1993.

along with another single "San Francisco" in a musical officially became the city song of San Francisco. On April 25, 2020, when the COVID-19 epidemic in California exacerbated, San Francisco residents sang the song on the Internet, with 94-year-old Bennett as the leading singer. The song brought George Cory and Douglas Cross not only the success, fame, and of course wealth that they longed for, but also them back to their hometown by the bay. The song and Bennett also helped each other succeed, and in the following decades, he continuously recorded the single. Due to the song, his statue was built outside the Fairmont Hotel to celebrate his 90th birthday.

Completely different from "New York, New York," the melancholy in "I Left My Heart in San Francisco" was lingering like the cold mist of that city:

The loveliness of Paris
seems somehow sadly grey
The glory that was Rome is of another day
I've been terribly alone and forgotten in Manhattan
I'm going home to my city by the Bay
I left my heart in San Francisco
High on a hill it calls to me
To be where
little cable cars climb halfway to the stars
The morning fog may chill the air I don't care
My love waits there in San Francisco
Above the blue and windy sea
When I come home to you San Francisco
Your golden sun will shine for me[2]

It sketches the contours of the nature in San Francisco, the saltwater bay where ships from all corners of the world are moored are embraced by low hills and small ranges. The location of the city, which faces the vast Pacific Ocean and distant Asia, is one of the most ideal natural ports in the world. The surrounding ranges and hills provide shelter for the bay from storms in the Pacific, and leave only an exit, one and a half mile wide, which is the famous Golden Gate in the world. The marine climate makes the city cool and mild all year round, and also shrouded in thick or

[2] George Cory & Douglas Cross, "I Left My Heart in San Francisco," 1953, translated by Hou Wenhui.

light milky white mist every morning. In the twentieth century, San Francisco deeply embedded the romantic imagination in Americans, natives, or foreigners living elsewhere. What lingers in the singer's memory are the undulating landforms of the city which are sometimes steep and sometimes continuous, the cold stars twinkling in the hillsides, the mist in the morning, and the "the blue and windy sea." In his memory of the hometown, that city, the place of romance and love, is completely different from Manhattan's indifference. It has a small cable car and warm human relationships, and it remains a fresh and modern city that makes Rome's glory become the past. However, it is not New York, Boston, Pittsburgh, Kansas City, or Las Vegas and Los Angeles, which are also located in the west of the Rocky Mountains. The city has a peculiar charm that inspires people's romantic imagination of nature.

There is no doubt that every big city is polyphonic, they don't have fixed main melody or invariable sound accompaniment. When all melodies are played together, they don't necessarily constitute perfect and smooth harmony, but always heighten a certain open and outstanding ethos. And so is San Francisco. The small town, with a population of only 850 people on March 15, 1848, a year before the gold rush, grew into a city with a population of nearly 900,000 in just 172 years, among which the white only accounted for 40%, Asian people was up to 34%, in addition, there were 15% Latino and 5% African American. Its evolutionary history was full of diverse melodies, either intertwined, contradictory, or independent.[3] Different listeners with different ears may hear a completely different melody, even for environmental historians who try to listen to the collision of nature and culture in that city, the melodies they hear are quite different. Gray Brechin heard in his book *Imperial San Francisco: Urban Power, Earthly Ruin* not only greedy capitalists noisily planning, arguing, and sharing profits in the skyscrapers in the city, but

[3] For the population of San Francisco in 1848, see Frank Soulé, John H. Gihon, and James Nisbet, The Annals of San Francisco; Containing a Summary of the History of the First Discovery, Settlement, Progress, and Present Condition of California, and a Complete History of All the Important Events with It Great City. Originally published in 1854, with Introduction by Richard Dillon and index by Charles Goehring, Berkeley: Berkeley Hills Books, 1999: p. 200. The yearbook, published in 1854, provides the full, accurate and comprehensive record of the early history of San Francisco and California. For other population and ethnic composition, see United States Census Bureau, https://www.census.gov/quickfacts/sanfranciscocitycalifornia, which was logged in on September 20, 2020.

also machines constantly extracting minerals and water from mountains and rivers day and night.[4] While Richard Walker heard in his book *The Country in the City* the appeal made by men and women from different races and classes over the last century to build a green, livable, and healthy city.[5] Joanna Dyl heard about not only the major disasters caused by the Earth's own movement on natural fault zones in *Seismic City: An Environmental History of San Francisco's 1906 Earthquake*, but also people's confidence and disillusionment in the safety brought by technology, in addition to complaints, denunciations, and protests from different races, classes and groups due to their respective interests, demands and perceptions.[6] All the voices are not the auditory hallucinations of historians,

[4] *The Imperial San Francisco: Urban Power, Earthly Ruin* is an important work on the study of urban history of San Francisco. In the book, urban geographer Brechin tries to break with the romantic imagination of San Francisco and reveal how a small portion of capital controllers used remote control technology to grab energy and resources from their "empire" sprawl during the expansion of San Francisco. He believes that San Francisco and the entire urbanization process were both the process in which a few people accumulated a large amount of wealth at the cost of sacrificing hinterland and the well-being of their future generations. In the process, these people controlled the voice of the city, reshaped their self-image through mass media, and hidden their greed under the construction of myths. Grey Brechin, *Imperial San Francisco: Urban Power, Early Ruin*. Berkeley: University of California Press, 1999.

[5] In the book *The Country in the City*, historian Walker observes the evolution of the city from a perspective completely different from the *Imperial San Francisco: Urban Power, Earthly Ruin*. He believes that the city began to reflect on its own development trace and established a new urban culture after the crazy plunder during half a century from its birth. Walker believes that the urban culture of San Francisco, compared to all other big cities in the United States, is the greenest and most environmentally friendly. More importantly, the environmental culture was not created by a few intellectual elites like John Muir, but it was the result of the joint efforts of people from different classes and races in the city. In Walker's view, these people fought against the ambition and power of capital, limited the unregulated development of the city, and worked together to build it into a city and metropolitan area where the countryside was fully integrated into the city. Richard A. Walker, *The Country in the City: The Greening of the Bay of San Francisco Area*. Foreword by William Cronon, Seattle: University of Washington Press, 2007.

[6] Compared to Imperial San Francisco: *Urban Power, Early Ruin* and *The Country in the City*, the famous works concerning the environmental history of San Francisco, *Seismic City: An Environmental History of San Francisco's 1906 Earthquake* published in 2019 truly makes the nature of San Francisco play its role in the narrative of environmental history once again. Nature, in Brechin's works, means that resources and energy are the objects of plunder, while nature, in Walker's works, means landscape, which is the object to be designed, constructed and protected. Although their analysis perspectives are

but powerful voice of recovery over the city and a certain reality of class conflict. The polyphonic scores performed by them in coordination are full of unexpected changes, as well as contingency and breakdown. It is sometimes a response to the aesthetic traditions of the eastern region on the other end of the continent, sometimes an imagination of the ancient East Asian civilization on the other side of the Pacific, sometimes a thirst for natural wealth here, and sometimes a discovery of the beauty of nature here. What all the listeners of its history capture is just a piece of movement, or different notes on the polyphonic cross-section at a moment of separation. In the chapter, we hope to listen to the interpretation of the mode of urban existence that result from unique natural environment and poetic contemplation of San Francisco, which has been constantly evolving in the history of the city since 1849. In the new ideas brought by the sea breeze of the Pacific Ocean and the changeful natural world of California, people's reflections on beauty, cities and nature have finally broken free from the localization of cities on the East Coast of the United States, and inspired people to make new attempts.

2 Passing through the Golden Gate

The birth of the city has nothing to do with any poetic imagination or moral demands. It is not the model of global ethics, the "City upon a Hill" that Winthrop hoped to create. It is not the development of natural agricultural wonderland in the eye of the boosters of Kansas City, by means of private yet organizational capital. Its initial formation stemmed from the desire of innumerable people for a precious metal that was once deeply hidden within the Earth's body. The California Gold Rush, officially launched in 1849, was the most significant event of the entire nineteenth century in the west of the United States and western coasts. It made California a state of the United States and initiated an ongoing

different, the nature in San Francisco, in both books, is a passive object, and human power holds an absolute dominant position. In *Seismic City: An Environmental History of San Francisco's 1906 Earthquake*, Dyl points out that since San Francisco is located above the fault line, the city has been constantly adjusting and adapting to its position in nature since its birth, and its confidence and aimlessness that are formed in the process of technological development are constantly being challenged by nature. The spontaneous forces of nature have been involved in the history of the city from beginning to end in the form of earthquakes, not only in its architectural history and landscape change, but also in the history of its social and political structure. Joanna Dyl, *Seismic City*.

population migration. Innumerable miners, mine owners, and merchants of the future flooded in from all over the country and the world. Most of them knew nothing about the place they were about to reach and had no knowledge of the mining of precious metals. Among them, even some intellectuals and artists came from the East of the United States and Europe. The strange place on the other side of the continent, in the eyes of those intellectuals and artists, not only stimulated a desire for wealth but also aroused a strong curiosity. All the people who came here during the gold rush had a common motivation—greed, but after they came here, the majority of them did not find gold, but discovered a sea and land landscape they had never foreseen. The broad water surface, known as Bay of San Francisco, which is mixed with tides flowing from the Pacific and rivers flowing from Sierra Nevada Mountains, formed the largest bay in the Pacific on the American side.

The "Golden Gate" is the narrow entrance demarcated by the coastal cape at the north and south ends of the bay. In people's cognition, the name "Golden Gate" seems to be naturally associated with the gold rush. However, as Col. Fremont, the famous Western explorer and presidential candidate nominated by California Republican Party later, put it, he named the place "Chrysopyrae, or Golden Gate" in 1846, inspired by the Chrysoceras, or Golden Horn during the Byzantine period. The *Annals of San Francisco* published in 1854, which is recognized as the authoritative work on early California history, states, "The term [the Golden Gate] was descriptive, not of the literal golden regions within, then as yet undiscovered, but merely of the rich and fertile country which surrounded the shores of the bay, and of the wealth which the commerce of the Pacific, passing through the strait, would certainly give to the future great city of the place. The name was probably suggested by the Golden Horn of Constantinople." The *Annals* goes on to write that "Since the discovery of the auriferous character of the country, the title has become of a still more happy nature; and its bestower must surely have had a prophetic soul, though he himself knew it not."[7] In the eyes of San Franciscans in 1854, the reason why the Golden Gate became true to its name was the gold mines that still attracted people to arrive at one after another, and Fremont named the place only by chance. However, it is precisely because Fremont named it that the future city was able to become a world-famous

[7] Frank Soulé, John H. Gihon, and James Nisbet, *The Annals of San Francisco*. p. 149.

metropolis, and it was ranked first in the West before Los Angeles caught up from behind. Through "the Golden Gate," wealth, immigrants, and ideas came from the direction of the Pacific in a continuous stream, thus the city and country would be redefined as a result. It is precisely because the soil in the interior of the bay could still be built by those who came here to realize the "American Dream" into a thriving agricultural empire through irrigation after the gold mine was exhausted and the dream of finding gold was broken that natural wealth was transformed into food and fruits and flew into San Francisco and the whole world from "Golden Cornucopia."

On passing through the Golden Gate, the rocky shores and hills surrounding it slowly improve on both sides. The future San Francisco is located to the south winding and undulate sand hills are about three to four hundred feet high. To the north of the bay stands Mount Tamalpais, which is 2572 feet tall. By Western American standards, it is just a hill, yet it is clear and towering. The dense mangrove of the bay makes its summit in a dark green. The *Annals* records that "On both sides [north and south] they are quite bare and barren. The strong winds and heavy fogs which constantly assail them, and their own sandy or rocky nature, have effectually prevented trees or luxuriant vegetation of any kind from growing. On the very summit, however, of the mountains on the northern side of the strait, there happens to be a solitary group of redwood trees, whose tall forms make a striking landmark to the mariner at sea. As he approaches the strait from the south, the voyager has seldom perhaps seen so dismal a looking place. A multitude of low, bleak sand hills on the seashore, often swept over by flying clouds of dense mist, first greet his eyes."[8] However, the author of the book continuously writes that once entering the bay and looking west, the shore in the interior becomes much more pleasant, and there are some beautiful green islands.

Sharper observers can see from the eyes of the author of the *Annals* wild flowers swaying in the spring breeze on the desolate sand dunes, and the golden light that withered and yellow grass gleams under the summer sun, they will also be shocked at rough and hard rocks that are constantly hit and scoured by roaring waves in millions of days and months on the westward side. Possibly, more navigators on deck, who eagerly anticipated the wealth of the continent, eastwards fixed their eyes on the Sierra

[8] Frank Soulé, John H. Gihon, and James Nisbet, *The Annals of San Francisco*. pp. 150–1.

Nevada Mountains in the distance; Carl Meyer was the one among them. He came from the German-speaking region of Switzerland and was obviously a highly educated man. In 1849, like thousands of gold-seekers, he left his homeland and set sail from New Orleans along the Atlantic Ocean, crossed the Isthmus of Panama on a mule, and then again sailed north from the Pacific Ocean. After a period of gold rush experience, he arrived in San Francisco in 1851. As he walked through the Golden Gate, he wrote, "Once again I looked far into the distance at the Sierra Nevada's snow-clad peaks towering in the infinite where the rose-colored sun set on the last day of my first mine pilgrimage." However, after a long journey, Meyer yearned for civilization far more than nature: "Then my eye glided past dark Monte Diabolo of the coast range to the shores of the bay before us … I had renounced the beautiful mountains and the glorious slopes and the more I breathed the fresh sea air which blew across to us from the harbor city, the more I longed for the comforts of civilization."[9] Meyer cheerily stepped foot on the muddy land of San Francisco, but didn't feel down-hearted due to the simplicity of the new city. He excitedly thought that five years ago (1847), the place was still in "its original state," and those "natives" with do-nothing spirit did not think about the greater future of the place. Now, it has become the third largest port in the United States, and he called it the "freshly blooming 'Queen of the West'" and wrote:

Here lies one of the greatest and most astonishing demonstrations of the enterprise of the Americans who build new states and cities from materials which have been thrown together as Romulus and Remus built Rome.

You can call it materialism, greed, selfishness, or egoism, but American projects are everywhere crowned with rich and enviable fruits, significant proofs that America progresses completely with the times in which a nation must be assured physical freedom in order that there be a successful mental uplift of the people; the golden present in which knowledge and truth are sought but in which man, bound to the physical world, does not hesitate to expose his varied material aspirations. Time is different from style in that it produces no renaissances, but it produces development, culture, and completion; thus the present is the past to a higher power.

[9] Carl Meyer, *Bound for Sacrament: Travel-Pictures of a Returned Wanderer*. Trans. by Ruth Frey Axe. Originally published in German in 1855, republished in English in 1938. Claremont, Calif.: Saunders Studio Press, 1938, p. 114.

The material activity of the American nation has made it great. It is the youngest and it will be the most powerful of all nations. Commercial ambition and greed drove the Americans to new discoveries and new sources of profit. To satisfy these desires this nation offered defiance both to wild nature and to her wild aborigines.

In the end, he declared, "San Francisco is a site of happiness and is indestructible; its space was allotted to it by fate, and it must occupy it."[10]

However, compared to most people who came to the newly emerging city at the same time, Meyer's enthusiasm is rare. For example, James Ayers from St. Louis, also a gold-seeker during the gold rush of 1849, wrote a few years later, "When one looks back and recalls the topographical features of San Francisco at that time, he is amazed at its selection as a site for a great city. It had, indeed, nothing to recommend it for that purpose except its magnificent harbor. There was absolutely no level ground beyond the narrow rim that formed the crescent beach against which the tides rose and fell. Where it was not shut in by almost precipitous and rugged acclivities it was obstructed by formidable sand hills. There was here and there a little sheltered valley, but immense sand dunes covered what is now the fairest part of the city."[11] Another gold-seeker, Hiram Pierce, a blacksmith, who came from New York after leaving his wife and son, lamented, "San Francisco is a miserable dusty dirty town of some 5000, out of every kindred tongue & people under Heaven."[12] Even the author of the *Annals*, who firmly believes in the future of San Francisco, admits that San Francisco's "situation happens to be about the most barren part of the district; and the immediate vicinity consists chiefly of low sand hills, covered with coarse shrubs and scattered patches of grass."[13]

Before the vast Pacific World opened up by "the Golden Gate" revealed its potential magnificent future, or before the discovery of what

[10] Carl Meyer, Bound for Sacrament:Travel-Pictures ofa Returned Wanderer. pp. 115–6.

[11] James Ayers, Gold and Sunshine, Reminiscences of Early California. Boston, R. G.

[12] Hiram Dwight Pierce, A Forty-Niner Speaks; A Chronological Record of a New Yorker and His Adventures in Various Mining Localities in California, His Return Trip across Nicaragua, Including Several Descriptions of the Changes in San Francisco and Other Mining Centers from March 1849 to January 1851. Oakland, Calif.: Keystone-Inglett printing Co., 1930, p. 33.

[13] Frank Soulé, John H. Gihon, and James Nisbet, *The Annals of San Francisco*. p. 157.

the Chinese later referred to as "goldmine," few people would indeed combine the place with a great city. The Ohlones, who lived here for a long time by fishing and hunting, did not seek wealth from the land, instead searched for food in the ocean, such as oysters, shellfish, whales, and various sizes of marine fish. The Spaniards who arrived in 1769, like those who came here during the gold rush, also found the land on the coastal cape barren and useless. However, at that time, they could not predict the New World that would emerge here, so they arrived and left in haste, only to leave a small stronghold and establish a small village named Yerba Buena after the peppermint that grew everywhere. The fur trade that covered the entire continental North America also extended here. Michael White, one of the earliest white settlers in San Francisco, recalled in an interview with famous historian Hubert Howe Bancroft in 1877 that he first arrived in San Francisco in 1828 (which Bancroft believed was 1829), which was not then a port of entry. He "bought two fine otter skins in exchange for a barrel of whiskey," and within a day, he saw at least 10 aborigines drunk.[14] Obviously, at that time, the aborigines on the peppermint grassland and the ecosystem they relied on for survival had been partially involved in the broader commercial world, but such trade was still extremely limited at that time.

The Spanish Catholic missionaries who came here also gave the place another more religious name, San Francisco, in memory of Francis of Assisi, the famous Catholic saint in the thirteenth century. Like the later name of the Golden Gate, it is also a name with the flavor of precognition by accident. In the demands of the Catholic Church to conquer nature, Francis of Assisi became eminent and preached to all living beings created by God. The image that he preached later became a symbol of Catholicism caring for nature, and he himself was also regarded as a saint protecting nature. In the city named after him, which is far away from mountains and rivers, the wave of the most thorough and fierce nature conservation was also set off, but it was unable to find out the missionary who named it at that time. In the coming decades, people were still accustomed to calling the place Yerba Buena, the more vivid and down-to-earth name, until January 30, 1847, when it was officially renamed

[14] Michael C. White, *California All the way back to 1828*. Dictated by Michael C. White and written by Thos. Savage for the Bancroft Library 1877. Republished in Los Angeles, G. Dawson, 1956, p. 26.

San Francisco.[15] At that time, California switched the allegiance in the previous year and became a territory of the United States. However, the real change still needed to wait for the golden moment in the early spring of 1848. Then thousands of people flooded in from all directions around the world by means of various modes of transportation in the following year. Most of them left again through the Golden Gate in haste, heading deep into the mining areas among hills along the Sacramento River and its tributaries, where gold flew downstream.

In the early months of the gold rush, San Francisco fell into a depression. The new adventurers were unwilling to stay in the small town that yet didn't present a clear outline and go on with the previous monotonous work of meager salaries. People rushed to the hills without hesitation and used all available tools, from butchers' knives to shovels and spades, to dig, with wicker baskets to clay pots and old hats as containers. Of course, the most important mixing tools were their hands. People could earn US $ 10–15 a day under normal circumstances, but everyone always believed that they would be the next person to earn US $ 8000 the next day. In the meantime, prices were skyrocketing. An egg was sold at a price of US $ 1, US $ 2, or even US $ 3, and one pound of tea or sugar, US $ 4. Even worse, there happened to be a "sickly season" in the autumn at the mines when miners arrived. The severe kind of daily labor, insufficient shelter, and the climate for them to adapt to caused various diseases to take hold here. But all the hard work couldn't stop people's determination to make a fortune. The author of the *Annals* wrote, "about the end of May [1849] we left San Francisco almost a desert place, and such it continued during the whole summer and autumn months."[16] Ships continuously arrived in the Bay of San Francisco, but as soon as they reached the shore, the seamen deserted their ships and rushed to the mining areas. The merchants sold goods actually arrived at high prices but could get no hands to assist them in removing and delivering the articles, thus only letting them rot on the beach. The common laborer, who had formerly been content with his dollar a day, now proudly refused ten for people's hopes were in the mining areas in the interior.

[15] Frank Soulé, John H. Gihon, and James Nisbet, *The Annals of San Francisco*: pp. 178–9.

[16] Frank Soulé, John H. Gihon, and James Nisbet, *The Annals of San Francisco*: chapter VI.

However, the circumstance didn't last too long. Subsequently immigrants began to arrive in numerous bands, from New York and Chicago in the east, Hamburg and Lyons in Europe, the mountains in Latin America, and the Pearl River on the other side of the Pacific Ocean. In a few months, the supply–demand ratio for labor turned upside down. It is inevitable to build a new city, although no preparations had hitherto been made for it. The *Annals* says: "Building lots had to be surveyed, and streets graded and planked—hills leveled—hollows, lagoons, and the bay itself piled, capped, filled up and planked—lumber, bricks, and all other building materials, provided at most extraordinarily high prices—houses built, finished and furnished—great warehouses and stores erected—wharves run far out into the sea—numberless tons of goods removed from shipboard, and delivered and shipped anew everywhere—and ten thousand other things had all to be done without a moment's unnecessary delay."[17] Before the houses were built, tents of various materials had already been set up on the sand dunes and open spaces, and then were inhabited with in the true sense of human beings from every corner of the universe: different races, cultures, languages and social classes, but they shared a common desire—making a fortune and suddenly growing rich. Nearly 120 years later, tents of all sizes set up again in the same place, by a group of young men and women who wore flowers on their heads. They came to San Francisco in the "Summer of Love" in 1967, in hopes of changing the mainstream American culture dominated by materialism. However, the people who came here in 1849 had no idea what kind of city they were going to build, nor did they know how their fate and that of the city would change.

3 Construction of "Gold Mountain"

Most miners eventually returned with no money in their pockets but a sickly body. Of course, some people made money hand over fist in the gold rush by investing in new mining techniques and hiring more labor to dig deep into the hills for the purpose searching for deposits, the gold deposits formed during the Earth evolution of billions of years. The era of unregulated free mining, which was spontaneously initiated by thousands of people, quickly came to an end. Soon, people needed stronger

[17] Frank Soulé,John H. Gihon, and James Nisbet, *The Annals of San Francisco*: pp. 214 & 215.

technology to make the gold deposits hidden beneath the riverbed to be exposed, thus required more capital activities in an organized manner. Only those who controlled the gold deposits, rather than those who obtained only a few pieces of nugget, could become true tycoons.[18] The wealth excavated from natural hills was left in the locality, but most of it flew to the city under construction to purchase equipment, clothing, and food, the most important supply, for miners. What's more, when the wealth from gold mines flew into San Francisco, the city became a true gold mountain. The tycoons, like the Big Four of the Central Pacific Railroad, did not actually enter the mining areas to mine for gold, but instead acquired wealth in the supply business of the gold rush by targeting business opportunities, investing in railroads, and obtaining contracts to build the first intercontinental railway in the West. In 1869, they used railroads to connect the city and bay with distant Omaha in Nebraska, Kansas City in Missouri and Chicago in Illinois.[19] These people became the Nabobs of San Francisco. The word was used to refer to the British who feathered their own nest in India through the British East India Company. At that time, another group of people found new wealth on the west coast of North America and became the new Nabobs, or Nobs. In 1873, they built their magnificent mansions and hotels on the top of the hill in the center of the city after the cable car system in San Francisco was opened to the public. Then the community became known as the Nob Hill, which is located at the center and peak of the geography and wealth in the city.

Compared with cities on the East Coast like Boston or central cities (for instance, Kansas City) established at the same time as it, San Francisco rose at a surprising rate. In 1848, its population was only 800 persons, but by 1849 it had surged to 25,000 persons. The Frenchman

[18] There are too many treatises on the California Gold Rush to recount, the best work on exploring the gold rush from the perspective of environmental history is *Mining California: An Ecological History* by Andrew Isenberg. New York: Hill and Wang, 2005. For a comprehensive history of the gold rush, see: Malcolm Rohrbough, *Days of Gold: The California Gold Rush and the American Nation and the American Nation*. Berkeley: University of California Press, 1998.

[19] The "Big Four" are Leland Stanford, the founder of Stanford University later on, Collis Potter Huntington, whose nephew donated the famous Huntington Library, Mark Hopkins and Charles Crocker. The "Big Four" all came from ordinary families on the East Coast, who were typical cases of realizing their American Dream of starting from scratch. For an environmental narrative about capital and expansion in San Francisco, see Gray Brechin, *Imperial San Francisco*.

Ernest de Massey, who came here that year, claimed to be an Argonaut of 1849 and went overseas to search for his golden fleece. He described in vivid words, "Two years ago this spot was almost a wilderness. Now, it is crowded with wooden and sheet-iron houses of every kind, shape, and description, and with tents of every color forming an amphitheater. These house a population of adventurers, vagabonds, bankrupts, refugees from justice, merchants, deserting sailors, gamblers, and vagabonds who have no home or country. Interspersed among them are some honest men, workmen, and speculators, who have come here from all over the world. This is what we see about two kilometers ahead of us—a great city in the making."[20] By 1860, on the eve of the American Civil War, the city still rapidly expanded, with a population of 56,000 persons, 13 times that of Kansas City, although the gold rush began to shrink five years ago. Chemist William H. Brewer, who arrived in the city at the end of 1859, marveled: "a city only about ten years old, it seems at least half a century. Large streets, magnificent buildings of brick, and many even of granite, built in the most substantial manner, give a look of much greater age than the city has." He added, "The weather is perfectly heavenly."[21]

A striking contrast to the magnificent architecture of the city was its "evil temperament." At that time, San Francisco was not the sweet city that people could leave their hearts stay there. Although a small cable car climbed halfway up the mountain and the city was shrouded in thick mist, it was neither romantic nor warm. It was a tumultuous, brutal, and competitive city which was determined to become New York on the West Coast. Corruption, violence, degeneration, and all other ethos of metropolises and frontier cities were fully displayed in the city. Racial discrimination and persecution were omnipresent, lynches, and murder frequently occurred. The *Annals* repeats the evils of the city throughout the book. Four years after the gold rush, it states: "during 1853, most of the moral, intellectual, and social characteristics of the inhabitants of San Francisco were nearly as already described in the reviews of previous years. There was still the old reckless energy, the old love of pleasure,

[20] Ernest Massey, *A Frenchman in the Gold Rush: The Journal of Ernest de Massey, Argonaut of 1849*. Trans. by Marguerite Eyer Wilbur. San Francisco: California Historical Society, 1927, p. 12.

[21] William H. Brewer, *Up and Down California in 1860–1864: The Journal of William H. Brewer*. ed. by Francis P. Farquhar, with a preface by Russell H. Chittenden. New Haven: Yale University Press; London: H. Milford, Oxford University press, 1930, p. 9.

the fast making and fast spending of money, the old hard labor and wild delights, jobberies and official and political corruption, thefts, robberies and violent assaults, murders, duels and suicides, gambling, drinking, and general extravagance and dissipation." Every person who came to the early city was deeply impressed by the ubiquitous gambling-houses in the city. For people, the gold rush itself was a huge gamble, and most people were losers, but everyone believed they could become the next winner. When they bad luck in the mining areas, they returned to the city. Their fate might be changed the moment the beautiful female dealers threw the dice in the warm salon. "The cards are often still dealt out and the wheels turned, or dice thrown, by beautiful women, well skilled in the arts calculated to allure, betray and ruin the unfortunate men who become their too willing victims."[22] In 1877, Mrs. Miriam Squier, a wealthy madame who came to the Golden Gate from New York, seated in the first-class cabin, along the intercontinental railway which was built less than ten years. She recorded that local police officers told her, "THERE were about one hundred and fifty gambling-houses in China Town a year ago [1876], and I suppose there are fully half as many now, and I wish I knew the precise address of one of them."[23]

Behind the evil and corrupt, yet rapidly-growing and charming city lay the unconstrained private capital that spun like dices in casinos. Although violence and discrimination occurred within and between every social class and race in the city. In 1858, Henry George, who was only 19 years old, left Philadelphia, his hometown, and came to San Francisco. Like millions of newcomers, he once fantasized about going to the mining areas to try his luck. But by various coincidences, he took up journalism and settled down in San Francisco, where an inch of land was an inch of gold at that time. He became a journalist and started his own newspaper in the 1870s, but he had difficulty in carrying out his cause and fell into abject poverty several times, even begging in the streets. The huge wealth gap and insatiable accumulation of capital in the emerging city caused him to contemplate the entire society. When he went to New York, he, in the city with a history of over 200 years, was surprised to find that poverty

[22] Frank Soulé, John H. Gihon, and James Nisbet, *The Annals of San Francisco*: pp. 500–2.

[23] Miriam Squier, California: A Pleasure Trip from Gotham to the Golden Gate, April, May, June, 1877. Nieuwkoop [Netherlands] B. De Graaf, 1972, p. 167.

was not an early problem of a city, but an intractable disease of the civilization in which he lived. In 1879, he published *Progress and Poverty; An Inquiry into the Cause of Industrial Depressions, and of Increase of Want with Increase of Wealth—The Remedy*, the most important book in the publishing history of California. The book was sold over 3 million copies at that time and opened up the progressive era in America in terms of ideology. Since then, many people became Georgianists or were deeply influenced by him, especially advocates of "back to the land" after the Great Depression of the 1930s, such as Ralph Borsodi and Scott Nearing. In the book, George deems that the biggest mystery of the era lies in why technology has advanced so much, while so many people live in poverty. His ultimate conclusion is that the key to all problems lies in land monopoly. A few people monopolize the land and lease it to the poor at extremely high prices. The key to solving these problems is to impose a single heavy taxation on private property. Only in this way can social justice be rebuilt.[24]

Like most progressives, George's care is limited to social justice. Although he saw that the land and everything on it were not created by any individual, and no one had legitimate rights to monopolize it, he did not see the fact that nature was deeply damaged due to social poverty. The creation of progress is not exclusive to the technology invented by the great human mind. In all emerging big cities such as San Francisco, New York, Boston, Chicago, and Kansas City, human wealth and progress are first and always the acquisition and transformation of natural wealth. If in San Francisco, which was founded 30 years ago, violence constituted one of the basic forms of interpersonal relationships, and the same kind of violence was also imposed by various social classes and races on nature. The only common topographic feature shared by Boston, Kansas City and San Francisco is that their cities don't have boundless plain as Manhattan in New York. So during the establishment process, all three cities adopted a common method of bulldozing some hills. Unlike Boston and Kansas City, San Francisco's hills and sand dunes stretch in an unbroken chain and are almost impossible to completely level. However, from the beginning of the establishment of the city, people began such work without hesitation. The *Annals* states: "While the position of San Francisco on the

[24] Henry George, Progress and Poverty; An Inquiry into the Cause of Industrial Depressions, and of Increase of Want with Increase of Wealth—The Remedy. San Francisco: W. M. Hinton & Co., 1879.

shores of the bay was undoubtedly the best that could have been selected for maritime purposes, there certainly have been sad drawbacks to the extension of the place on the land side. The want of sufficient level space on which to found so great and growing a city, has been partially rectified, at an enormous expense, by taking building ground from the waters, and by lowering, and in many cases absolutely removing bodily the multitude of sand hills, by which the place is immediately surrounded. What with digging out and filling up, piling, capping and planking, grading and regrading the streets, and shifting, and rebuilding, and again rebuilding the houses, to suit the altered levels, millions upon millions of dollars have been spent."[25] What the author of the *Annals* did not mention was that the place, which was bleak and desolate in the eyes of early Spanish colonists but still covered with wild peppermint, soil, animals, and plants, was completely buried under the rising city during the process.

If sand dune leveling is an environmental cost that has to be paid to build a city, then the landfill in the Bay of San Francisco has created more unexpected results in the conquest of environment. Miners digging for gold in the mountainous areas of the upper reaches of the Sacramento River used a variety of tools and techniques to create a large amount of gravel. When gold was separated, the gravel rolled down the river and, together with other sediments, flowed into the Bay of San Francisco, thus burying various aquatic organisms that once thrived there. In the entire mining story, the wildest operation is "hydraulic mining," which is a great technological invention in the California mining area. It not only caused the most serious environmental disruption in the upstream mountainous areas, but also brought huge ecological disasters to the downstream Bay of San Francisco. Millions of tons of rocks, grit, and soil on the hills of California were washed down by powerful water currents, then flew into rivers and ultimately into the Bay of San Francisco. Together with the planned landfill in San Francisco, the waters in the entire Bay ultimately decreased by one-third.[26] The entire ecosystem of Bay of San Francisco was constantly disturbed by the ongoing landfill, but the victims were not just migratory birds, oysters, or crabs in the bay. A large number of houses

[25] Frank Soulé, John H. Gihon, and James Nisbet, *The Annals of San Francisco*: p. 161.

[26] In his book *Down by the Bay: San Francisco's History between the Tides*, Matthew Booker tells the story of the environmental changes in Bay of San Francisco, which are full of unexpected results, and all gains come at the cost of some kind of huge sacrifice. Although Booker believes that nature demonstrates its power in earthquakes, Bay of San

were built on both the new land formed by landfilling along the coast and the construction land covered by garbage, grit, and gravel in the salt swamps in the city. At the point, San Francisco, as a bay, exactly resembles New York, which is also located in a bay and has a lot of swamps.[27] But the difference is that San Francisco is located on the most dangerous fault zone in the entire North America. Once an earthquake occurs, the ground, which looks solid, will collapse into a puddle of mud, and become the area with the heaviest losses in the earthquake.[28]

The entire process of building gold mountain synchronized with the process of destroying the natural mountain. In the western city by the bay, people's desires, behaviors, pursuit of money, and conquest of nature seem to gain unprecedented freedom. As Carl Sandburg's poetry puts it, it seems to be "new cities and new people," only an ocean of to-morrows, a sky of to-morrows. However, is it really the case?

4 Invisible Wall

In 1911, San Francisco was already a wealthy and elegant mature metropolis with famous universities such as Stanford University, University of California-Berkeley, and University of San Francisco. It remained the largest city on the West Coast, but in the next decade, Los Angeles to its south caught up and surpassed it, and became the first largest city on the West Coast. However, at that time, San Francisco, after experiencing the destructive earthquake five years ago, was rapidly rebuilt, and permeated with the optimism and confidence that were shared by the countries on both sides of entire North Atlantic World before the outbreak of World War I. Progressive reforms also fundamentally changed the social mood of the city, for example, the label like the evil and rough frontier city was thrown into the Bay of San Francisco. In that year, George Santayana, a well-known pragmatic philosophy and philosophy professor at Harvard University, was invited to the University of California-Berkeley across from San Francisco to deliver a speech titled "The Genteel Tradition

Francisco itself has become a hybrid landscape, which reflects the subsequent environmental, political and social conflicts. Mathew Booker, *Down by the Bay: San Francisco's History between the Tides*. Berkeley: University of California Press, 2013.

[27] For information on the landfill and ecological consequences of Manhattan swamps, see Theodore Steinburg, *Gotham Unbound*.

[28] For specific analysis, see Joanna Dyl, *Seismic City*.

in American Philosophy" to the Philosophical Union of the University of California. As a Hispanic philosopher who immigrated to the United States during his youth, Santayana always observed the country with the calm attitude of an outsider. His own non-Anglo-Saxon background also made him cautiously suspect the Transcendentalist tradition in the New England. At the beginning of the speech, Santayana said, "America is a young country with an old mentality; it has enjoyed the advantages of a child carefully brought up and thoroughly indoctrinated; it has been a wise child. But a wise child, an old head on young shoulders, always has a comic and an unpromising side." He went on to say that America was not simply a young country with an old mentality. "It is a country with two mentalities, one a survival of the beliefs and standards of the fathers, the other an expression of the instincts, practice, and discoveries of the younger generations."[29] When Santayana made the speech, he was in one of the youngest cities in the young country, and as he pondered his philosophy, he faced a young city growing up in two different ways of thinking.

Like all modern cities in the United States, San Francisco doesn't have city walls, which declares its clear break with outdated conventional ideas, clearly expresses its openness and inclusiveness, and also pronounces its determination to expand. Compared with cities on the East Coast and Midwest cities, the city faces the Pacific Ocean. After the Age of Discovery took the Pacific world by storm, San Francisco, due to its geographical location, was doomed to face a world greatly different from the one embraced by Boston and Kansas City, and San Francisco was. In contrast, the multiculture of cities in the eastern and Midwestern regions ceaselessly grew and expanded with the continuous arrival of new migrants, while San Francisco, since its establishment, had no choice but to become the multicultural center. The sensitive Frenchman Massey had already noticed the ethos in 1849, and wrote: "In the midst of this babel you can distinguish the jargon of many languages. Ask a question in English and your reply may be in German; if you speak in French you may perhaps be answered in Spanish, Italian, Russian, Polish or Chinese."[30] People who

[29] George Santayana, "The Genteel Tradition in American Philosophy," *Address Delivered before the Philosophical Union of the University of California, August 25, 1911*, p. 2, 3. https://monadnock.net/santayana/genteel.html, which was logged in on September 10, 2020.

[30] Ernest Massey, A Frenchman in the Gold Rush, p. 15.

spoke each language brought their unique culture into the city. However, a new culture was born when they gathered together. In its evolution of the first 50 years, San Francisco boasted the ethos of wallless cities: innovative technology, tough minds, the ambition of strong vitality rather than the truculence of laissez-faire capitalism, and the conviction of the freedom and self-reliance of individualism. But when we truly arrived at San Francisco, went into the real city and knew about the evolution of its thought, we found that its builders didn't actually tear down the invisible wall in their thought. Boston, New York, Pittsburgh, Chicago, and Kansas City all didn't have such ethos. They are built within an invisible wall in some kind of thought. If the Boston and Cambridge where Santayana lived were clever children carefully cultivated, then San Francisco was a wild child growing up without discipline. In its growth experience, it yearned for the admission of that respectable and noble tradition on its bright brothers' model.

As mentioned above, San Francisco is a mountainous place with many sand dunes. The common ground in terrain that it shares with Manhattan is the large number of swamps formed by the bay, but the great difference from that distant eastern city is that Manhattan is flat and wide, while San Francisco is rugged and narrow. When Manhattan city was planned, grid blocks were not only an old tradition brought by the urban planners from Europe, but also a cheap option for land trade and management. Although it still disregarded the local ecosystem, the rationality of economy and management could be at least seen from it. But San Francisco made the same choice in its undulating narrow space—grid blocks. The author of *Annals* describes: "Upon this limited space stands the most valuable and business portion of San Francisco, and its most substantial and magnificent buildings. However, over all the hills, and much of the country on every side beyond them, the ground has been regularly laid out into building lots, lying upon long straight streets, crossing each other at right angles; and many excellent houses, together with a scattered multitude of an inferior description, have been erected on them."[31] The Swiss Meyer has a similar record: "like those of all modern American cities, are laid out at right angles. The long streets are laid out parallel with the meridian and form with the broad streets which lead to the harbor

[31] Frank Soulé, John H. Gihon, and James Nisbet, *The Annals of San Francisco*: p. 159.

a 360-foot square which is crossed by a communicating street."[32] The entire design completely disregards natural features and lines, rigidly adds the traditional urban aesthetics inherited from Europe to the fresh city which is quite different from them in geography. It is believed that only in this way can the scientific feature of a modern city be distinguished. Therefore, much greater financial and environmental cost is bound to be paid than that of the design combining nature.

The authors of the *Annals* undoubtedly have their own judgment of this. As early as 1854, before the construction of Central Park in New York began, Olmsted, who still knocked around plantations in the southern United States, contemplated the survival status of the land and slaves there.[33] The *Annals* had already begun issuing its own warnings for San Francisco, which had been built for five years. They wrote, "Indeed the eye is wearied, and the imagination quite stupefied, in looking over the numberless square—all square—building blocks, and mathematically straight lines of streets, miles long, and every one crossing a host of others at right angles, stretching over sandy hill, chasm, and plain, without the least regard to the natural inequalities of the ground." In the eyes of these three authors, the city needs curves, arcs, decorations, or in their words, "a public park—the true 'lungs' of a big city."[34]

The voice of 1854 had a huge echo after the end of the American Civil War. Aster Central Park in New York was built, Olmsted left California's "savage mining areas" and returned to the "civilized world" of the East, and officially embarked on his career of using nature to civilize American cities. San Francisco, like other western cities in the United States, was anxious for new development and recognition. Just as they designed the basic pattern of the city with grid blocks, they found that in order to break free from the rough and uncivilized identity of frontier cities, they

[32] Carl Meyer, Bound for Sacrament: Travel-Pictures of a Returned Wanderer: p. 121.

[33] From 1852 to 1857, Olmsted was hired by The New-York Daily Times (predecessor of The New York Times) to investigate the slave-owning economy in the southern United States and Texas, and wrote for the special column of The Times. Later, these column was included in three works and regarded as the most comprehensive and objective analysis of southern society, economy and environment from the perspective of northern America before the American Civil War. His roam from 1853 to 1854 was included in this book: Frederick Law Olmsted, *A Journey in the Back Country in the Winter of 1853–4* New York: G. P. Putnam's Sons, 1907.

[34] Frank Soulé, John H. Gihon, and James Nisbet, *The Annals of San Francisco*: pp. 160–1.

should update their urban landscape with new urban concepts. Therefore, an urban park movement began in the city.

On the surface, the park movement was closely connected with urban naturalization movements in New York, Boston, and other places. Their fundamental objective is to recognize the necessity of nature for urban civilization, attempt to mingle the beauty of nature with industrial civilization, and bridge the emotional gap between urban people. However, if we reread the proposal and park practices of San Francisco at that time, we will find that the city once again confined itself to the intangible walls built by tradition. As Olmsted was about to leave the West, he was invited to offer suggestions for the construction of parks in the initial stage of the city. He surveyed the entire city and its surrounding water sources, water quality, terrain, soil, and vegetation, and submitted a quasi-design report in 1866. He bluntly pointed out, "The special conditions fixed by natural circumstances, to which the plan must be adapted ... it must, I believe, be acknowledged, that, neither in the beauty of greensward, nor in great umbrageous trees, do these special conditions of the topography, soil, and climate of San Francisco allow us to hope that any pleasure ground it can acquire, will ever compare in the most distant degree, with those of New York or London."[35] Olmsted sensitively saw that San Francisco's sandy and flowing soil, arid climate, and the invasion of sea breeze prevented arbors from growing naturally into huge trees with beautiful crown, and local species formed low shrubs and vegetation featuring low water consumption during their evolution over millions of years. It differs from New York or New England, where has abundant rainfall and fertile soil. Although there is no good soil on the original site of Central Park, the soil and plants that were transported from New Jersey are also native species to New York and can make plants easily grow and even thrive. The lush lawns of England and Western Europe are fully adapted to the climate of New England, New York, and other places, and can be easily maintained without the need for substantial investment and water. As a landscape designer, Olmsted always adheres to the basic principle of combining design with nature. Although his own aesthetic interests fully

[35] Frederick Law Olmsted, Preliminary Report in Regard to a Plan of Public Pleasure Grounds for the City of San Francisco. New York: W. C. Bryant, 1866. The translation is referenced in San Francisco, Local Attraction, Civilizing American Cities: A Selection of Frederick Law Olmsted's Writings on City Landscape by F L. Olmsted, translated by Wang Sisi et al., Nanjing: Yilin Publishing House, 2013, p. 87.

agree with England's rural landscapes, he is well aware that such aesthetic concept cannot be mechanically applied to the natural environment which are quite different. More importantly, when he trembled at the grandeur and wildness of Yosemite and Niagara Falls, called for the protection from California and New York, he clearly expressed a fundamental idea: the beauty of nature was diverse and ever-changing, it was not just a beautiful form of picturesque British countryside. The colors of nature were far richer than the spectrum that human beings could imagine, green was just one of them. When people try to apply green to all environments, they fall into another ossified tradition.

The elites of San Francisco in 1866 had the idea of building a nature park in an emerging city, as it was a symbol of civilization. But they didn't have the courage to break with the tradition of walls, and even lacked a profound understanding of the ever-changing nature itself. They were deeply disappointed with Olmsted's advice, as in their original ideas, Olmsted's Central Park was exactly what they wanted to imitate and surpass, just like their attitude toward New York. The quasi-design sheets were thrown into paper piles by them, and Olmsted left the West. Although he was later hired by Stanford University established in 1885 to design the campus, his proposal to use native species for design was rejected by Stanford University once again. The Board of Trustees of Stanford university was the same group of people as the elite of San Francisco, they yearned for open green lawns and tall and unfolding shrubs, and the aesthetic concept transplanted from the East Coast and European homeland.

Hence, in 1870, the city of San Francisco hired engineer William Hammond Hall to design Golden Gate Park on the sand dunes along the bay in the west of the city, but in reality, the most important role was John McLaren, a Scottish immigrant.[36] His design philosophy was completely different from that of Olmsted for San Francisco, whether getting its bird's-eye view or walking along its alameda, the Golden Gate Park created an illusion that people came to Central Park. The only difference is that it covered a larger area than Central Park, whose area is 843 acres, while the Golden Gate Park covered an area of 1017 acres. In

[36] The Golden Gate Park here belongs to City of San Francisco and is the urban park of the city. It is named after the Golden Gate Strait nearby. After the construction of the Golden Gate Bridge, both sides of it were opened up as a national park and managed by the National Park Service.

order to build deep shade on the flowing dry sandy land, McLaren transplanted a large number of tree species from around the world, including various types of evergreen trees such as eucalyptus, pines, and cypresses. Everyone loves the park deeply. When in the park, people will not only forget the scenes of craving personal fame and gain in the industrial and urban civilization, but also don't truly perceive the ecological reality of the place—San Francisco is a water-deficient city! In order for these trees to survive and eventually grow up into a wood, a large amount of artificial irrigation should be used. However, where did irrigation water come from? It could come from the Hetch Hetchy Valley within Yosemite National Park, located 180 miles away. The Golden Gate Park required water for irrigation, a large amount of water, which became an important argument for the San Francisco side in the debate centering on the Hetch Hetchy Valley.[37]

In the winter of 1873, John Muir walked out of the cold Sierra Nevada Mountains and came to San Francisco, where he chose to settle. Although Muir always referred to wilderness as his "true home," in the eyes of Donald Worster, author of his biographer, such a declaration was somewhat self-deceiving. He no longer refused the city because "the city promised not only an outlet for his writing talents but also a home place, a circle of friendship, and a richness of culture that the Valley in the off-tourist season lacked."[38] At that time, no one could predict what kind of change this passionate, argumentative, bearded middle aged man who loafed about the wilderness all day would bring to the city and country. Muir had a completely different understanding of the city and nature from that of Olmsted, and even more so from many of his contemporaries. Although he shows deepest affection for the majestic and magnificent mountains in California, he respects all forms of life, natural colors, and the power of nature. When he settled in the city, he began his writing

[37] For research on the Golden Gate Park, see Terence Young, *Building San Francisco's Parks, 1850–1930*. Baltimore: Johns Hopkins University Press, 2004. In 2015, when San Francisco was hit by an even greater drought, Young wrote a excellent short essay for *Los Angeles Times* about the deadlock of the Olmsted's plan and the high water consumption for the irrigation of the vegetation in the Golden Gate Park. Terrence Young, "The Great Park San Francisco Needed; But Rejected," *Los Angeles Times*, Sept. 20, 2015. https://www.latimes.com/opinion/op-ed/la-oe-young-olmsted-drought-golden-gate-park-20150920-story.html. which was logged in on December 24, 2020.

[38] Donald Worster, A *Passion for Nature: The Life of John Muir*. New York: Oxford University Press, 2008: p. 216.

career, using his pen to depict an artificial world or the parks in the city that completely differed from people's daily lives. Such place is no doubt beautiful to Muir. Each place had its own unique beauty, but beauty was not the only reason to cherish and preserve it. In Muir's view, it is its power, the vigorous, wild, and unruly power of life itself that is the reason why it is most formidable. He hoped that people living in cities could go out to discover and observe this world, and establish spiritual connections with the world created by non-culture. His shout was heard by more and more people, and more and more urban dwellers coming from different directions went to the high mountains in the west to experience the world that Muir had experienced. Although in his time, almost no one truly understood Muir's desire to redefine the position of human beings in nature with the democratic spirit of liberalism in his heart of hearts, people could still feel another kind of beauty in the mountains in California that was different from the soft and bright rural scenery in the UK, a cold beauty which is free from traditional aesthetic limitations.

In 1892, several professors from University of California-Berkeley and Stanford University, government scientists and lawyers established the Sierra Club in San Francisco, and voted for Muir as the first president of the Sierra Club. Initially, the club advocated walking and also called for a ban on logging and grazing in the Sierra Nevada Mountains. However, the famous Hetch Hetchy Dam Debate, the first environmental debate in American history, changed not only the nature of the club and but also the vision of the American environmental protection movement. At the heart of the debate was whether to construct a dam in the Hetch Hetchy Valley within Yosemite National Park to provide cheap fresh water for the development of the growing city of San Francisco. At that time, the city was already nagged by the limits of nature, and the local water sources were completely unable to provide the city with a steady water supply for its further expansion. Then the city sought to transform all the nature around it into its own hinterland, especially by establishing a new water hinterland to create a great city.[39] The focus of the debate is not whether San Francisco should get water, but whether it should get water from national parks that are not allowed by law to be developed. In fact, San Francisco was not without other options at that time, but the water source in the national park was far cheaper than those that were already

[39] For the extraction of water in San Francisco, see Grey Brechin, *Imperial San Francisco*.

privately owned. Driven by interests rather than survival, San Francisco set its sights on Hetch Hetchy Valley. In the debate, an important viewpoint was the irrigation needs of the Golden Gate Park. Therefore, the urban park built to meet traditional aesthetic discipline became one of the legitimate reasons for destroying the Hetch Hetchy Valley which formed through a long evolution process.

The debate ultimately ended with the failure of the Sierra Club and Muir. In existing interpretations of the debate, historians see it as a confrontation between the development views of utilitarianism and aesthetics-centered conservation views. There is no problem with such interpretation, but if we put ourselves in their shoes, it is also a conflict between two aesthetic views, and a contradiction between two interpretations of the relationship between nature and civilization. The San Francisco side sincerely believed that after the dam was built, the picturesque reservoir was far more beautiful than the rough, disorderly, rocky, overgrown wild canyon, just as they believed that a nature park with Central Park as prototype was far more pleasant than a nature park with local species used but exposure to the nature that the city dried up. For them, the existence and protection of nature are only necessary for the development of civilization. But it can be precisely seen in the debate that a new understanding of nature, cities, and civilization was emerging. People began to realize that there were several traditional, invisible huge walls that constrained the new city's understanding of itself, restricted the city from thinking about the relationship between it and its natural environment through instinct, practice, and discovery. A better city was not necessarily a larger city, but the one that knew better how to adapt to its environment.

Obviously, Santayana had a profound understanding of it. As he emerged from his review of America's genteel traditions, he discovered a revolution in American thought that was set off by Whitman's poetic naturalistic emotions and William James's philosophical pragmatic spirit. Facing his audience, a group of philosophers surrounded by Bay of San Francisco and California mountains, he said.

This revolution, I should think, might well find an echo among you, who live in a thriving society, and in the presence of a virgin and prodigious world. When you transform nature to your uses, when you experiment with her forces, and reduce them to industrial agents, you cannot feel that nature was made by you or for you, for then these adjustments would have been pre-established [without your efforts]. Much less can you feel it when she destroys your labor of years in a momentary spasm. You must feel, rather, that you are an offshoot of her life; one brave little force among her immense forces. When you escape, as you love to do, to your forests and your sierras. I am sure again that you do not feel you made them, or that they were made for you. They have grown, as you have grown, only more massively and more slowly. In their non-human beauty and peace they stir the sub-human depths and the superhuman possibilities of your own spirit. It is no transcendental logic that they teach; and they give no sign of any deliberate morality sealed in the world. … A Californian whom I had recently the pleasure of meeting observed that, if the philosophers had lived among your mountains their systems would have been different from what they are. Certainly, I should say, very different from what those systems are which the European genteel tradition has handed down since Socrates; for these systems are egotistical; directly or indirectly they are anthropocentric, and inspired by the conceited notion that man, or human reason, or the human distinction between good and evil, is the center and pivot of the universe. That is what the mountains and the woods should make you at last ashamed to assert. … They allow you, … simply, humbly, for what you are, and to salute the wild, indifferent, non-censorious infinity of nature.[40]

Obviously, when Santayana spoke these words, his mind was filled with scenes of San Francisco's evolution in nature: mountains were shattered, rivers blocked, rivers shrinking, miners in mining areas completely exhausted in California, as well as the scenes that swamps were filled, sand dunes eliminated, and hills leveled in San Francisco. Five years ago (1906), the gentle shaking of the Earth's crust caused a collapse, the "sturdy city" built for half a century was destroyed in seconds. Once-proud technologies like power lines became an important reason for the raging fire that could not be put out after the earthquake, and the water supply system broke down. When these scenes rose one by one, they forced Santayana not to consider nature as the background existence of the city or human thought. He realized that human thought was not generated out of thin air; on the contrary, nature had a powerful shaping effect on it. Meanwhile, several scenes rose before his eyes: as he looked eastward from San Francisco, the enormous Sierra Nevada Mountains and its forests as well as those natural forms that had slowly evolved long

[40] George Santayana, "The Genteel Tradition in American Philosophy," pp. 30–2.

before the emergence of any human had their own power and their own history. The ultimate perceptions that people fled toward them were not the moral truths of transcendentalists or the aesthetic discipline of traditionalists, but a thorough reflection on humanity itself and a reverence for a far greater power not created by humanity, even if the power is not, or not always the tender affection of the Mother Earth. Santayana believed that in the face of such nature, San Franciscans had the best driving force to break with the genteel traditions of American philosophy, and confine their cities and the invisible walls of thought.

5 The Home of Life

If Santayana sees infinity, vastness, beauty, and power in nature, and how nature allows the thinking of America, the young country, to transcend its genteel traditions, what he did not see, or at least not spoken of, is the limitations that nature imposes on the young city and country. For American cities, the most difficult intangible wall to depart from is not some inherent and rigid tradition, but nature itself. Nature uses its own boundaries and limits to tell the city and every city what they can and cannot do. In the USA's urban evolution history of two hundred years, the most unquestionable effort is to challenge the boundaries of nature. However, nature, in the end, also used its way, beauty and life, to respond to the limited and unbridled power of human desires, reshape people's thinking patterns of challenging it and allow them to discover the significance of the existence of the invisible wall of nature. San Francisco is the birthplace of the city for rediscovery.

Undoubtedly, the reason that millions of people were attracted to San Francisco from all over the world is largely the extraordinary economic opportunities that people saw in its mining, finance, trade, fishery, and forestry sectors. The retreat of the gold rush did not cool the craze of people dreaming of the Gold Mountain. The original meaning of the Golden Gate became increasingly clear, and it became the largest port on the west coast of the United States, and a gateway to the Pacific and Asia. In the rising metropolis, people could find food from all over the world, hear local dialects and tittle-tattles in various languages, shuttle among people wearing various clothing and having different skin colors, and constantly feel the surging business opportunities and wealth in the fresh city.

However, the reason why San Francisco attracted people was not only the pursuit of wealth, but also nature. On so many aspects, the unique charm of the city is a masterpiece of natural forces. Its location is one of the most magnificent environments in the world, with a mild and pleasant climate throughout the year. The thick fog in the morning casts a mysterious veil over it, and when the bright sunshine penetrates the mist in the afternoon, the entire city and the bay which the city is situated in are glittering. There are so many plants growing on its sand dunes, perhaps not tall and lush, but colorful and vibrant. In the interior, there are various aquatic lives, whether they are permanent residents or annual visitors, each can find their suitable place in the bay. The bay guarded by the Golden Gate Strait has never been an unchanging monotone blue. Every turbulence, every undulation, every shadow under the sunlight, and every change projected by floating clouds is a new color and shape. The rocks on both sides of the strait are lofty and rugged, stretching southward toward the Big Sur and northward toward the eye-catching redwood trees. To the east is the spiritual home of innumerable San Franciscans and Americans, the mountains in California as mentioned by Muir.

When people's aesthetics finally climbed over the invisible wall built by the moist pastoral scenery of New England, the natural beauty of San Francisco became the source of attracting a continuous stream of artists, poets, and writers, and also the reason why more ordinary people who yearn for a spiritual world beyond material abundance built their homes here. Writers Mark Twain, Jack London, Mary Austin, John Muir, Dashiell Hammet, poet Ina Coolblith, painter William Keith, and photographer Edward Muybridge could be found in San Francisco during the decades from the late nineteenth century to the first half of the twentieth century. They, with a perspective different than traditional one, each examined the nature and civilization there, observed every change of nature and civilization, and conveyed their new interpretations with different characters and forms of records. After photography became a new art of the twentieth century, San Francisco once again attracted the lenses of the best photographers in the United States. Ansel Adams, one of the greatest landscape photographers of the twentieth century, was born in San Francisco in 1902. He used sharp and contrasting black and white light and shadow to capture every aspect of the nature in California, especially his beloved Yosemite. However, Adams was not only a recorder of the nature in Yosemite, but also figured out and pondered over the city where he was born and brought up. Like all San Francisco artists who are extremely sensitive to beauty, Adams did not overlook the changes in the Golden Gate Strait in his eyes. He took two photos before and after the completion of the Golden Gate Bridge. The bay fluctuated under the same wind and clouds, but the difference was that a bridge that became the future symbol of San Francisco appeared on the second photo. Between clouds and the bay, the bridge stood silently but not abruptly, where Adams saw the beauty jointly built by civilization and nature. Adams' love for nature made him a true environmentalist, and he was once one of the leaders of the Sierra Club. Along with many others, he successfully made their ideas visualized, heard, understood, and accepted by more people, and San Francisco the most environmentally friendly city in the United States.

Almost all of these artists in the twentieth century had a posture of refusing to compromise with mainstream culture, a bohemian ethos different from the normal social values that are natural and right. They interpreted it as the freedom which broke free from the genteel American tradition and the discipline of industrial civilization, and transcended the freedom of the morality and value that were born from the philosophy of human egocentrism. What allows such freedom to be freely expressed was the rich and wild nature of the West. Thought-provokingly, the vast majority of them still yearned for the collision of the flame of civilization with intelligent souls, just like Muir who walked out of Yosemite in the snowy winter. But for them, such a civilization should be a new one, with less traditional arrogance but more humility and introspection. The American culture on the East Coast is too scholarly and mainstream, even the wandering poets and artists in Greenwich Village were completely captured by the small world of human self, they were full of complaints but willingly put themselves in a highly technological urban civilization. So when these Bohemians looked around in search of a place to establish a new civilization on the continental North America, they discovered San Francisco. Robinson Jeffers, an American modernist poet, was one of the pioneers. He was born in Pennsylvania, but immigrated to the West. What attracted him was not the opportunity to acquire wealth there, but the extraordinarily beautiful rocks and coastline in his eyes. Eventually, he settled in Carmel-by-the-Sea, the important town in the cultural circle of San Francisco. There gathered poets, writers, musicians, and painters, who searched for the middle ground between city and nature. Jeffers became just right one of them, and an idol poet of post-war environmentalists in the United States in the collision of sea breeze.

Compared with mainstream East Coast art of the same time, the artists who came to San Francisco at that time had a certain special insight into nature. In their view, the poetry and art circles in the eastern region had lost their fresh and original creativity, but are complacent about the arrangement of rhetoric. So they abandoned the mainstream world there and came to San Francisco, where they began the San Francisco Renaissance in American poetry in the 1940s. Its founder, Kenneth Rexroth, like other major characters, once again stepped into nature to seek new inspiration. But unlike before, they rediscovered East Asian culture in San Francisco, a city where multiculturalism had been shaping its fundamental ethos from the beginning. No previous American thinkers didn't read ancient Chinese or Japanese culture. Emerson, Thoreau, or pragmatic thinkers or anti-modernists in the late nineteenth century had read the writings of Confucius and Lao Tseu, heard of Chuang Tzu and Sun Tzu. However, for them, East Asian culture was just a foreign embellishment representing erudition in their complete Western ideological training, and they had never truly defined their thoughts, whether it was their reflection on society or their appreciation of nature. Although Santayana was well aware of the rigidity of genteel thinking, he himself remained stuck in the thread of Western thought. Therefore, he found the power of nature to impact, shock, and even shatter the thread. Muir, Austin, and Keith at the turn of the nineteenth and twentieth centuries all made similar attempts, and they indeed opened up a new understanding of nature and civilization. It is precisely the new understanding that gives rise to a new understanding of other cultures: since people can find their own value in non-human nature, discover the diversity of beauty beyond the existing aesthetic admonitions of the West, and since the ethics of equality shall extend to all species, there shall be a reverence for the power that is not created by us, shall we not hold the same attitude toward cultures that differ from ourselves?

When these poets and artists who yearned to break free from the shackles of genteel traditions came to San Francisco for the beauty of nature in the West and the freedom against tradition, they saw a quite different world. They found that they should also allow for the freedom of various cultures to coexist in addition to the spiritual freedom that individuals experienced in the wilderness. Such freedom was thriving in the city mingled with international community. There, there is the North Beach established and settled by Italian, and Chinatown, the oldest community in the city. It was the latter that inspired their new ideas.

How could they ignore the existence of East Asian culture represented by China and Japan in cities like San Francisco? When it opened its history with the development of the Golden Gate, it was destined to be a city facing the Pacific. As a matter of fact, at that time, in the new social and cultural context after World War II, many artists and writers came here precisely because it was the gateway to Asia—China, Japan, India—especially their philosophy and art. The disillusionment of Western civilization, especially the lifestyle of the middle class, led them to seek a new possibility of survival in the city. The "Beat Generation" emerged in the wave of thought and society.

The term "Beat Generation" first appeared on the other side of the continent in New York in 1948. Jack Kerouac, born in Massachusetts, was the author of the novel *On the Road*, the Bible of the world's arty persons. Famous figures in the Beat Generation also include another genteel idol of arty persons, the Jewish poet Allen Ginsberg, and novelist William Burroughs, who was born in St. Louis. At first, they also chose to settle in Greenwich Village, New York, which was once a holy land in Bohemia. However, in the end, they left one after another. Although their settlement locations were different, they stayed in San Francisco for a long time and formed their own literary associations, thus fundamentally redefining the culture of the city. Their intellectual home was the City Lights Bookstore located at the intersection of Chinatown and North Beach. It was founded and managed by poet Lawrence Ferlingetti. A group of poets singing about the light of nature and wildness gathered in City Lights Bookstore, expressing new expectations and reflections on the city.

In the 1960s, they truly formed the "Beat Generation," and these "Beatniks" gave birth to their cultural descendants, the "hippies" in counterculture, especially the "flower children" of the "Summer of Love" who came to San Francisco. The "hippies" that are called by the "Beaters" were not just wanderers wearing long hair, dressed in wide robes, and always singing with a guitar held in arms. They should have a profound insight into life and nature, break free from the blindness of traditional American materialism, and turn around to enter the earth and universe. However, the acquisition of such deep consciousness required some kind of assistance, while too many people sought physical indulgence in alcohol and marijuana, and hoped for freedom of thought. Thus, it was not fair to simply define them as a group of youths who grew up in a middle-class life, freed of worries about food and clothing, but were idle and indulgent. They all rebelliously thought about the relationship between persons in a civilized society, the relationship between human beings and self-awareness, and the relationship between human beings and the rest of nature, and hoped to find ways to free America from new material restrictions. Or perhaps most of them did not think deeply and the ways they found were too extreme. They at least tried to resist various disciplines and dogmas which were imposed by society and civilization, reflected on human pursuit of growth, faith in material progress, and the rationality of plundering nature, all of which were traits that were not present in previous social movements. Before the movement, such thoughts only struck sporadic individuals, but at that time, they truly became a universal questioning power. Although those youths wearing long hair eventually grew up and many returned, the questioning power they sparked did not disappear with their long hair, but forced the entire American society with high urban civilization to rethink their relationship with nature.

Gary Snyder, the prototype of the Dharma Bums in Kerouac's writing, was one of the thinkers among them. In the mid-1950s, he ended his wander and settled in Mill Valley, about 15 miles away from San Francisco. Unlike most of the "Beat Generation" who were born in cities, Snyder grew up in a poor rural environment and was the "son of nature" among them. During his studies in San Francisco, he was exposed to Chinese and Japanese poetry and haiku, and deeply moved by the ethereal Zen philosophy of self-congratulation in all things. However, when he actually went to Japan and China, he began to realize that "I got interested in China for the wrong reasons." In his collection of essays called *The Great Clod: Notes and Memoirs on Nature and History in East Asia*, which borrowed the meaning of Zhuangzi's "The Great Clod burdens me with form, labors me with life, eases me in old age, and rests me in death. So if I think well of my life, for the same reason I must think well of my death." he explained this sentence. He wrote that the reason why his interest in China for the wrong reasons was "That is, I thought I had come onto a fully engaged civilization that maintained a respectful and careful regard for the land itself, and the many other beings who already lived there. It turned out that I was wrong, but in a very complex and challenging way."[41] He realized that the natural sentiments chanted in

[41] Gary Snyder, The Great Clod: Notes and Memoirs on Nature and History in East Asia. Los Angeles: Counterpoint Books, 2016. The translation is referenced in The Great Clod: Notes and Memoirs on Nature and History in East Asia by Gary Snyder, translated by Wu Yue and Li Jing, Beijing: People's Literature Publishing House, 2018, p. 6.

East Asian poetry were only a small part of the vast civilization of East Asia. The part is a real idea, but it cannot represent the relationship between the entire East Asian civilization and nature, which is also full of slaughter, destruction, and conquest. However, no matter how much bewilderment the new understanding brought to him, that small fraction of true thoughts had a profound and lasting influence on him. He began to combine the traditions of his own Western wilderness culture with the insights of East Asian Zen meditation to perceive the close connections between the subtlest changes in all things and thought. He wrote, "All change/in thoughts/as well as things."[42]

In 1975, Snyder went back to San Francisco once again at the age of 45 and gave a speech at an event called "North Beach as Habitat." This community, known for its restaurants of all kinds, Bocce ball and Catholic Church, is his favorite place to wander in the city. But what he deeply loved here was not only the community culture full of human touch or the Italian cuisine tickling his taste buds, but also a "small watershed" with several small creeks that once flowed through the place and eventually merged into bays and to oceans. However, when he returned, those creeks were already built into roads and houses. What flew in those creeks was the winter rainstorm that blew from the sea—"the dark running sea," and that sea was also one of the sources of the city's unique spirit. There were more things in the habitat where that watershed passed through, such as food, cable cars, bookstores, steep slopes up and down, and the lights of boats entering and exiting the Golden Gate. In Snyder's view, San Francisco is a habitat where nature and culture blend together, it provides refuge for artists and revolutionaries, anyone seeking freedom, and also a habitat for innumerable people, as well as numerous other lives respected by Snyder. In order to repay the rainfall brought by the sea, the artists of the city gave the world a new idea about urban life. "Like the sub-Aleutian storms, pulse after pulse came out of North Beach from the fifties forward that touched the lives of people around the world." He recalled that in the city, "there is the rich soil of much beauty, and the

[42] Gary Snyder, "Riprap," https://www.poetryfoundation.org/poems/47178/riprap, which was logged in on November 20, 2020. The translation is referenced in the *Riprap and Cold Mountain Poems* by Gary Snyder, translated by Liu Xiangyang, Beijing: People's Literature Publishing House, 2018, p. 51.

good work of hatching something *else* in America; pray it cracks the shell in time. Gratitude to the Spirits of the Place; may all Beings flourish."[43]

Like all other cities, San Francisco is a city created by human beings, where they use their imagination, knowledge, technology, and traditions to construct a new habitat. However, every city is created within its own natural environment, nurtured, shaped, and constrained by the ecology in which it is located. San Francisco is like that, Boston and Kansas City are no exception. In the city, the sea, the water, the land that emerges from the water, the plants, animals, insects, and bacteria that live on the land and in the water, the "big clods" that carry the miners, poets, immigrants, and merchants who come here, are not just ideas constructed by human beings. The nature here, like nature anywhere else, is the real power that strongly changes genteel traditions and promotes new ideas. Those who come to San Francisco in search of the power can not only see, hear, and feel it, but also change the city they live in with such ideas. While it allows for the coexistence and integration of all cultures, it also hopes that it will not only be defined by the beauty that human beings expect, but also become a habitat for all life forms to bloom together.

[43] Gary Snyder, "North Beach," *A Place in Space: Ethics, Aesthetics, and Watersheds*. Los Angeles: Counterpoint Books, 2008: p. 3.

Epilogue: Fragile Balance

American cities grow within the ecological paradox they create themselves. In their evolution, there are two kinds of fundamental power—nature and thought—that interact intermittently to shape the pattern and ethos of cities, as well as their relationship with the environment. On the surface, it seems to have become an existing paradigm in the narratives of American environmental history: nature continuously falls into decay while environmental thought always becomes progressively more enlightened. However, on a deeper level, it narrates the complex process of environmental adaptation.[1] Whether it is decline or progress, it is a linear process with an inherent trajectory that points toward a fixed endpoint. However, the evolution process of American cities did not follow such a linear trajectory. On the contrary, as stated in the "Introduction" about the ecological paradox of American cities, the process is full of twists and turns, as well as contradictions, contingency, and unexpected consequences, and it always in an open and unfinished state. It is precisely

[1] Paul Sutter, in his article reviewing the development of environmental history over the past thirty years, believes that there are two basic narrative patterns of American environmental history, namely the continuous decline of the environment under human destruction, and the constant progress of environmental thought. He believes that the research on environmental history shall break with the narrative path. However, his summary still has the suspicion of too much simplicity. Paul Sutter, "The World with Us: The State of American Environmental History"; *The Journal of American History*. 100 (June 2013): pp. 94–119.

S. Hou, *Wallless Cities*, https://doi.org/10.1007/978-981-97-7827-0

because nature and thought are both in the ecological paradox of cities, and ceaselessly change.

For environmental historians, nature is a term worth a lot of thinking. As a concrete existence, undoubtedly, since the intelligent species changed the entire Earth through agriculture, there has been no "nature" without cultural participation. It has become a "hybridity" in the words of Bruno Latour and Richard White, or the "second nature" in the words of William Cronon.[2] Hybridity refers to the product of the interaction between natural and cultural forces. On the level, cities, hinterland, and even seemingly pure "wilderness" are all a certain degree of hybridity. However, it does not mean that nature, as an independent, spontaneous, and non-cultural force, no longer exists. In fact, throughout the evolution of cities, the power of nature has never been absent.

On the one hand, it is a biological force, such as the reproductive and survival needs of humanity itself, the ubiquitous microorganisms that exist in the air, water, soil, and all organisms including humans; various plants, whether they are called wild grasses by human beings, or exotic and strangely scented flowers carefully cultivated and built in greenhouses; animals except human beings, various texas cockroaches, free creatures drifting in the vast sea and sky, domestic animals (birds) that are used for human labor and consumption, rare birds and animals that are kept in cages to satisfy human curiosity, and cats and dogs that are dressed in thick woolen sweaters in winter. They together constitute the biological force of nature, and tend to be objects that are intentionally or unintentionally damaged or destroyed during the process of urbanization. A city is a simplified ecosystem, and people hope to only preserve the species that meet their own needs, economic, physiological, intellectual, aesthetic, or psychological need, and to eliminate those organisms that they believe are dangerous, harmful, unhealthy, and disruptive. It is a long confrontation process, and the reduction of biodiversity is the biggest ecological crisis that the Earth is currently facing. However, cities have never achieved an ultimate victory. Various living beings, including humans with their powerful fertility, often escape human rational control, transcend the ecological niches allocated by human beings, and disrupt the ecological order that human beings attempt to construct in cities. These living beings are always in the process of evolution, and some species, such as

[2] Richard White, "From Wilderness to Hybrid Landscapes: The Cultural Turn in Environmental History," *Historian*, 66 (Fall 2004), pp. 557–64.

microorganisms, insects, and weeds, are evolving at a pace that human culture cannot keep up with, thus forcing the latter to constantly make adjustments to adapt to new ecosystems.

On the other hand, nature is a physical force, more ancient and vast than the biological force.[3] Sunspot activity and continental drift are fundamental factors that constitute its physical force. The influence of hydrology, terrain, climate, soil, and geology on a city is never limited to the beginning of its establishment although they are often key factors why a city is established and grows. However, even in modern cities with highly-developed industrial civilization, seemingly natural rhythms and limitations have been broken through due to modern transportation and technology which are supported by fossil energy. In essence, they still regulate human economy, politics, and culture in various ways. Even in terms of human dependence on fossil energy, the limitations of the physical force of nature on modern cities have surpassed that of traditional agricultural civilization. Urban dwellers often only face the existence of natural force during major geological disasters, such as the San Francisco Earthquake, Kansas City Floods, or Hurricane Katrina. However, the affluence and scarcity of nature, the warmth and coldness of the climate, the ebb and flow of tides, and the exhaustion and overflow of rivers are the most common yet enduring natural force that constantly shapes the form of cities and individual thoughts in the daily life of cities.

It is a force that modern cities around the world face together, not the one which is unique to American cities. Since the birth of the modern world, the seemingly regional differences in global ecology have been eliminated through species dispersal and the creation of global ecological hinterland. However, just as technology has helped us build indestructible cities, it is largely another illusion that modernity has given. The phenomenon of ecological teleconnection, it is true, has become more prominent in the establishment of global ecological hinterland, such as the Irish Potato Famine which became a direct cause of the demographic, economic, cultural, and political fission in the United States after the mid-nineteenth century, the outbreak of dust storm on the U.S. Great Plains which was closely related to the expansion of European markets in the

[3] The idea of defining nature as two forces, biological and physical forces, comes from "Nature Rules Butt Which Nature," a talk given by Donald Worster in Trinity College, December 1, 2020, on Zoom. He defined "nature" as two forces: "climate" and "biology" in a broad sense.

twentieth century, and the continuous wildfire in the Amazon rainforest which was reflected in the beef consumption of urban populations in the United States and China in the twenty-first century. Global warming, the disappearance of biodiversity, marine and atmospheric pollution, and the global spread of COVID-19 pandemic are not the crises that a country or a city faces alone. However, regional natural systems still play their own role, as Mumford said, cities cannot exist without their regional ecosystems.

The ecological paradox of American cities occurs in their unique natural environment. Natural abundance, which has long disappeared in the Old World while is easily available in the United States, gave rise to wallless cities, which transform the thought of the definition of a city from security to freedom, openness, and growth. Such a city is a unique creation of the United States in the New World, ultimately spreads to cities around the world. The paradox is that during the pursuit of new urban themes, the gradual decline of natural abundance leads to new ecological realities and inspires various new ideas that once again begin to create new ecological changes. Like the response of it to nature, the ideological power that shapes cities is also not uniform or unchanging. On the contrary, in any era, there are various sounds echoing in the air, but people can still recognize the loudest one and call it the mainstream. The clearest sound over the new cities constantly emerging in America in the nineteenth century was the conquest of nature, which became a essential task of achieving freedom, democracy, and growth. However, the new ecological reality inspired a challenge to sounds that were once taken for granted, and it made suppressed whispers gradually stronger and the legitimacy of conquering nature less solid. More and more people were yearning not for larger and wealthier cities, but for cleaner and better cities, not just cities with cultural diversity, but cities that embrace biodiversity. More and more people realized that modern cities were not as safe as they imagined, and nature never stopped confronting the order imposed on humanity.

Could the new ideas help American cities, or general modern cities in today's context, resolve their ecological paradoxes? I'm afraid as history comes to the present day, a clear answer has not yet emerged. If the first two ecological paradoxes in American urban history have gradually been resolved under the dual impact of nature and thought, the third paradox is becoming increasingly prominent in the shrinking Earth. The sprawl of American cities today is not the expansion in terms of population and area, but fundamentally an expansion of consumption in the urbanization. In Boston, the scenery remains its charm, Pittsburgh reproduces the blue sky, Kansas City rises again, and San Francisco strives to become a habitat for all kinds of life forms. At the same time, like other big cities in the world, they are consuming energy, soil, air, water, and creatures from other parts of the planet. However, since in the environmental history of American cities, environmental pollution within cities and wilderness landscapes outside cities can give rise to waves of ideological change and lead American cities out of the first two ecological paradoxes as the abundance of the continental North America ends. Or perhaps our cognition of living on the same planet can ultimately hatch new ideas, and awaken the resilient vitality shared by cities and their ecosystems to help the modern world establish a new balance between shrinking planets and expanding cities even if it is even more fragile than ever before.

Postscript

At the end of May 2020, I returned to my home in Beijing from Qingdao after a four-and-a-half-month absence. At that time, summer had set in Beijing, but the air conditioner at home broke down. When the air conditioner repairman opened the cabinet of the outer unit of the air conditioner that is mounted on the exterior wall, he suddenly discovered a pair of eyases staring at the intruder, whose half of the body was hanging high in the 23rd floor, with their round eagle eyes in fear. The repairman was a young man from the county town of Shandong who had been working in Beijing for two years. He was obviously as excited and at a loss as I was about such a discovery. We had no experience of living in wild, and had no knowledge of how to coexist with animals other than dogs, cats, cockroaches in the kitchen, or mosquitoes in the bedroom. After calling for help, I learned that the eyases were kestrels (Falco tinnunculus). I decided not to disturb their lives, move them or feed them; otherwise their parents would abandon the old nest and birdlings.

For the next half month, I woke up every day in the hot and still air without the regulation of modern technology, and constantly heard the chirping of birds outside the window, or the sound of flapping wings as big birds entered and exited. That was a fresh and unique perception. In the sky of 23rd floor, the familiar sound to me is the rolling sound of cars on the crisscross roads, like a tidal surge, day and night, as well as the monotonous and endless operation of the outer unit. On a morning about half a month later, when I woke up, I suddenly had

S. Hou, *Wallless Cities*, https://doi.org/10.1007/978-981-97-7827-0

an unusual feeling. Outside the window was only the sound of cars, not the chirping of eyases. The fence of the cabinet of the outer unit was half open, opening and closing in the gentle breeze. I leaned out of the window, finding no eyas. The air conditioner repairman came and went again, and a familiar and reassuring buzzing sounded outside my window. The indoor air became cool and pleasant, and I had no excuse for not writing due to the hot weather. However, looking at the blue sky cut by high-rise buildings every day, I could still see the traces of the wings of the kestrels, or their high-spirited figures on the opposite building. These high-rise buildings resemble deep canyons for them. During the evolution of the giant city, they found their own ecological niche. These kestrels are not hybrid, our cultural power doesn't invent them or shape them. It is their own biological instincts that help them become the fittest to survive in new environments, and they that perhaps can find the hope of new redemption for the huge and crowded city.

I am very grateful to the cruise of the pair of kestrels in the "valley" made up of tall buildings and to the unknown wild grass growing in the cracks of the cement road. Their existence convinces me that the city is always a habitat for all kinds of life forms, yet I am just a member of the vast ecological community. Similarly, I am fortunate that I have always been in a vibrant cultural community. The most important force that helped me find my own niche in the cultural community is my tutor Mr. Donald Worster. If, during my six years when I studied on the Great Plains of the United States, Don completely shattered the wall of my former thoughts, and made me see a quite different yet diverse world of intelligence and nature. In the past eight years, Don came to Beijing every May to remind me with his unique observation perspective and sharp questioning attitude not to stay within the knowledge and thinking that I am already familiar with. He always asked about the rationality of deserved values, thoughts, and knowledge to let my imagination to break with all tangible and intangible boundaries and touch the fresh and wild power of thought. Although the book was ultimately completed in Chinese, there was a collision with Don's ideas from the initial outline to the writing of each chapter.

Another part of the cultural community is the discipline of environmental history as we call. I first knew the term more than thirty years ago. My mother invited Mr. Susan Flader, a Professor of the University of Missouri in the United States to give a lecture at Lanzhou University in

1986 for the Lanzhou University offered a course on American environmental history. It was at that time that I, as a completely naive child, came to understand the term. Although I had no idea of its meaning, I could not predict that my future career would be closely related to it. Susan, one of the founders of environmental history, combined knowledge with environmental activism and devoted her life to practicing Leopold's land ethics. She has a unique meaning for the book. When COVID-19 swept the world, she went to the Archives of the Missouri Historical Society on my behalf to take photos of the files I had failed to complete due to lack of time. In the community of environmental history, William Cronon is another figure who has had a profound impact on my thought about cities. At the top of Mount Tianshan in Xinjiang, there are two beautiful rainbows in the sky, and before my eyes are entire hillsides with yellow flowers in full bloom. Bill patiently listened to me explaining the structure of the book, and finally convinced me to replace Los Angeles with San Francisco as my third case of cities. He also persuaded me not to stay within the confines of *Nature's Metropolis: Chicago and the Great West* by him, but to seek the possibility of challenging his pattern. Similarly, in the community, I was fortunate to have met teachers and schoolmates such as Xia Mingfang, Mei Xueqin, Fu Chengshuang, Gao Guorong, Christof Mauch, Bjorn-Ola Linner, Beth LaDow, Lisa Brady, Mark Hersey, Wang Lihua, Fei Sheng, Cao Mu, Qiao Yu and others. Especially Teacher Xia, Teacher Fu, and Guo Rong spared no trouble to give advices on the first draft of my book and read the entire book. I admire the knowledge of these colleagues in environmental history, and I am even happier to be with them not only in terms of academic knowledge, but also as friends in daily life. The editor-in-chief of the series, Tan Xufeng, my junior fellow, hopes that the book will not fall short of his expectations.

Another part of the cultural community is the School of History, Renmin University of China, where I have lived and worked for the past decade. It is a group that does not have strict hierarchical cultures, but is a vigorous intellectual force. Although my colleagues at the Youth Salon of History, Chen Hao, Wu Tingting, Gu Liwei, Hu Heng, and others are different from me in the majors and interests, we share our curiosity about new knowledge and experience the same joy in thought and analysis. I also express heartfelt thanks to the senior teachers in the School of History, who are lenient and open. They allow the interests of future students to grow freely. My graduate students are a new type of intellectual force. They are young, intelligent, and ambitious, but their thinking gradually grows mature. The communication with them often gives me unexpected inspirations.

In addition, in the cultural community, I have benefited greatly from the predecessors and colleagues in the historian circles of China and the United States. The American History Research Association of China has a democratic and equal academic atmosphere. With the efforts of teachers such as Wang Xi, Wang Xu, Li Jianming, Liang Maoxin, Han Yu, and Zhang Yong'an, the Association has created a large number of opportunities for young scholars to maintain close communication with American colleagues. As a researcher of urban history, I am also deeply inspired by teachers and schoolmates such as Wang Xu, Han Yu, Yang Changyun, and Li Wenshuo.

Libraries and archives are the most important institutions that maintain the effective operation of an academic and cultural community. The writing of the book greatly benefited from the help of a large number of professional librarians in the libraries and archives in the United States, who made the process of searching and checking files an experience of enjoyment and learning. These libraries and archives include The Bancroft Library (University of California, Berkeley), Boston Public Library, East Asian Library (UC Berkeley), Francis Loeb Library (School of Design, Harvard University), Kansas City Public Library, Kenneth Spencer Research Library (University of Kansas), the Library of Congress, Massachusetts Historical Society, Missouri Historical Society, and San Francisco Public Library. Compared with the period when I traveled to Boston, New York, and Washington D.C. to search for information for my doctoral dissertation more than a decade ago, the accessibility of digital archives is now much easier. Especially during the outbreak of COVID-19, I was unable to return to the United States to continue archival research, and the existing files on San Francisco were not enough to support the final writing of the chapter. The digital archives of the Library of Congress and the San Francisco Public Library became important sources of historical data. Such convenience is deeply touching, but I always believe in the significance of physical libraries and archives, and I also believe that before he/she writes about a place as an environmental historian, he/she shall go to that place, experience that place directly with his/her eyes and taste buds, touch the carrier of words related to that place with both hands, such behaviors will increase his/her power to write.

Similarly, in the cultural community, there are many of my family members, relatives, and friends. In our career, we each have our own choices. But in our perception of beauty and expression of emotions, we are always in a community. My elder brother Hou Chen, his wife Sara, and their son Xin Han let me know that I always have a home in a foreign land far away. However, my elder brother's significance to me goes beyond that. His professional knowledge enabled me, as a student of liberal arts, to truly understand evolutionism. Whenever needed, My aunt, Ms. Tang Huilan, would surely save my messy life like an angel during my stay in Beijing for twelve years as a guest. My teenage friend Xing Yanying and her mother are my ecological hinterland. Every few weeks, they send steamed stuffed bun made by my aunt from Qingdao, thousands of li away from Beijing, to entertain me, the glutton, and help me cope with my homesickness. Of course, there are also Meng Yin, Wang Ling, Wang Jihong, Feng Yurong, Wang Zhaoyong, Dai Chong, Liu Zhe, Li Aixia, Hou Dasheng and so on. The list can be up to several lines. By now, I realize my luck.

For me, whether as a biological life form or a cultural life form, the most important people in my life are my parents. It is they who brought me into the world that, despite setbacks, pain, problems and crises, is still beautiful, abundant, and full of deep love, and allows me to always grow in love and beauty. It is they who taught me the first word, recited the first poem, opened the first page of books, and they who together led me into the world of history, thus I wander in it, settle down, and live in peace. At the time of completion of the book, my father had already left the world he loved so much for over two years. I dedicate the book to my father, thanking him for teaching me the value of life and the power of thought through his life, and showing me the importance of thought and knowledge to everything we cherish.

References

Primary References

The Library of Congress:

Frederick Law Olmsted Papers, Correspondences, 1838–1928
Digital Collections:
California as I Saw It: First-Person Narratives of California's Early Years, 1849 to 1900
Chronicling America (Historic American Newspapers)
Cities and Towns (Images, maps, and photos)
The Evolution of the Conservation Movement
The Nineteenth Century in Print: Periodicals:
The Atlantic Monthly,
The Century,
Garden and Forest,
Harper's New Monthly Magazine

On Boston:

Boston Public Library:
City Documents Archives
Massachusetts Historical Society:
Commonwealth of Massachusetts Senate Reports
Map Collection
Francis Loeb Library, Graduate School of Design, Harvard University
Charles Eliot Papers, 1859–1897, Special Collection

S. Hou, *Wallless Cities*, https://doi.org/10.1007/978-981-97-7827-0

On Kansas City:

Kansas City Public Library:
Archives of *Kansas City Star*
Missouri Valley Special Collections:
Charles P. Deatherage Book Manuscripts
George Kessler Collection
Kansas City Stockyards Collections
Map Collection
Missouri Historical Society:
Reports of the Trade and Commerce of Kansas City, 1878–1939
Kenneth Spencer Research Library, University of Kansas:
Rare Books on Kansas City's Early History

On San Francisco:

Bancroft Library, University of California, Berkeley:
Western Americana Collection
The Gold Rush,
Westward Migration,
Urban Communities: Emergence and Growth
East Asian Library, University of Californian, Berkeley:
The Him Mark Lai(麦礼谦)Collection
San Francisco Public Library:
SF Historical Photos

Books:

Adams, Charles F. *An Autobiography, 1835–1915*. Boston & New York: Houghton Mifflin Company, 1916.

Addams, Jane. *Twenty Years at Hull House*, with Autobiographic Notes. New York: The Macmillan company, 1937.

Ayers, James. *Gold and Sunshine, Reminiscences of Early California*. Boston, R. G. Badger, 1922

Bailey, Liberty Hyde. *The Country Life Movement in the United States*. New York: Macmillan Co., 1911.

Bailey, Liberty Hyde. *The Holy Earth*. New York: Charles Scribner's Sons, 1915. Bellamy, Edward. *Looking Backward: From 2000 to 1897*. Boston: William Ticknor, 1888.

Borsodi, Ralph. *Flight from the City*. New York: Harper & Brothers, 1933.

Borsodi, Ralph. *This Ugly Civilization*. New York: Simon and Schuster, 1929.

Brewer, William H. *Up and Down California in 1860–1864: The Journal of William H. Brewer*. ed. by Francis P. Farquhar, with a preface by Russell H.

Chittenden. New Haven: Yale University Press; London: H. Milford, Oxford University press, 1930.

Deatherage, Charles P. *Early History of Greater Kansas City, Missouri and Kansas, the Prophetic City at the Mouth of the Kaw*, Vol. 1. Kansas City: Interstate Pub. Co., 1927.

Eliot, Charles W. *Charles Eliot, Landscape Architect; A Lover of Nature and of His Kind, Who Trained Himself for a New Profession, Practised It Happily and through It Wrought Much Good*. Boston and New York: Houghton, Mifflin and Co., 1902.

Farnham, Charles Haight. *A Life of Francis Parkman*. Boston: Little, Brown and Company, 1900.

George, Henry. *Progress and Poverty; An Inquiry into the Cause of Industrial Depressions, and of Increase of Want with Increase of Wealth—The Remedy*. San Francisco: W. M. Hinton & Co., 1879.

Hamilton, Alice. *Industrial Poisons in the United States*. New York: McMillian, 1925.

George, *Henry. Industrial Toxicology*. 2nd ed. New York: Scribner, 1934.

Hill, W. R. *The Great Flood of 1903: Being a Graphic Story of How Two Mad Rivers, the Missouri and the Kaw, Deluged Kansas City and Its Suburb*. Kansas City: Enterprise Publishing Company, 1903.

Howard, Ebenezer. *Garden Cities of To-morrow*. London: S. Sonnenschein & Co., Ltd. 1902.

Kains, Maurice Grenville. *Five Acres and Independence*. New York: Greenberg, 1935.

Massey, Ernest. *A Frenchman in the Gold Rush: The Journal of Ernest de Massey, Argonaut of 1849*. Trans. by Marguerite Eyer Wilbur. San Francisco: California Historical Society, 1927.

Meyer, Carl. *Bound for Sacrament: Travel-Pictures of a Returned Wanderer*. Trans. by Ruth Frey Axe. Originally published in German in 1855, republished in English in 1938. Claremont, Calif.: Saunders Studio Press, 1938.

Muir, John. *The Mountains of California*. New York: Century Co., 1894.

Nearing, Helen and Scott. *Living the Good Life: How to Live Sanely and Simply in a Trouble World*. New York: Schocken Books, 1954

Nearing, Scott, *The Making of a Radical: A Political Autobiography*. New York: Harper and Row, 1972.

Olmsted, Frederick L. *Civilizing American Cities: Writing on City Landscapes*. Edited by S. B. Sutton. New York: Da Capo Press, 1997.

Olmsted, Frederick L. *Journeys and Explorations in the Cotton Kingdom of America: A Traveller's Observations on Cotton and Slavery in the American Slave States, based upon three former volumes of journeys and investigations by the same author*. 2 vols. London: Sampson Low, Son and Co., 1861.

Olmsted, Frederick L. *The Papers of Frederick Law Olmsted*. 6 vols. & supplementary vol. 1. Edited by Charles Capen McLaughlin, Charles E. Beveridge, David Schuyler, Jane Turner Censer, and Victoria Post Ranney. Baltimore: Johns Hopkins University Press, 1977–1997.

Olmsted, Frederick L. *Walks and Talks of an American Farmer in England*. 2 vols. New York: G. P. Putnam and Co., 1852.

Pierce, Hiram Dwight. A Forty-Niner Speaks; A Chronological Record of a New Yorkerand His Adventures in Various Mining Localities in California, His Return Trip across Nicaragua, Including Several Descriptions of the Changes in San Francisco and Other Mining Centers from March 1849 to January 1851. Oakland, Calif., Keystone-Inglett printing Co., 1930.

Pinchot, Gifford. *Breaking New Ground*. New York: Harcourt, Brace, and Co., 1947.

Richards, Ellen and Alpheus Woodman, *Air, Water, and Food: From a Sanitary Standpoint*. 1st edition, New York: John Wiley and Sons, 1900.

Richards, Ellen. *Euthenics, the Science of Controllable Environment; A Plea for Better Living Condition As the First Step toward Higher Human Efficiency*. Boston: Whitcomb and Barrow, 1910.

Riis, Jacob, *How the Other Half Lives: Studies among the Tenements of New York*. New York: Charles Scribner's Sons, 1890.

Schofield, John. *Forty-Six Years in the Army*. New York: Century Publisher, 1897.

Sloan, Walter B. *History of Kansas and Nebraska: Describing Soil, Climate, Rivers, Prairies, Mounds, Forests, Minerals, Roads, Cities, Villages, Inhabitants, and Other Subjects Relating to that Region, with a Correct Map*. Galesburg, ILL: Boishel, Kuhn, & Co., 1857.

Snyder, Gary. *The Great Clod: Notes and Memoirs on Nature and History in East Asia*. Los Angeles: Counterpoint Books, 2016.

Snyder, Gary. *A Place in Space: Ethics, Aesthetics, and Watersheds*. Los Angeles: Counterpoint Books, 2008.

Soulé, Frank, John H. Gihon, and James Nisbet, The Annals of San Francisco; Containing a Summary ofthe History of the First Discovery, Settlement, Progress, and Present Condition of California, and a Complete History of All the Important Events with Its Great City. Originally published in 1854, with Introduction by Richard Dillon and index by Charles Goehring, Berkeley: Berkeley Hills Books, 1999.

Squier, Miriam. *California: A Pleasure Trip from Gotham to the Golden Gate*, April, May, June, 1877. Nieuwkoop [Netherlands] B. De Graaf, 1972.

Urban League of Kansas City, *The Negro Worker of Kansas City: A Study of Trade Union and Organized Labor Relations*. Kansas City: Urban League of Kansas City, 1939.

White, E. B. *One Man's Meat*. New York: Harper & Brothers, 1942.

Whitehill, Walter Muir. *A Life of Francis Parkman*. Boston: Little, Brown, and Company, 1901.

Whitman, Walt, *Leaves of Grass*. Philadelphia: D. McKay, 1884.

Winthrop, John. *A Model of Christian Charity,* Collections of the Massachusetts Historical Society. Boston: 1838.

Wood, William. *New England Prospect: A True, Lively, and Experimental Description of That Part of America Commonly Called New England....* London: Printed by Tho. Cotes for John Bellamie and are to be sold at his shop..., 1634.

Research References

Works (in English):

Abbott, Gordon Jr. *Saving Special Places: A Centennial History of the Trustees of Reservations: Pioneer of the Land Trust Movement*, Ipswich, Mass, 1993.

Albert, Fein. *Frederick Law Olmsted and the American Environmental Tradition*. New York: G. Braziller, 1972.

Allerfeldt, Kristofered. *The Progressive Era in the USA, 1890–1921*. Aldershot, England; Burlington, VT: Ashgate, 2007.

Armitage, Kevin. *The Nature Study Movement: The Forgotten Popularizer of America's Conservation Ethic*. Lawrence: University of Kansas, 2009.

Barns, David. *The Great Stink of Paris and the Nineteenth-Century Struggle against Filth and Germs*. Baltimore: Johns Hopkins University Press, 2006.

Bender, Thomas. *Toward an Urban Vision: Ideas and Institutions in Nineteenth-Century America*. Lexington: The University Press of Kentucky, 1975.

Bernhardt, Christophed. *Environmental Problems in European Cities of the 19th and 20th Century*. New York/ Muenchen/ Berlin: Waxmann, Muenster, 2001.

Biehler, Dawn Day. *Pests in the City: Flies, Bedbugs, Cockroaches, and Rats*. Seattle: University of Washington Press, 2015

Binfor, Henry C. *The First Suburbs: Residential Communities on the Boston Periphery, 1815-1860*. Chicago: The University of Chicago Press, 1985.

Birnbaum, Charles A. and Robin Karson, eds. *Pioneers of American Landscape Design*. New York: McGraw-Hill, 2000.

Booker, Mathew. *Down by the Bay: San Francisco's History between the Tides*. Berkeley: University of California Press, 2013.

Boyer, Paul. *Urban Masses and Moral Order in America*. Cambridge, Mass.: Harvard University Press, 1978.

Brady, Lisa. *War Upon the Land: Military Strategy and Southern Landscapes during the American Civil War*. Athens: University of Georgia Press, 2012.

Brechin, Grey. *Imperial San Francisco: Urban Power, Early Ruin*. Berkeley: University of California Press, 1999.

Brenner, Neil ed. *Implosions/ Explosions: Towards a Study of Planetary Urbanization.* Berlin: Jovis, 2014.

Brewer, Richard. *Conservancy: The Land Trust Movement in America.* Lebanon, New Hampshire: Dartmouth College Press, 2004.

Brosnan, Kathleen. *Uniting Mountain & Plain: Cities, Law, and Environmental Change along the Front Range.* Albuquerque: University of New Mexico Press, 2002.

Brown, Frederick. *The City Is More Than Human: An Animal History of Seattle.* Seattle: University of Washington Press, 2019.

Buell, Lawrence. *The Environmental Imagination: Thoreau, Nature Writing, and the Formation of American Culture.* Cambridge, Mass.: Belknap Press of Harvard University Press, 1995.

Burke, Diane Mutti, Jason Roe, and John Herron eds. *Wide-Open Town: Kansas City in the Pendergast Era.* by, Lawrence: University of Kansas Press, 2018.

Burnstein, Daniel Eli. *Next to Godliness: Confronting Dirt and Despair in Progressive Era New York City.* Urbana: University of Illinois Press, 2006.

Burrows, Edwin and Mike Wallace. *Gotham: A History of New York City to 1898.* New York: Oxford University Press, 1999.

Carlson, Allan. *The New Agrarian Mind: The Movement toward Decentralist Thought in Twentieth Century America.* New Brunswick, N.J.: Transaction Publishers, 2000.

Carson, Rachel. *Silent Spring.* Boston: Houghton Mifflin, 1962.

Castonguay, Stéphane. *Metropolitan Nature: Environmental Histories of Montreal.* Pittsburgh: University of Pittsburgh Press, 2011.

Chudacoff, Howard. *The Evolution of American Urban Society.* Englewood Cliffs, NJ: Prentice-Hall, 1975.

Castongua y, Stéphane. Judith E. Smith, and Peter C. Baldwin. Castonguay, Stéphane. 7th ed. Upper Saddle River, NJ: Prentice Hall, 2010.

Clifford-Napoleone, Amber R. *Queering Kansa City Jazz: Gender, Performance, and the History of a Scene.* Lincoln: University of Nebraska Press, 2018.

Contosta, David. *Suburb in the City: Chestnut Hill, Philadelphia, 1850-1990.* Columbus: Ohio State University Press, 1992.

Conzen, Kathleen. *Immigrant Milwaukee, 1836-1860: Accommodation and Community in a Frontier City.* Cambridge: Harvard University Press, 1976.

Cranz, Galen. *The Politics of Park Design: A History of Urban Parks in America.* Cambridge: MIT Press, 1982.

Cronon, William. *Changes in the Land: Indians, Colonists, and the Ecology of New England.* New York: Hill and Wang, 1983.

Cronon, William. *Nature's Metropolis: Chicago and the Great West.* New York: W. W. Norton, 1991.

Cronon, William, ed. *Uncommon Ground: Toward Reinventing Nature.* New York: W. W. Norton, 1995.

Crosby, Alfred. *Ecological Imperialism: The Biological Expansion of Europe, 900-1900*. New York: Cambridge University Press, 1986.
Culver, *Lawrence. The Frontier of Leisure: Southern California and the Shaping of Modern America*. New York: Oxford University Press, 2012.
Davis, Allen F. *Spearheads for Reform; The Social Settlements and the Progressive Movement, 1890–1914*.
New York: Oxford University Press, 1967.
Davis, Mike. *City of Quartz: Excavating the Future in Los Angeles*. London: Verso, 1990.
Davis, Mike. *Ecology of Fear: Los Angeles and the Imagination of Disaster*. New York: Metropolitan Books, 1998.
Davis, Mike. *Late Victoria Holocausts: El Nino Famines and the Making of the Third World*. London, New York: Verso: 2001.
Dawley, Alan. *Class and Community: The Industrial Revolution in Lynn*. Cambridge: Harvard University Press, 1976.
Detwyler, Thomas R. and Melvin G. Marcus, eds. *Urbanization and Environment: The Physical Geography of the City*. Belmont,
California: Dexbury Press, 1972.
Doell, M. Christine Klim, *Gardens of the Gilded Age: Nineteenth-Century Gardens and Homegrounds of New York State*. Syracuse: Syracuse University Press, 1986.
Domosh, Mona. *Invented Cities: The Creation of Landscape in Nineteenth-Century New York & Boston*. New Haven: Yale University Press, 1996.
Donahue, Brian. *The Great Meadow: Farmers and the Land in Colonial Concord*. New Haven: Yale University Press, 2004.
Dyl, Joanna L. *Seismic City: An Environmental History of San Francisco's 1906 Earthquake*. Seattle: University of Washington Press, 2019.
Elkind, Sarah, *Bay Cities and Water Politics: The Battle for Resources in Boston and Oakland*. Lawrence, KS: University of Kansas Press, 1998.
Falck, Zachary J. S. *Weeds: An Environmental History of Metropolitan America*. Pittsburgh: University of Pittsburgh Press, 2016.
Fisher, Irving D. *Frederick Law Olmsted and the City Planning Movement in the United States*. Ann Arbor, Mich.: UMI Research Press, 1986.
Forman, Richard T. T. and Michel Gordon, *Landscape Ecology*. New York: John Wiley and Sons, 1986.
Foster, John Bellamy. *Marx's Ecology: Materialism and Nature*. New York: Monthly Review, 2000.
Fox, Stephen. *The American Conservation Movement: John Muir and His Legacy*. Madison: University Wisconsin Press, 1985.
Fraser, Derek and Anthony Sutcliffe, eds. *The Pursuit of Urban History*. London: Edward Arnold, 1983.

Gandy, Matthew. *Concrete and Clay: Remaking Nature in New York City*. Cambridge, Mass.: MIT Press, 2002.

Giesen, James C. *Boll Weevil Blues: Cotton, Myth, and Power in the American South*. Chicago: University of Chicago Press, 2011.

Glaab, Charles N. ed. *The American City, a Documentary History*. Homewood, Ill.: Dorsey Press, 1963.

Glaab, Charles N. *A History of Urban America*. New York: Macmillan, 1967.

Glaab, Charles N. *Kansas City and the Railroads: Community Policy in the Growth of a Regional Metropolis*. Lawrence, KS: University Press of Kansas, 1993.

Gotham, Kevin Fox. *Race, Real Estate, and Uneven Development: The Kansas City Experience, 1900- 2010*. Albany: Sunny Press, 2014.

Gottlieb, Robert. *Forcing the Spring: The Transformation of the American Environmental Movement*. Washington, D. C.: Island Press, 1993.

Green, James R. *Death in the Haymarket: A Story of Chicago, the First Labor Movement and the Bombing that Divided Gilded Age America*. New York: Pantheon Books, 2006.

Griffin, G. S. *Racism in Kansas City: A Short History*. Traverse City, MI: Chandler Lake Books, an imprint of Mission Point Press, 2015.

Griswold, Mac and Eleanor Weller. *The Golden Age of American Gardens: Proud Owners Private Estates, 1890- 1940*. New York: H. N. Abrams, 1991. Haglund, Karl. *Inventing the Charles River*. Cambridge, Mass.: MIT Press, 2003.

Hall, *Lee. Olmsted's America: An "Unpractical" Man and His Vision of Civilization*. Boston: Little, Brown, 1995.

Hannickel, Erica. *Empire of Vines, Wine Culture in America*. Philadelphia: University of Pennsylvania Press, 2013.

Havlick, Spenser W. *The Urban Organism: The City's Natural Resources from an Environmental Perspective*. New York: Macmillan Co., 1974.

Hays, Samuel P. *Conservation and the Gospel of Efficiency: The Progressive Conservation Movement, 1890-1920*. Cambridge: Harvard University Press, 1959.

Hays, Samuel P. *Beauty, Health, and Permanence: Environmental Politics in the United States, 1955–1985*. New York: Cambridge University Press, 1987.

Hoffmann, Richard. *An Environmental History of Medieval Europe*. New York: Cambridge University Press, 2014.

Horowitz, Andy. *Katrina: A History, 1915–2015*. Cambridge, Mass.: Harvard University Press, 2020.

Hou, Shen. *The City Natural: Garden and Forest Magazine and the Rise of American Environmentalism*. Pittsburgh: University of Pittsburgh Press, 2013.

Hurley, Andrew ed. *Common Fields: An Environmental History of St. Louis*. St. Louis: Missouri Historical Society Press, 1997.

Hurley, Andrew ed. *Environmental Inequalities: Class, Race, and Industrial Pollution in Gary, Indiana, 1945-1980*. Chapel Hill: University of North Carolina Press, 1995.

Irland, Lloyd. *The Northeast Changing Forests*. Hanover: University of New England Press, 1999.

Isenberg, Andrew ed. *The Nature in Cities: Culture, Landscape and Urban Space*. New York: University of Rochester Press, 2006.

Isenberg, Andrew ed. *The Destruction of the Bison: An Environmental History, 1750-1920*. Cambridge, U. K.: University of Cambridge Press, 2000.

Isenberg, Andrew ed. *Mining California: An Ecological History*. New York: Hill and Wang, 2005.

Jacoby, Karl. *Crimes against Nature: Squatters, Poachers, thieves, and the Hidden History of American Conservation*. Berkeley, California: University of California Press, 2001.

Jackson, Kenneth. *Crabgrass Frontier: The Suburbanization of the United States*. New York: Oxford University Press, 1985.

Kalfus, Melvin. *Frederick Law Olmsted: The Passion of a Public Artist*. New York: New York University Press, 1990.

Kazin, Michael. *The Populist Per- suasion: An American History*, Ithaca and London: Cornell Univ. Press, 1998.

Kelman, Ari. *A River and Its City: The Nature of Landscape in New Orleans*. Berkeley Cal.: University of California Press, 2nd ed. 2006.

Kennedy, Lawrence W. *Planning the City upon a Hill: Boston since 1630*. Amherst: The University of Massachusetts Press, 1992.

Kiechle, Melanie. *Smell Detective: An Olfactory History of Nineteenth-Century Urban America*. Seattle: University of Washington Press, 2019.

Klingle, Matthew. *Emerald City: An Environmental History of Seattle*. New Haven: Yale University Press, 2007.

Kowsky, Francis R. *Country, Park, and City: The Architecture and Life of Calvert Vaux, 1824–1895*. New York: Oxford University Press, 1997.

Krech, Shepard. *The Ecological Indian: Myth and History*. New York: W. W. Norton & Co. 1999.

Langston, Nancy. *Toxic Bodies: Hormone Disruptors and the Legacy of DES*. New Haven: Yale University Press, 2010.

Lasch-Quinn, Elisabeth. *Black Neighbors: Race and the Limits of Reform in the American Settlement House Movement, 1890- 1945*. Chapel Hill: University of North Carolina Press, 1993.

Lears, T. J. Jackson. *No Place of Grace: Antimodernism and the Transformation of American Culture, 1880-1920*. New York: Pantheon Books, 1981. Reprint, Chicago: The University of Chicago Press, 1994.

Lees, Andrew. *Cities Perceived: Urban Society in European and American Thought, 1820–1940.*

Manchester: Manchester University Press, 1985.

Lemon, James. *Liberal Dreams and Nature's Limits: Great Cities of North America since 1600*, New York: Oxford University Press, 1996.

Leopold, Aldo. *A Sand County Almanac, With Other Essays on Conservation from Round River*. New York: Oxford University Press, 1966.

Levy, John M. *Urban America: Processes and Problems*. Upper Saddle River, NJ: Prentice Hall, 2000.

Liner, Marc and Lawrence S. Zacharias. *Of Cabbages and Kings County: Agriculture and the Formation of Modern Brooklyn*. Iowa City: University of Iowa Press, 1999.

Logan, Michael F. *Desert Cities: The Environmental History of Phoenix and Tucson*. Pittsburgh: University of Pittsburgh Press, 2006.

Lowenthal, David. *George Perkins Marsh: Prophet of Conservation*. With a forward by William Cronon. Seattle: University of Washington Press, 2000.

Katz, Michael. *The People of Hamilton, Canada West: Family and Class in a Mid-nineteenth-century City*. Cambridge: Harvard University Press, 1975

Machor, James L. *Pastoral Cities: Urban Ideals and the Symbolic Landscape of America*. Madison: University of Wisconsin Press, 1987.

Major, Judith. *To Live in the New World: A. J. Downing and American Landscape Gardening*. Cambridge, Mass: MIT Press, 1997.

Mallea, Amahia. *A River in the City of Fountains: An Environmental History of Kansas City and the Missouri River*. Lawrence: University of Kansas Press, 2018.

Mauch, Christof and Tom Zeller eds. *Rivers in History: Perspectives on Waterways in Europe and America*. Pittsburgh: University of Pittsburgh Press, 2008.

Max, Leo. *The Machine in the Garden: Technology and the Pastoral Ideal in America*. New York: Oxford University Press, 1964.

McBride, Paul. *Culture Clash: Immigrants and Reformers, 1880- 1920*. San Francisco: R and E Research Associates, 1975.

McClelland, Linda F. *Building the National Parks: Historic Landscape Design and Construction*. Baltimore: Johns Hopkins University Press, 1998.

McCammack, Brian. *Landscapes of Hope: Nature and the Great Migration in Chicago. Cambridge*: Harvard University Press, 2017.

McNeill, John. *Something New Under the Sun: An Environmental History of the Twentieth Century World*. New York: Norton, 2000.

McNeur, Catherine. *Taming Manhattan: Environmental Battles in the Antebellum City*. Cambridge, Mass.: Harvard University Press, 2014.

McShane, Clay and Joel Tarr, *The Horse in the City: Living Machines in the Nineteenth Century*. Baltimore: Johns Hopkins University Press, 2007.

Melosi, Martin. *Fresh Kills: A History of Consuming and Discarding in New York City*. New York: Columbia University Press, 2020.

Melosi, Martin. *Garbage in the Cities: Refuse, Reform, and the Environment: 1880-1980*. College Station, TX: Texas A & M University Press, 1981.

Melosi, Martin. *The Sanitary City: Urban Infrastructure in America from Colonial Times to the Present*. Baltimore: Johns Hopkins University Press, 2000.

Merchant, Carolyn. *The Death of Nature: Women, Ecology and the Scientific Revolution*. San Francisco: Harper & Row, 1980.

Merchant, Carolyn. *Ecological Revolutions: Nature, Gender, and Science in New England*. Chapel Hill: University of North Carolina Press, 1989.

Meyers, Amy R. ed. *Art and Science in America: Issues of Representation*. San Marino, Calif: Huntington Library Press, 1998.

Miller, Char and Hal Rothman, eds. *Out of the Woods: Essays in Environmental History*. Pittsburgh: University of Pittsburgh Press, 1997.

Miller, Perry. *Errand into the Wilderness*. Cambridge: Belknap Press of Harvard University Press, 1956.

Minteer, Ben. *The Landscape of Reform: Civic Pragmatism and Environmental Thought in America*. Cambridge, Mass: MIT Press, 2006.

Miller, Zane and Patricia M. Melvin. *The Urbanization of Modern America: A Brief History*, 2nd e d. San Diego: Harcourt Brace Jovanovich, Publishers, 1987.

Monkkonen, Eric. *The Dangerous Class: Crime and Poverty in Columbus, Ohio, 1860- 1885*. Cambridge: Harvard University Press, 1975.

Mumford, Lewis. *The Brown Decades: The Study of the Arts in America, 1865-1895*. New York: Harcourt, Brace, and Co., 1931.

Mumford, Lewis. *The City in History: Its Origins, Its Transformations, and Its Prospects*. New York: Harcourt, Brace, and World, 1961.

Nash, Roderick. *Wilderness and the American Mind*. New Haven: Yale University Press, 1967, and subsequent editions. Needham, Andrew. *Power Lines: Phoenix and the Making of the Modern Southwest*. Princeton University Press, 2015.

Newman, William A. and Wilfred E. Holton. *Boston's Back Bay: The Story of America's Greatest Nineteenth- Century Landfill Project*. Boston: Northeastern Univ. Press, 2006.

Nicolson, Marjorie Hope. *Mountain Gloom and Mountain Glory: The Development of the Aesthetics of the Infinite*. Seattle: University of Washington Press, 1997.

O 'Connor, Thomas H. *The Hub: Boston Past and Present*. Boston: Northeastern University Press, 2001.

O Grada, Cormac. *The Great Irish Famine*. Cambridge: Cambridge University Press, 1989.

Pauketat, Timothy R. *Cahokia: Ancient America's Great City on the Mississippi*. New York, N. Y.: Viking, 2009.

Penna, Anthony N. and Conrad Edick Wright ededs. *Remaking Boston: An Environmental History of the City and Its Surroundings*. Pittsburgh: University of Pittsburgh Press, 2009.

Peterson, Jon A. *The Birth of City Planning in the United States, 1840- 1917*. Baltimore: Johns Hopkins University Press, 2003.

Philippon, Daniel J. *Conserving Words: How American Nature Writers Shaped the Environmental Movement*. Athens: The University of Georgia Press, 2004.

Piott, Steven L. *American Reformers, 1870–1920: Progressives in Word and Deed*. Lanham, Md.: Rowman & Littlefield Publishers, 2006.

Platt, Harold L. *Building the Urban Environment: Visions of the Organic City in the United States, Europe, and Latin America*. Philadelphia: Temple University Press, 2015.

Postel, Charles. *The Populist Vision*. New York: Oxford University Press, 2007.

Potter, David. *People of Plenty: Economic Abundance and the American Character*. Chicago: University of Chicago Press, 1954.

Pratt, Joseph A., Martin V. Melosi, and Kathleen A. Brosnan eds. *Energy Capitals: Local Impacts, Global Influence*. Pittsburgh: University of Pittsburgh Press, 2014.

Pregill, Philip and Nancy Volkman. *Landscapes in History: Design and Planning in the Eastern and Western Traditions*, 2nd ed. New York: John Wiley, 1999.

Price, Jennifer. *Flight Maps: Adven- tureswith Nature in Modern America*. New York: Basic Books, 1999.

Prud' homme, Alex. *The Ripple Effect: The Fate of Fresh Water in the twenty-first Century*. New York: Simon & Schuster, 2011.

Punch, Walter T. ed. *Keeping Eden: A History of Gardening in America*. Massachusetts Horticultural Society. Boston: Bulfinch Press, 1992.

Rawson, Michael. *Eden on the Charles: The Making of Boston*. Cambridge: Harvard University Press, 2010.

Richardson, Robert. *Henry Thoreau: A Life of the Mind*. Berkeley: University of California Press, 1986.

Richardson, Robert. *Emerson: The Mind on Fire: A Biography*. Berkeley: University of California Press, 1995.

Risjord, Norman K. *Populists and Progressives*. Lanham: Rowman & Littlefield, 2005.

Robison, Charles Mulford. *The Improvement of Towns and Cities*. New York & London: G. P. Putnam's sons, 1901.

Robison, Charles Mulford. *Modern Civic Art, or, the City Made Beautiful*. New York: G. P. Putman, 1903.

Rodger, Daniel. *Atlantic Crossings: Social Politics in a Progressive Age*. Cambridge, Mass.: Harvard University Press, 1998.

Rogers, Jedediah S. and Matthew C. Godfrey eds. *The Earth Will Appear As the Garden of Eden: Essays on Mormon Environmental History*. Salt Lake City: University of Utah Press, 2019.

Rohrbough, Malcolm. *Days of Gold: The California Gold Rush and the American Nation*. Berkeley: University of California Press, 1998. Rome, Adam. *The Bulldozer in the Countryside: Suburban Sprawl and the Rise of American Environmentalism*. New York: Cambridge University Press, 2001.

Roper, Laura Wood.*FLO: A Biography of Frederick Law Olmsted*. Baltimore: The Johns Hopkins Univ. Press, 1973. Rosenzweig, Roy. *Eight Hours for What We Will: Workers and Leisure in an Industrial City, 1870–1920*. New York: Cambridge University Press, 1983.

Rosenzweig, Roy. and Elizabeth Blackmar. *The Park and the People: A History of Central Park*. Ithaca: Cornell University Press, 1992.

Rothman, Hal. *Neon Metropolis: How Las Vegas Started the Twenty-first Century*. New York: Routledge, 2002.

Runte, Alfred. *National Parks: The American Experience*, 3rd ed. Lincoln: University of Nebraska Press, 1997.

Rybczynski, Witold. *A Clearing in the Distance: Frederick Law Olmsted and America in the Nineteenth Century*. New York: Scribner, 1999.

Sachs, Aaron. *The Humboldt Current: Nineteenth-century Exploration and the Roots of American Environmentalism*. New York: Viking, 2006.

Scheese, Don. *Nature Writing: The Pastoral Impulse in America*. New York: Twayne, 1996.

Schirmer, Sherry Lamb. *A City Divided: The Racial Landscape of Kansas City, 1900–1960*.

Columbia: University of Missouri Press, 2002.

Schmitt, Peter J. *Back to Nature: The Arcadian Myth in Urban America*. With a forward by John R. Stilgoe. New York: Oxford University Press, 1969.

Reprint, Baltimore: Johns Hopkins University Press, 1990.

Schultz, *Stanley K. Constructing Urban Culture: American Cities and City Planning, 1800–1920*.

Philadelphia: Temple University Press, 1989.

Schuyler, David. *The New Urban Landscape: The Redefinition of City Form in Nineteenth-Century America*. Baltimore: Johns Hopkins University Press, 1986.

Schuyler, David. *Apostle of Taste: Andrew Jackson Downing, 1815- 1852*.

Baltimore: Johns Hopkins University Press, 1996.

Scott, James C. *Seeing like a State: How Certain Schemes to Improve the Human Condition Have Failed*. New Haven: Yale University Press, 1998.

Seasholes, Nancy S. *Gaining Ground: A History of Landmaking in Boston*. Cambridge: MIT Press, 2003.

Sellars, Richard. *Preserving Nature in the National Parks: A History*. New Haven: Yale University Press, 1997.

Shortridge, James R. *Kansas City and How It Grew, 1822-2011.* Lawrence: University of Kansas Press, 2012

Simo, Melanie. *Forest and Garden: Traces of Wildness in a Modernizing Land, 1897-1949*. Charlottesville: University of Virginia Press, 2003.

Smith, Carl. *The Plan of Chicago: Daniel Burnham and the Remaking of the American City*. Chicago: University of Chicago Press, 2006.

Smith, Henry Nash. *Virgin Land: The American West as Symbol and Myth*. Cambridge, Mass.: Harvard University Press, 1950.

Smith-Howard, Kendra. *Pure and Modern Milk: An Environmental History Since 1900*. New York: Oxford University Press, 2013.

Soluri, John. *Banana Cultures: Agriculture, Consumption, and Environmental Change in Honduras and the United States*. Austin: University of Texas Press, 2005.

Spain, Daphne. *How Women Saved the City*. Minneapolis: University of Minnesota Press, 2001.

Steinberg, Theodore. *Acts of God: The Unnatural History of Natural Disaster of America*. New York: Oxford University Press, 2006.

Steinberg, Theodore. *American Green: The Obsessive Quest for the Perfect Lawn*. New York: W. W. Norton, 2007.

Steinberg, Theodore. *Down to Earth: Nature's Role in American History*. New York: Oxford Press, 2002.

Steinberg, Theodore. *Gotham Unbound: The Ecological History of Greater New York*, New York: Simon and Schuster, 2014.

Steinberg, Theodore. *Nature Incorporated: Industrialization and the Waters of New England*. New York: Cambridge University Press, 1991.

Stevenson, Elizabeth. *Park Maker: A Life of Frederick Law Olmsted*. New Brunswick, N. J.: Transaction Publishers, 2000.

Stilgoe, John R. *Borderland: Origins of the American Suburb, 1820-1939*. New Haven: Yale University Press, 1988.

Stock, Catherine M. *Rural Radicals: Righteous Rage in the American Grain*, Ithaca and London: Cornell Univ. Press, 1996.

Stoll, Mark. *Protestantism, Capitalism, and Nature in America*. Albuquerque: University of New Mexico Press, 1997.

Stradling, David. *Making Mountains: New York City and the Catskills*. Seattle: University of Washington Press, 2007.

Stradling, David. *Smokestacks and Progressives: Environmentalists, Engineers and Air Quality in America, 1881–1951.*

Baltimore: Johns Hopkins University Press, 1999.

Sutter, Paul & Christopher Manganiello eds. *Environmental History and the American South: A Reader*. Athens, GA: University of Georgia Press, 2009.

Sutter, Paul. *Driven Wild: How the Fight against Automobiles Launched the Modern Wilderness Movement*. Seattle: University of Washington Press, 2005.

Tarr, Joel A. ed. *Devastation and Renewal: An Environmental History of Pittsburgh and Its Region*. Pitts- burgh: University of Pittsburgh Press, 2005.

Tarr, Joel A. ed. *The Search for the Ultimate Sink: Urban Pollution in Historical Perspective*. Akron: of Akron Press, 1996.

Teaford, Jon C. *The Unheralded Triumph: City Government in America, 1870-1900*. Baltimore: The Johns Hopkins University Press, 1984.

Thernstrom, Stephan. *Poverty and Progress: Social Mobility in a Nineteenth Century City*. Cambridge: Harvard University Press, 1964.

Thernstrom, Stephan. *The Other Bostonians: Poverty and Progress in the American Metropolis, 1880- 1970*. Cambridge, Mass.: Harvard University Press, 1973.

Thomas, John, *Alternative America: Henry George, Edward Bellamy, Henry Demarest Lloyd, and the Adversary Tradition*. Cambridge, Mass.: Belknap Press of Harvard University Press, 1983.

Thomas, Keith. *Man and the Natural World: Changing Attitudes in England, 1500-1800*. London: Allen Lane, 1983.

Turner, Frederick. *The Frontier in American History*. New York: Henry Holt and Company, 1920.

Uekoetter, Frank. *The Age of Smoke: Environmental Policy in Germany and the United States, 1880-1970*. Pittsburgh: University of Pittsburgh Press, 2009.

von Hoffman, Alexander. *Local Attachments: The Making of an American Urban Neighborhood, 1850 to 1920*. Baltimore: The Johns Hopkins University Press, 1994.

Walker, Peter and Melanie Simo, *Invisible Gardens: The Search for Modernismin the American Landscape*. Cambridge, Mass: MIT Press, 1994.

Walker, Richard A. *The Country in the City: The Greening of the San Francisco Bay Area*. Foreword by William Cronon, Seattle: University of Washington Press, 2007.

Warner, Sam Bass. *Streetcar Suburbs: The Process of Growth in Boston, 1870 1900*. Cambridge: Harvard University Press, 1962.

Washington, Sylvia. *Packing Them In: An Archaeology of Environmental Racism in Chicago, 1865–1954*. Lanham, Maryland: Lexington Books, 2005.

Webb, Walter Prescott. *The Great Plains*. Boston: Ginn, 1931.

Weber, Michael P. *Social Change in an Industrial Town: Patterns of Progress in Warren, Pennsylvania, from Civil War to World War I*. University Park: Pennsylvania State University Press, 1976.

White, Dana F. and Victor A. Kramer. *Olmsted South: Old South Critic/New South Planner*. Westport: Greenwood Press, 1979.

White, Dana F. *The Urbanists, 1865-1915*. New York: Greenwood Press, 1989.

White, Morton and Lucia. *The Intellectual versus the City: From Thomas Jefferson to Frank Lloyd Wright*. Cambridge, Mass: Harvard University Press/M. I. T. Press, 1962.

White, Richard. *Railroaded: The Transcontinentals and the Making of Modern America*. New York: W. W. Norton, 2011.

Whitehill, Walter Muir. *Boston: A Topographical History*. 2nd edition, enlarged, Cambridge, Mass.: The Belknap Press of Harvard University Press, 1959.

Wiebe, Robert H. *The Search for Order, 1877-1920*. New York: Hill and Wang, 1967.

Wilkie, Richard & Jack Tager, *Historical Atlas* of *Massachusetts*. Amherst: University of Massachusetts Press, 1991.

Williams, Raymond. *The Country and the City*. London: Chatto & Windus, 1973.

Wilson, William H. *The City Beautiful Movement*. Baltimore: Johns Hopkins University Press, 1989.

Wilson, William H. *The City Beautiful Movement in Kansas City*. Columbia: University of Missouri Press, 1964.

Worster, Donald. *Dust Bowl: The Southern Plains in the 1930s*. New York: Oxford University Press, 1979.

Worst er, Donald. ed. *The Ends of Earth: Perspectives of Modern Environmental History*. New York: Oxford University Press, 1988.

Worster, Donald. *Nature's Economy: A History of Ecological Ideas*, 2nd ed. New York: Cambridge University Press, 1994.

Worster, Donald. *A Passion for Nature: The Life of John Muir*. New York: Oxford University Press, 2009.

Worster, Donald. *A River Running West: The Life of John Wesley Powell*. New York: Oxford University Press, 2001.

Worster, Donald. *Shrinking the Earth: The Rise & Decline of American Abundance*. New York: Oxford University Press, 2016.

Worster, Donald. *The Wealth of Nature: Environmental History and the Ecological Imagination*. New York: Oxford University Press, 1993. Young, Terence. *Building San Francisco's Parks, 1850–1930*. Baltimore: Johns Hopkins University Press, 2004

Zaitzevsky, Cynthia. *Frederick Law Olmsted and the Boston Park System*. Cambridge: Belknap Press, 1982.

Works (in Chinese, including translations):

Neil Brenner, *City, Territory, Planet: Studies in Critical Urban Theory*, Translated by Li Zhigang, et al., Beijing: The Commercial Press, 2019.

Robert Darnton, *The Great Cat Massacre and Other Episodes in French Cultural History*, translated by Lv Jianzhong, Beijing: The Commercial Press, 2014.

Christophe Defeuilley, *L'Entrepreneur et le Prince: La création du service public de l'eau, LONDREA, NEW YORK, PARIS RACONTÉS, AU FIL DE L'EAU*, translated by Tang Jun, Beijing: Social Sciences Academic Publishing House, 2019.

Emily Dickinson, *Selected Poems of Emily Dickinson*, translated by Jiang Feng, Beijing: Foreign Language Teaching and Research Press, 2012.

Johann Heinrich von Thünen, *Der isolierte Staat in Beziehung auf Landwirtschaft und Nationalökonomie*, translated by Wu Hengkang, The Commercial Press, published for the first time in 1986, reprinted in 2010.

Eric Foner, *The Fiery Trial: Abraham Lincoln and American Slavery*, translated by Yu Liuzhen, Beijing: The Commercial Press, 2017

付成双:《动物改变世界:海狸、皮毛贸易与美洲开发》,北京:北京大学出版社,2016年。

付成双:《美国现代化中的环境问题研究》,北京:高等教育出版社,2017年。

Gao Guorong: *American Environmental History: A Historiography*, Beijing: China Social Sciences Press, 2014.

侯文蕙:《征服的挽歌:美国环境意识的变迁》,北京:东方出版社,1995年。

Walt Whitman, *Leaves of Grass*, translated by Zou Zhongzhi, Shanghai: Shanghai Translation Publishing House, 2019.

William M. Cavert, *Smoke of London: Energy and Environment in the Early Modern City*, translated by Wang Qinghua et al., Beijing: Social Sciences Academic Press, 2019.

William Cronon, *Nature's Metropolis: Chicago and the Great West*, translated by Huang Yanjie et al., Nanjing: Jiangsu People's Publishing House, 2020.

Alfred W. Crosby, *Ecological Imperialism: The Biological Expansion of Europe, 900–1900*, translated by Zhang Suguo, Beijing: The Commercial Press, 2017.

Aldo Leopold, *A Sand County Almanac*, translated by Hou Wenhui, Beijing: The Commercial Press, 2016.

梁茂信:《都市化时代:20世纪美国人口流动与城市社会问题》,长春:东北师范大学出版社,2002年。

Ruth Rogaski, *Hygienic Modernity: Meanings of Health and Disease in Treaty-Port China*, translated by Xiang Lei, Nanjing: Jiangsu People's Publishing House, 2007.

Daniel Rodgers, *Atlantic Crossings: Social Politics in Progressive Age*, translated by Wu Wanwei: Nanjing: Yilin Publishing House, 2011. 19. Adam Rome, *The Bulldozer in the Countryside: Suburban Sprawl and the Rise of American Environmentalism*, translated by Gao Guorong et al., Beijing: China Social Sciences Press, 2011.

Kenneth T.Jackson, *Crabgrass Frontier: The Suburbanization of the United States*, translated by Wang Xu, Beijing: The Commercial Press, 2017.

J. R. McNeill & William H. McNeill: *The Human Web: A Bird's-Eye View of World History*, translated by Wang Jinxin, Song Baojun, et al., Beijing: Peking University Press, 2007.

Lewis Mumford, *Technics and Civilization*,translated by Chen Yunming et al, Beijing: China Architecture Industry Press, 2006.

Lewis Mumford, *The Culture of Cities*,translated by Song Junling Yunming et al, Beijing: China Architecture Industry Press, 2009.

梅雪芹:《环境史研究叙论》,北京:中国环境科学出版社,2011年。

Roderick Nash, *Wilderness and the American Mind*, translated by Hou Wenhui et al., Beijing: China Environmental Science Press, 2015.

Ted Gioia, *The History of Jazz*, translated by Li Jianmin, and reviewed by Li Wan, Beijing: The Commercial Press, 2020.

Oswald Spengler, translated by Wu Qiong: *The Decline of the West*, Perspectives on World History, Volume II, Shanghai: SDX Joint Publishing Company, 2006.

James C. Scott, *Weapons of the Weak: Everyday Forms of Peasant Resistance*, translated by Zheng Guanghuai et al., Nanjing: Yilin Publishing House, 2011

James C. Scott, *Seeing Like a State: How Certain Schemes to Improve the Human Condition Have Failed*, translated by Wang Xiaoyi, Beijing: Social Sciences Academic Press, 2019.

David Stradling, *Smokestacks and Progressives: Environmentalists, Engineers, and Air Quality in America, 1881-1951*, translated by Pei Guangqiang, Beijing: Social Sciences Academic Press, 2019.

孙群郎:《美国城市化郊区研究》,北京:商务印书馆,2005年。

Alexis-Charles-Henri Clérel de Tocqueville: *De la démocratie en Amérique*, translated by Dong Guoliang, Beijing: The Commercial Press, 1996.

王旭:《美国城市发展模式:从城市化到大都市区化》,北京:清华大学出版社,2006年。

Raymond Williams, *The Country and the City in the Modern Novel*, translated by Han Ziman, et al., The Commercial Press, Beijing: 2013.

Donald Worster, *Nature's Economy: A History of Ecological Ideas*, translated by Hou Wenhui, Beijing: The Commercial Press, 1999.

Donald Worster, *Dust Bowl: The Southern Plains in the 1930s*, translated by Hou Wenhui, Nanjing: Jiangsu People's Publishing House, 2020.

Donald Worster, *Rivers of Empire: Water, Aridity, and the Growth of the American West*, translated by Hou Shen, Nanjing: Yilin Publishing House, 2018.

夏明方:《文明的"双相":灾害与历史的缠绕》,桂林:广西师范大学出版社,2020年。

Joseph Alois Schumpeter, *Capitalism, Socialism, and Democracy*, translated by Wu Liangjian, Beijing: The Commercial Press, 1999.

Mark Elvin, *The Retreat of the Elephants: An Environmental History of China*, translated by Mei Xueqin et al., Nanjing: Jiangsu People's Publishing House, 2014.
赵毅衡编译:《美国现代诗选(上)》,北京:外语教学与研究出版社,2019年。
The theses in Chinese and English and doctoral dissertations are omitted.

GPSR Compliance

The European Union's (EU) General Product Safety Regulation (GPSR) is a set of rules that requires consumer products to be safe and our obligations to ensure this.

If you have any concerns about our products, you can contact us on ProductSafety@springernature.com

In case Publisher is established outside the EU, the EU authorized representative is:

Springer Nature Customer Service Center GmbH
Europaplatz 3
69115 Heidelberg, Germany

The manufacturer's authorised representative in the EU is Springer Nature Customer Service Centre GmbH, Europaplatz 3, 69115 Heidelberg, Germany. If you have any concerns regarding our products, please contact ProductSafety@springernature.com

Printed and bound by CPI Group (UK) Ltd, Croydon, CR0 4YY
01/10/2026
02225924-0009